Molecular Biotechnology and its Applications

Molecular Biotechnology and its Applications

Edited by **Oscar Watson**

New York

Published by Callisto Reference,
106 Park Avenue, Suite 200,
New York, NY 10016, USA
www.callistoreference.com

Molecular Biotechnology and its Applications
Edited by Oscar Watson

International Standard Book Number: 978-1-63239-467-5 (Hardback)

Contents

Permissions

List of Contributors

Preface

Every book is initially just a concept; it takes months of research and hard work to give it the final shape in which the readers receive it. In its early stages, this book also went through rigorous reviewing. The notable contributions made by experts from across the globe were first molded into patterned chapters and then arranged in a sensibly sequential manner to bring out the best results.

This book focuses on the significance of biology at the molecular level as a means of biotechnology for advancement in human life conditions. One of the engaging issues in this field is the identification of organisms producing bioactive secondary metabolites. This book also includes how to structure a plan for use and the preservation of species depicting a potential source for new drug development, particularly those acquired from bacteria. The book also presents new uses of biotechnology like the therapeutic applications of electroporation; enhancing value, microbial safety of fresh-cut vegetables; production of synthetic PEG hydro gels which can be used as an additional cellular matrix mimic for tissue engineering applications, and other innovative uses.

It has been my immense pleasure to be a part of this project and to contribute my years of learning in such a meaningful form. I would like to take this opportunity to thank all the people who have been associated with the completion of this book at any step.

Editor

Part 1

Molecular Studies

Ammonia Accumulation of Novel Nitrogen-Fixing Bacteria

Kenichi Iwata[1], San San Yu[2],
Nik Noor Azlin binti Azlan[1] and Toshio Omori[1]
*[1]College of Systems Engineering and Science, Department of Bioscience and Engineering,
Shibaura Institute of Technology,
[2]Department of Biotechnological Research, Ministry of Science and Technology,
[1]Japan
[2]Myanmar*

1. Introduction

Nitrogen is an essential element for many biological processes, including those occurring in plants (Ogura *et al.*, 2006). Despite the abundance of atmospheric nitrogen, production of nitrogen fertilisers by the Harber–Bosch process is increasing annually due to the deficiency of ammonia produced by biological nitrogen fixation – the enzyme-catalyzed reduction of nitrogen gas (N_2). Concern over 'greenhouse' gasses emitted by the Harber–Bosch process has resulted in a research focus on nitrogen-fixing bacteria, and in particular, their genetic modification to excrete excess ammonia for agricultural purposes (Terzaghi, 1980; Saikia & Jain, 2007).

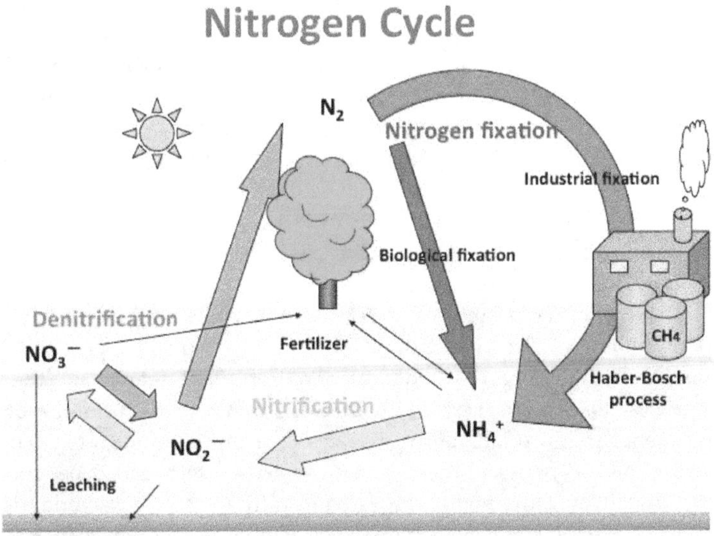

Fig. 1. The nitrogen cycle.

There are three main biological processes in the natural cycle of nitrogen (Fig. 1): fixation, nitrification and denitrification, which involve nitrogen-fixing, nitrifying and denitrifying bacteria, respectively.

Blue arrows indicate nitrogen fixation, including biological and industrial processes. Green arrows indicate microbial nitrification processes involving nitrifying bacteria, and pink arrows indicate microbial denitrification processes involving denitrifying bacteria. Black arrows indicate the flow of each compound in soils. The NH_3 produced by nitrogen fixation may be assimilated into amino acids and thence to protein and other N compounds, or it may be converted by nitrifying bacteria to NO_2^- and NO_3^-. In turn, NO_3^- may enter metabolism through reduction to NH_4^+ and subsequent assimilation to amino acids by bacteria, fungi and plants or can serve as an electron acceptor in denitrifying bacteria when oxygen is limiting. Losses from the nitrogen pool occur physically, when nitrogen (especially nitrate) is leached into inaccessible domains in the soils, and chemically, when denitrification releases N_2.

2. Biological nitrogen fixation

Decomposers use several enzymes to break down proteins in dead organisms and their waste, releasing nitrogen in much the same way as they release carbon. Proteinases break large proteins into smaller molecules. Peptidases break peptide bonds to release amino acids. Deaminases remove amino groups from amino acids and release ammonia.

According to Kneip *et al.* (2007), during biological nitrogen fixation (BNF), molecular nitrogen is reduced (Formula 1) in multiple electron-transfer reactions, resulting in the synthesis of ammonia and release of hydrogen. Ammonium is then used for the subsequent synthesis of biomolecules. This reduction of molecular nitrogen to ammonium is catalysed in all nitrogen-fixing organisms via the nitrogenase enzyme complex in an ATP-dependent, highly energy-consuming reaction (Fig. 2). The nitrogenase complex is composed of two main functional subunits, dinitrogenase reductase (azoferredoxin) and dinitrogenase (molybdoferredoxin). The structural components of these subunits are the Nif (nitrogen fixation) proteins: NifH ($\gamma2$ homodimeric azoferredoxin) and NifD/K ($\alpha2\beta2$ heterotetrameric molybdoferredoxin). Three basic types of nitrogenases are known based on the composition of their metal centres: iron and molybdenum (Fe/Mo), iron and vanadium (Fe/V) or iron only (Fe). The most common form is the Fe/Mo-type found in cyanobacteria and rhizobia. Electrons are transferred from reduced ferredoxin (or flavodoxin) via azoferredoxin to molybdoferredoxin. Each mole of fixed nitrogen requires 16 moles ATP to be hydrolysed by the NifH protein. The NH_3 produced is utilised in the synthesis of glutamine or glutamate for N-metabolism. NifJ: pyruvate flavodoxin/ferrodoxin oxidoreductase, NifF: flavodoxin/ferredoxin). An important feature of the nitrogenase enzyme complex is its extreme sensitivity to even minor concentrations of oxygen. In aerobic environments and in photoautotrophic cyanobacteria, in which oxygen is produced in the light reaction of photosynthesis, nitrogenase activity must be protected. This protection is mediated by different mechanisms in nitrogen-fixing bacteria, depending on their cellular and physiologic constitutions. Aerobic bacteria (like *Azotobacter*) prevent intracellular oxygen concentrations from reaching inhibitory levels by high rates of respiratory metabolism in combination with extracellular polysaccharides that reduce oxygen influx.

$$N_2 + 8H^+ + 8e^- + 16ATP \rightarrow 2NH_3 + H_2 + 16ADP + 16Pi \qquad (1)$$

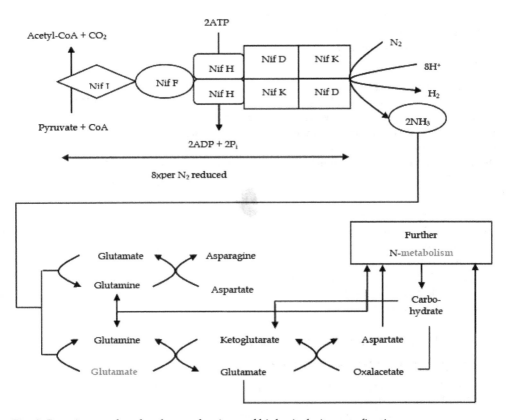

Fig. 2. Reactions and molecular mechanisms of biological nitrogen fixation.

General reaction of molecular nitrogen fixation. Schematic of the structure and operation of the nitrogenase enzyme complex and subsequent metabolism of nitrogen.

Azotobacter vinelandii, *Azotobacter beijerinckii* and *Klebsiella pneumoniae* are nitrogen-fixing bacteria commonly used for genetic modification. Metabolic mutants of *A. vinelandii* were first isolated over 50 years ago, but the mutants were unstable and some researchers were unable to mutate this bacterium. However, whether *Azotobacter* was itself difficult to mutate or the selection procedures were inadequate has remained unclear. Such failures have contributed to the continuing studies of this strain mutation.

Ultraviolet mutagenesis, the most easily controllable method of mutation, was thus often the first choice. Ultraviolet irradiation was used to modify *A. vinelandii* and *Azomonas agilis*, but the problems of segregation and mutant stability remained, despite their nitrogen-fixation activity. Several years later, it became clear that nitrogen fixation by *Klebsiella pneumoniae* is complicated by the presence of biochemically and genetically distinct nitrogenase enzymes, each of nitrogenase enzymes is synthesized under different conditions of metal supply. However, experiments continued and Bali and colleagues (1992) generated the mutant MV376 of *A. vinelandii*, which excreted about 9.3 mM of ammonium in stationary phase cultures. No excretion by the wild type was reported (Bali *et al.*, 1992). Another improvement was achieved by Brewin and colleagues (1999), resulting in production of greater quantities of ammonium. Again, the wild type did not excrete ammonium (Brewin *et al.*, 1999).

The same results arose from mutation of *A. beijerinckii* by chemical mutagens such as *N*-methyl-*N'*-nitro-*N*-nitrosoguanidine (NTG) and ethylmethane sulphonate (EMS), together with UV radiation (Owen & Ward, 1985). The same group generated some mutants by means of transposon-insertion mutagenesis several years after the study using chemical mutagens. However, no mention was made of their ammonia-excreting activities, and again, the abnormal growth and instability of putative transposition isolates precluded routine use of the method.

With regard to the carbon sources used to culture these two species, most of the studies described above used Burk's medium, which contains 2% sucrose, or modified Burk's medium (0.5% or 2% glucose) as carbon sources. The latest researches on *A. vinelandii*, *A. beijerinckii* and a new nitrogen-fixing *Lysobacter sp.* have demonstrated that cultures grown in nitrogen-free medium with ≤0.7% glucose resulted in excrete ammonia. This suggests that no modification of these nitrogen-fixing bacteria is required. Even though the mechanisms remain unclear, further research on this topic will contribute greatly to the agriculture industry development (Iwata *et al.*, 2010).

3. Screening and identification of nitrogen-fixing bacteria

3.1 Screening of nitrogen-fixing bacteria

To screen for nitrogen-fixing bacteria, 1 g of soil was suspended in 10 mL of sterilized dH_2O in a 15-mL Eppendorff tube that was left to stand until the soil solution settled. A 1-mL aliquot of supernatant was then added to 200 mL of fresh NFMM or NFMM liquid medium and incubated for 1 week on a rotary shaker at 120 rpm and 30°C. Subculture was carried out twice by adding 2 mL of liquid culture to 200 mL of new C–NFMM medium and incubated as before. Single-colony isolation was performed on NFMM plates. Nitrogen-fixing activity was tested by growing the strains on glucose–NFMM plates substituted with BTB. From the 20 soil samples collected, we obtained four strains that showed a colour change in BTB-containing medium, suggesting excretion of ammonia. These strains were named C4, E4, G6 and G7.

3.2 Identification of nitrogen-fixing bacteria

DNA extraction was performed using a Miniprep DNA Purification Kit (TaKaRa). Bacterial 16S rDNA was amplified over 35 PCR cycles. Each cycle consisted of denaturation for 1 min at 94°C, annealing for 30 s at 60°C and extension for 4 min at 72°C. DNA purification was performed using the Agarose Gel DNA Extraction Kit (Roche Diagnostics GmbH). Ligation was conducted using the DNA Ligation Kit (TaKaRa) and the pT7 Blue T-vector (Novagen) as the plasmid. Transformation used *Escherichia coli* JM109, and plasmid purification was performed according to the manufacturer's protocols. Nucleotide sequences were analyzed using the ABI PRISM 310 Genetic Analyzer (Applied Biosystems) and Basic Local Alignment Search Tool (BLAST) on the National Center for Biotechnology Information (NCBI).

The nucleotide sequences of C4 and G7 showed high similarity (99%) to *A. beijerinckii,* and E4 and G6 were most similar to *Lysobacter enzymogenes* DMS 2043^T (99% identity), as recently described. We therefore concluded that E4 and G6 belong to this genus. Subsequently an experiment was performed to determine of ammonia accumulation by *Azotobacter* using the common species *A. beijerinckii, A. vinelandii* and *Lysobacter* sp. E4.

3.3 Classification of isolated strains

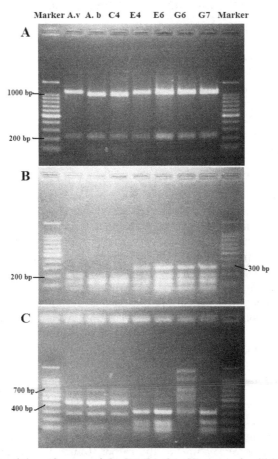

Fig. 3. RFLP analysis of the *nifL* gene of C4, E4, E6, G6, G7, *A. vinelandii* (A.v) and *A. beijerinckii* (A.b). (A) *Afa*I, (B) *Hae*III and (C) *Alu*I.

RFLP of the amplified *nifL* gene of C4, E4, G6 and G7 suggested that these may represent of nitrogen-fixing bacteria. Due to the similarities of strains C4, E4, E6, G6 and G7 to *Azotobacter* species and the amplification of the *nifL* gene from them, RFLP of the amplified *nifL* genes was conducted. Only strain C4 possessed the same restriction fragment pattern as *Azotobacter* species, showing the same length of fragments as both *A. vinelandii* and *A. beijerinckii* for *Hae*III and *Alu*I and as *A. beijerinckii* for *Afa*I (Fig. 3). From this result, it was assumed that the probability of this strain to belong to *A. beijerinckii* was high. E4, E6, G6 and G7 showed the same fragment lengths after digestion with *Afa*I and *Hae*III but these four strains were divided into two groups by *Alu*I digestion; G6 differed from the other three strains (C4, E4 and G7). Additionally, G6 and G7 showed different 16S rDNA RFLP fragment lengths; thus the data suggest that these represent different strains.

4. Mutation of *Azotobacter nif* genes for ammonia accumulation

Azotobacter is a free-living nitrogen-fixing microbial genus widely distributed in soil and rhizosphere (Martinez *et al.*, 1985; Kennedy & Tchan, 1992). Considering the possibility of replacing industrially produced ammonia fertilisers, many attempts to modify two species of this diazotroph—*A. beijerinckii* and *A. vinelandii*—were undertaken with the aim of producing an environmentally friendly bacterial fertiliser (Brewin *et al.*, 1999). Generally, regulation of ammonia producichtii by *Azotobacter*, especially *A. vinelandii*, is similar to that achieved by using *Klebsiella pneumoniae*, being regulated by *nifL* and *nifA*. The NifL protein binds to and inactivates NifA when ammonium is present where even at relatively low levels. At higher levels of ammonium, expression of the *nifLA* operon does not occur, and so NifA is not synthesized (Brewin *et al.*, 1999). An idea to mutate *nifL* for enhancing ammonia production by *Klebsiella pneumoniae* for agricultural purposes generated many studies to generate a mutant with a damaged *nifL* gene. Various methods of mutation were tested on *A. beijerinckii*, including UV radiation and chemical mutagenesis using *N*-methyl-*N'*-nitro-*N*-nitrosoguanidine (NTG) and ethylmethane sulphonate (EMS). However, no ammonia-excreting mutants were isolated, even using the mating approach (Owen & Ward, 1985). This may have been due to the production by *Azotobacter beijerinckii* of polysaccharide that surrounds the cell (Danilova *et al.*, 1992), rendering mutation problematic. However, for *A. vinelandii* a mutation in *nifL* (upstream of and regulatory to *nifA*) was successfully produced. This mutant was named MV376, and it secreted significant quantities of ammonium during diazotrophic growth (Bali *et al.*, 1992). According to Bali *et al.* (1992), the mutant strain MV376, but not the wild type, showed ammonium production up to 10 mM when grown in Burk's sucrose medium.

5. Accumulation of ammonia by wild-type strains

When wild-type *A. beijerinckii* and *A. vinelandii* were cultured in Glucose-Nitrogen Free Mineral Medium (G-NFMM) and Fructose-Nitrogen Free Mineral Medium (F-NFMM), respectively, both strains showed ammonium accumulation. This indicates that the concentration, as well as the nature, of the carbon source might influence ammonium accumulation; here we report a correlation of carbon source concentration with ammonium accumulation by both *Azotobacter* species.

6. Ammonia detection and estimation

Ammonia concentration was estimated using the Visocolor Alpha Ammonia Detection Kit (Macherey-Nagel). After centrifugation at 13,000 rpm for 10 min at room temperature (RT), supernatant (1 mL) was transferred into a test tube. Two drops of NH_4-1 were added to the sample and mixed well, after which one-fifth of a spoonful of NH_4-2 was added. After mixing, the sample was left at RT for 5 min. One drop of NH_4-3 was then added, mixed well and left at RT for 5 min.

Ammonia concentration was also estimated using ion chromatography. After centrifugation at 13,000 rpm for 10 min at RT, the supernatant was passed through a 0.2-μm filter and the ammonium concentration determined using an 861 Advanced Compact Ion Chromatography (Metrohm). The cation eluent used was 4 mM H_3PO_4 with 5 mM 18-crown 6-ether. The separation column was an IC YK-421 (Shodex) and the guard column was an IC-YK-G (Shodex). Standard ammonium solution was prepared from $(NH_4)_2SO_4$; the concentration was adjusted to 1000 parts per million (ppm) and diluted appropriately to obtain a standard curve. All experiments were performed in triplicate.

7. Cultivation of nitrogen-fixing *Lysobacter sp.*

A. beijerinckii, *A. vinelandii* and *Lysobacter* sp. were grown on 0.5% G-NFMM plates for 2 days and then inoculated into 6 mL G-NFMM or F-NFMM liquid media, respectively, containing various glucose and fructose concentrations. These species were then incubated for 2 (*Azotobacter*) or 3 (*Lysobacter*) days. Optical density (OD), pH and ammonium concentrations were then measured to examine the relationship between the carbon source concentration and ammonia accumulation. Best concentration was chosen for examining the correlations among incubation time, ammonia accumulation and carbon uptake. *A. beijerinckii*, *A. vinelandii* and *Lysobacter* sp. were pre-cultured in 6 mL of 0.5% G-NFMM and 0.25% F-NFMM, respectively, for 2 days and 2 mL was then transferred to 200 mL fresh media in 500-mL baffle flasks. Samples of cultures were taken at different times for measurement of OD, pH, ammonium ion and concentration of carbon source. All incubation periods were carried out aerobically at 30°C with shaking (200 rpm). Culture samples were centrifuged and filtered (0.2 μm) before being ammonium assayed by Nessler's reagent; ammonium concentration was estimated by ion chromatography. The cation eluent used for ion chromatography was 4 mM H_3PO_4 added to 5 mM 18-crown 6-ether. The residual carbon concentration in media was assayed by Somogyi-Nelson method. All experiments were performed in triplicate.

8. Effect of carbon concentration

The optimum carbon source concentration was used to determine the correlations among incubation time, ammonia accumulation and carbon uptake. *Azotobacter beijerinckii* and *A. vinelandii* were pre-cultured in 6 mL G-NFMM and F-NFMM, respectively, for 2 days and 2 mL was transferred to 200 mL fresh medium in 500-mL baffle flasks. The OD, pH, ammonium ion and residual sugar levels in cultures were determined. All incubation periods were carried out aerobically at 30°C on a rotary shaker at 200 rpm. Experiments were performed in triplicate. For *A. vinelandii*, almost no ammonium accumulation was detected in culture broth containing glucose as the carbon source (Table 1). However, ammonium accumulation was detected with fructose (Table 2). Similar to *A. beijerinckii*, ammonium accumulation started 16 h after incubation. At this time, the fructose level in the medium had decreased, and no fructose was detected using the Somogyi–Nelson method after 20 h incubation.

Glucose concentration		0.10%	0.25%	0.50%	0.70%	1.00%	2.00%
A. beijerinckii	OD	0.145	0.486	1.109	1.406	1.698	1.522
	pH	7.0 (7.0)*	7.0 (7.0)*	6.8 (7.1)*	6.6 (7.1)*	6.4 (7.1)*	6.3 (7.1)*
	NH_4^+	0.062	0.117	0.202	0.080	0.026	0.001
A. vinelandii	OD	0.189	0.478	0.950	1.391	1.710	1.948
	pH	7.1 (7.1)*	6.8 (7.1)*	6.1 (7.1)*	4.9 (7.1)*	4.7 (7.1)*	4.7 (7.0)*
	NH_4^+	0.010	0.024	0.020	0	0	0

OD: optical density (600 nm). *Figures in parentheses show the value before incubation.

Note: ammonium ion concentration is in mM. Presence of ammonium was primarily tested using Nesler's reagent before the concentration was determined by ion chromatography.

Table 1. OD, pH and ammonium accumulation by *A. beijerinckii* and *A. vinelandii* in G-NFMM liquid medium of various glucose concentrations after 2 days incubation.

		Glucose	Fructose	Galactose	Mannose	Sucrose	Citrate	Succinate
A. beijerinckii	OD	0.518	0.739	0.564	0.029	0.656	0.005	0.212
	pH	7.3 (7.0)*	7.2 (7.0)*	7.1 (7.1)*	7.1 (7.1)*	7.1(7.1)*	7.4	8.6 (7.2)*
	NH_4^+	0.296	0.315	0.201	0.041	0.192	(7.0)*	N. D.
							N. D.	
A. vinelandii	OD	0.442	0.704	0.573	0.122	0.655	0.361	0.361
	pH	7.0 (7.0)*	7.2 (6.9)*	7.1 (7.0)*	7.1 (7.1)*	7.2(7.0)*	8.4	8.8 (7.2)*
	NH_4^+	0.026	0.179	0.025	0.017	0.63	(7.0)*	N. D.
							N. D.	

N.D.: not determined, OD: optical density (600 nm). *Figures in parentheses show the value before incubation.

Note: ammonium ion concentration is in mM. Presence of ammonium was primarily tested using Nesler's reagent before the concentration was determined by ion chromatography.

Table 2. OD, pH and ammonium accumulation by *A. beijerinckii* and *A. vinelandii* in G-NFMM liquid medium containing various carbon sources after 2 days incubation.

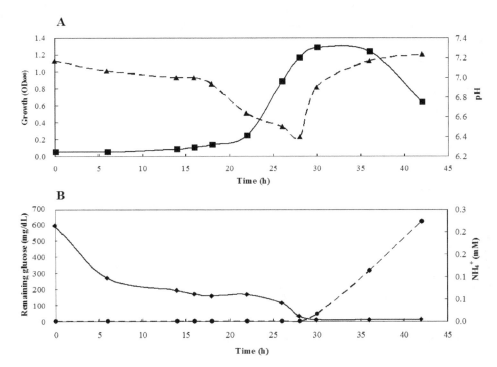

Fig. 4. A: Growth (■), pH (▲), ammonium concentration of *Azotobacter beijerinckii*. (●) B: remaining glucose concentration (◆) in cultures of *Azotobacter beijerinckii*. Samples were removed for analysis at the indicated times.

9. Time course of ammonia accumulation

As the *A. beijerinckii* population increased, medium pH decreased slowly due to production of acidic substances from glycolysis; a sharp decrease to pH 6.4 occurred after 16 h (Fig. 4A). Medium pH began to increase at the end of the log phase or early stationary phase due to production of ammonium around 30 h after inoculation. Medium pH remained steady at 7.1-7.2 beginning in the middle of stationary phase, whereas the amount of ammonium gradually increased to 0.46 mM after 54 h incubation (Fig. 4B).

10. Time course of ammonia accumulation by *Lysobacter* sp.

Time-course experiments suggested that ammonia accumulation began upon glucose depletion. In the 0.30% medium, no glucose remained after incubation for 3 days, resulting in ammonia accumulation. In media with higher glucose concentrations, residual glucose was present after 3 days. As a result, no ammonia accumulation occurred; longer incubation times may have resulted in production of detectable levels of ammonia (Fig. 5A).

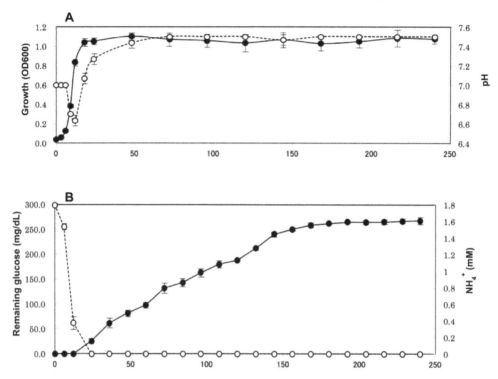

Fig. 5. A: Growth (■), pH (▲), ammonium concentration of *Lysobacter* sp. E4. (●) B: remaining glucose concentration (◆) in cultures of *Lysobacter* sp. E4. Samples were collected for analysis at every ten hours.

11. Effect of remaining sugar on ammonia accumulation

Residual sugar levels were determined using a glucose detection kit, according to the manufacturer's protocol (Miwa et. al., 1972). For *A. beijerinckii*, the concentration of glucose slowly decreased. Almost no glucose remained in the medium after 30 h incubation, at which point ammonia began to accumulate.

These data suggest that ammonia accumulation by strain E4 is dependent on sugar concentration. Glucose is required for bacterial growth until the middle of the logarithmic phase, and fixation of nitrogen during this period likely supports bacterial growth. Ammonia starts to accumulate when no more glucose remains in the culture, as shown by glucose and ammonia determinations after 14 h incubation (Fig. 5B).

For *A. vinelandii*, as for *A. beijerinckii*, bacterial growth and medium pH decreased slowly due to production of acidic substances from glycolysis; a sharp decrease to pH 6.4 occurred after 8 h. Medium pH began to increase at the end of log phase or early stationary phase due to production of ammonium approximately 16 h after inoculation. Medium pH remained neutral at 7.1–7.2 beginning in the middle of stationary phase, whereas ammonium levels gradually increased, reaching 0.1 mM after 28 h.

Thus, in both strains, ammonia began to accumulate at the end of log phase or in early stationary phase; no carbon source could be detected in the medium at this time. Higher ammonia levels in the medium will likely be detected after moreover 30 hours, longer incubation times, suggesting that the mechanism of nitrogen fixation might be influenced by sugar levels in the medium.

E4 strain grew well at pH 7.0 and produced the highest concentration of ammonia (~0.4 mM). Although media at pH 8.0 resulted in the greatest growth, ammonia accumulation was lower than at pH 7.0, suggesting that accumulated ammonia at the higher pH value may have been used for bacterial growth (Fig. 5B).

Ammonia was detected in E4 cultures incubated at 30°C, but not at 20°C. Ammonia may accumulate at 20°C after longer incubation times, since some glucose remained after 3 days incubation.

12. Conclusions

From the above, the following conclusions could be drawn. Firstly, the ammonium accumulation is clearly dependent on the carbon source concentration. Higher ammonium accumulation occurred in media with lower concentrations of the carbon source. Glucose was required for growth of *A. beijerinckii* until late logarithmic phase. Fixation of nitrogen during this time likely supports bacterial growth. Ammonium starts to accumulate after glucose depletion as determined by the Somogyi-Nelson method after 30 h incubation, which suggests that regulation of *nifL* and *nifA* genes might not be functioning when the medium contains less than 2.0% glucose. Normally, in the presence of excess ammonium or ammonia, *nifL* is expressed, resulting in repression of *nifA* and cessation of ammonia production. However, when glucose levels drop to 2.0% or less (0.5% for this experiment), we consider believe that the lowered glucose concentration renders the *nifL* system nonfunctional. This results in continuing *nifA*-mediated extracellular ammonium production and accumulation in the medium.

13. References

Bali, A., Blanco, G., Hill, S. & Kennedy, C. (1992). Excretion of ammonium by a *nifL* mutant of *Azotobacter vinelandii* fixing nitrogen. *Applied and Environmental Microbiology* 58, 1711–1718

Betty, E. Terzaghi. (1980). Ultraviolet sensitivity and mutagenesis of *Azotobacter*. *Journal of General Microbiology* 118, 271-273

Betty, E. Terzaghi. (1980). A method of isolation of *Azotobacter* mutants derepressed of Nif. *Journal of General Microbiology* 118, 275-278

Brewin, B., Woodley, P. & Drummond, M. (1999). The basis of ammonium release in *nifL* mutants of *Azotobacter vinelandii*. *Journal of Bacteriology* 181, 7356–7362

Danilova, V., Botvinko, I. V. & Egorov, N. S. (1992). Production of extracellular polysaccharides by *Azotobacter beijerinckii*. *Mikrobiologiya* 61(6), 950–955

Iwata, K., Azlan, A., Yamakawa, H. & Omori, T. (2010). Ammonia accumulation in culture broth by the novel nitrogen-fixing bacterium, *Lisobacter* sp. E4. *Journal of Bioscience and Bioengineering* 110 (4), 415-418

Kennedy, I. R. & Tchan, Y-T. (1992). Biological nitrogen fixation in non-leguminous field crops: Recent advances. *Plant and Soil* 141, 93–118

Kneip, C., Lockhart, P., Voβ, C., & Maier, U.-G. (2007). Nitrogen fixation in eukaryotes – New models for symbiosis. *BMC Evolutionary Biology* 7(55), 1471–2148

Martinez-Toledo, M. V., Gonzalez-Lopez, J. & Ramos-Cormenzana, A. (1985). Isolation and characterization of *Azotobacter chroococcum* from the roots of *Zea mays*. *FEMS Microbiology Ecology* 31, 197–203

Miwa, I., Okudo, J., Maeda, K. & Okuda. G. (1972). Mutarotase effect on colorimetric determination of blood glucose with –D-glucose oxidase. *Clinica Chimica Acta* 37, 538-540

Ogura, J., Toyoda, A., Kurosawa, A., Chong, L., Chohnan, S. & Masaki, T. (2006). Purification, characterization, and gene analysis of cellulose (*Cel8A*) from *Lysobacter* sp. IB-9374. *Bioscience, Biotechnology, and Biochemistry* 70, 2420–2428

Owen, D. J. & Ward, A. C. (1985). Transfer of transposable drug-resistance elements Tn5, Tn7, and Tn76 to *Azotobacter beijerinckii*: Use of plasmid RP4::Tn76 as a suicide vector. *Plasmid* 14, 162–166

Saikia, S. P. & Jain, V. (2007). Biological nitrogen fixation with non-legumes: An achievable target or a dogma? *Current Science* 92, 317–322

Terzaghi, B. E. (1980). Ultraviolet sensitivity and mutagenesis of *Azotobacter*. *Journal of General Microbiology* 118, 271–273

Bioactive Compounds from Bacteria Associated to Marine Algae

Irma Esthela Soria-Mercado[1], Luis Jesús Villarreal-Gómez[2],
Graciela Guerra Rivas[1] and Nahara E. Ayala Sánchez[3]
[1]*Facultad de Ciencias Marinas, Universidad Autónoma de Baja California, Ensenada, BC.,*
[2]*Centro de Ingeniería y Tecnología,*
Universidad Autónoma de Baja California, Tijuana, BC.,
[3]*Facultad de Ciencias, Universidad Autónoma de Baja California, Ensenada, BC.,*
México

1. Introduction

Since ancient times, humans have sought to satisfy their needs, one of which is, without a doubt, to stay alive. The fear of getting sick and dying, led man to study the organisms that surround him, discovering that the chemicals compounds present in some of them could be beneficial for treating illness. Thus; began the chemistry of the natural products; biotechnology area for human welfare. Several of these organisms produce secondary metabolites, which are part of a wide variety of natural compounds used by humans to combat diseases. Secondary metabolites are defined as organic compounds formed as bio products in organisms, not directly related to growth, development and normal reproduction of thereof. Some examples are fibers (cotton, silk, wool); fuels (oil and natural gas), and medicines (antibiotics, hormones, vaccines).

The importance of finding and using these secondary metabolites can be justified in two ways (1) to know the natural substances that can be beneficial for man and (2) to identify the organisms that produce these substances in order to make a rational exploitation of them, because they may be the only carriers of useful compounds to combat pathogenic microbes.

Marine organisms possess an inexhaustible source of useful chemical substances for the development of new drugs; among these organisms we find marine algae that are capable of biosynthesizing a broad variety of secondary metabolites and bacteria that live in the oceans and that are crucial organisms used in biotechnology in the discovery of new compounds from marine origin.

The discovery of new bioactive compounds necessarily involves previously diversity studies, because by knowing the type of microorganisms that reside in a certain environment, it is possible to design cultivation techniques adapted for all the microbial communities present in a certain ambience. That is why it is very important to identify the organisms that produce bioactive secondary metabolites, and to be able to structure a plan of use and preservation of those species that represent a potential source for new drug development, especially those obtained from bacteria, because of their own cultivation characteristics, have attracted attention on either a big quantity of investigators on a global scale in the search of new natural products with anticancer and antibiotic activity principally.

2. Anticancer activity

Cancer is an illness that comprises more than hundred types. This disease appears when old cells are not replaced by new cells and are accumulated in a mass of tissue known as tumor (Figure 1).

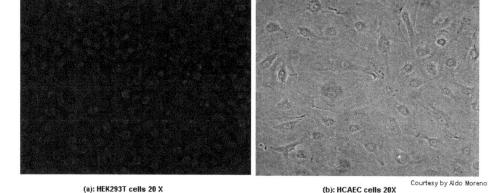

(a): HEK293T cells 20 X   **(b): HCAEC cells 20X**   Courtesy by Aldo Moreno

Fig. 1. Pictures of immortalized cells that resemble carcinogenic cells and cells of a normal tissue (20X). (a): immortalized cells of Human Embryonic Kidney Cells (HEK293T) visualized with 20X amplification. (b): Human Coronary Artery Endothelial Cells (HCAEC) visualized with 20X amplification. (A courtesy of Aldo M.Ulloa, UCSD, School of Medicine).

The incidence of cancer increases constantly constituting an enormous challenge for health institutions, in many of the cases, the medicines used in chemotherapy treatment provoke secondary toxicity or resistance (Isnard-Bagnis et al., 2005). For all of the above, there is an urgent need to discover new anticancer compounds from natural sources. Compounds like taxol, camptothecin, vincristine and vinblastine, are obtained from superior plants (Cragg & Newman, 1999), recently some medicines obtained from marine organisms have showed promising results when administered at different cancer stages.

The metabolic and physiological capacity that allows marine organism to survive in extreme conditions provides an enormous potential for the production of unique compounds that are not present in the terrestrial organisms. That is way marine organisms are an attractive source of compounds with pharmaceutical activity, (Faulkner, 2002). Seaweeds are recognized as one of the richest sources of new bioactive compounds of which reviews have been published recently on the biological activity of their derivative compounds (Blunt et al., 2006).

3. Resistance to antibiotics

The resistance to antibiotics is a phenomenon by which a microorganism stops being affected by an antimicrobial compound to which previously it was sensitive. It is a consequence of the capacity of certain microorganisms (bacteria, virus and parasites) to neutralize the effect of the medicine. The resistance can come from the mutation of the microorganism or from the acquisition of resistance genes. The infections caused by resistant microorganisms do not answer to the ordinary treatment, which result in a long illness and the risk of death (WHO, 2011).

Approximately 440 000 new cases of multiresistant tuberculosis produce at least 150 000 deaths every year. In South East Asia infections of *Plasmodium falciparum* that are late in disappearing after the beginning of the treatment with artemisinins are arising, which indicates resistance of the parasite to this specific medicine. Resistance has been found also to the antiretroviral medicines that are used in the treatment of the HIV's infection (WHO, 2011).

A high percentage of the infections contracted in the hospitals are caused by very resistant bacteria, like Methicillin Resistant *Staphylococcus aureus* (MRSA), *Enterococcus faecium* and several microorganisms Gram negative resistant to Vancomycin (WHO, 2011). New mechanisms can appear that can cancel completely the ability of antimicrobial drugs to act against bacteria. This could represent the last defense against multiresistant microorganism's strains. For example, a new β-lactamase, enzyme of the group of the carbapenemases that nowadays are named like NDM-1, gives resistance to the majority of the β-lactams medicines. The enzyme is linked to genes that are easily transferred between the common bacteria, and the infections caused by NDM-1's producer bacteria have no treatment or, if they have it, the therapeutic options are few (WHO, 2011).

4. Factors that enhance the resistance appearance to the antimicrobial effects

According to the World Health Organization (WHO) several factors exist that enhance the resistance to antibiotics. One problem is the lack of commitment of the government towards solving the problem, the bad definition of the responsibilities of the interested parts and the scarce participation of the consumers those results in the lack of coherent and coordinated methods to anticipate and to contain the resistance to the antimicrobial compounds. The improper and irrational use of antimicrobial drugs promotes conditions for the appearance of resistant microorganisms, which at once propagate. This happens, for example, when the patients do not take the complete treatment of a prescribed antibiotic or when the above mentioned medicine is of bad quality.

The nonexistence or weakness of the systems of alertness determines the lack   of information that can guide politicians to make recommendations and to closely continue to monitor the resistance to the existing antimicrobial compounds. Also, the scarcity of diagnosis means more medicines and vaccines for the prevention and treatment of illness, also, the shortcomings on the subject of research and development, debilitates the aptitude to combat the problem (WHO, 2011).

Right now, the WHO is focusing their efforts towards the regulation of the normatively by means of alertness, technical assistance, generation of knowledge and alliances, prevention and control of certain illnesses like tuberculosis, malaria, HIV, proper illnesses of the infancy, sexually transmitted diseases and hospitable infections; the quality, the supply and the rational use of essential medicines; the safety of the patients; and the guarantee of certified laboratories.

On the other hand, the struggle against the resistance to the antimicrobial compounds was the issue of the World Day of the Health of 2011 (April 7). On this occasion, the WHO called to contain the spread of the resistance to the antimicrobial compounds by means of a set of politics   that were recommended so that the governments can start to solve the problem (WHO, 2011).

5. Distribution and economic importance of marine algae

Marine algae are not only used in the discovery of new drugs, they are also used extensively as food on the Asian east coast (Japan, China, Korea, Taiwan and Vietnam), Indonesia, Peru, Canada, Scandinavia, Ireland, Wales, the Philippines and Scotland, among other places. From the economic point of view the marine algae represent an important resource of food and industrial input. The Caribbean Sea coast of Colombia contains innumerable species that have an economic value and are used as human food, medicinal products, fertilizers, fuel, and play an important role in the extraction of phycocolloids and hydrocolloids (Teas, 2007). All these products have a big industrial application. In spite of the speculation on the seaweed potential as a direct source of proteins and pharmaceutical products, the demand for phycocolloids will be the factor that will influence the future development of the marine algae world resources.

Many species have been exploited, but others like the genus *Sargassum* and *Codium* have been considered to be invasive   for their capacity of adaptation and their high growth rate. Due to the fact that it has been established that marine algae are a potential source for new drugs their study should become a priority. The chemical screening of all the seaweeds and their related organisms is necessary in order to establish which species can be exploited without consequences and those that must be protected.

6. Marine algae as producers of secondary metabolites

In 2010, Mexican investigators found that the marine algae *Codium fragile, Sargassum muticum, Endarachne binghamiae, Centroceras clavulatum* and *Laurencia pacifica* possess compounds that inhibit the growth of Gram negative bacteria *Proteus mirabilis* (Villarreal-Gómez et al., 2010), which provokes 90 % of the infections caused by *Proteus*. The bacteria causes the production of big levels of urease that hydrolyze the urea to ammonia increasing the pH and therefore the formation of glazing of struvite, carbonate of calcium and/or apatite, causing the formation of kidney stones.

In the South-west coast of India, a group of scientists studied 13 groups of marine algae to evaluate the cytotoxic, larvicide, nematicide and ichthyotoxic activities on *Artemia salina* larvae. This Indian region is the only marine habitat with great marine algae diversity. 13 algae extracts between them *Dictyota dichotoma* and *Hypnea pannosa* showed lethal effect against the root nematode *Meloidogyne javanica*. *D. dichotoma* and *Valoniopsis pachynema* showed an ichthyotoxic activity. *A. orientalis, Padina tetrastromatica* and *Centeroceras clavulatum* showed activity against the urban mosquito larvae *Culex quinquefasiatus* (Manilal et al., 2009). Another study done in the same country, found marine algae that belonged to the family *Chlorophyceae* (*Caulerpa racemosa* and *Ulva lactuca*) and Rhodophyceae (*Gracillaria folifera* and *Hypneme muciformis*) that showed antibacterial activity against the Gram negative bacteria *E. faecalis*, *K. pneumoniae* and *E. aerogens*, as well as in the Gram positive bacteria *S. aureus* (Kandhasamy & Arunachalam, 2008). In a work done in the Iberian Peninsula with 82 marine algae (18 Chlorophyceae, 25 Phaeophyceae and 39 Rhodophyceae) the antibacterial and antifungal activity was analyzed to evaluate their application as natural preservatives for the cosmetic industry. The raw extracts of every taxon, prepared from fresh material as well as from lyophilized one, proved to have the opposite effect on three Gram positive bacteria, two Gram negative bacteria and yeast, by means of the agar diffusion technique. Sixty seven % of the studied seaweeds did not show antimicrobial activity against opposite the tested microorganisms. The biggest percentage of active taxon were presented in the group of the Phaeophyceae (84 %) followed by the Rhodophyceae (67 %) and finally the Chlorophyceae (44 %). The red seaweed presented the highest activity, with the widest spectrum. Inside this group, the most active species were *Bonnemaisonia asparagoides*, *Bonnemaisonia hamifera*, *Asparagopsis armata* and *Falkenbergia rufolanosa* (*Bonnemaisoniales*). As for the microorganisms, *Bacillus cereus* was the most sensitive and *Pseudomona aeruginosa* the most resistant. Three taxonomic groups showed seasonal change in the production of antimicrobial substances, being autumn the station with major percentage of active taxon for the Phaeophyceae and Rhodophyceae, while for the Chlorophyceae it was summer (Salvador et al., 2007).

Now a day there is a database that contains all the known natural compounds derived from marine algae, creating a crucial tool in the scientific community dedicated to the search of new useful compounds for medicinal purposes. This database provides the user with more than 3,600 released articles that describe 3,300 secondary metabolites originated from seaweeds, and it is still considered to be insufficient even though is growing every day. According to the database, Phaeophytas and Rhodophytas present significantly more quantity of bioactive compounds than Chlorophytas. The red seaweed *Laurencia* (*Ceramiales, Rhodomelaceae*) is one of the most prolific algae in the production of secondary metabolites derived from the sea. Sesquiterpenes, diterpenes, triterpernes and acetogenins (characterized by the presence of halogens atoms in their chemical structures) have been found present on this seaweed (John Davis & Vasanthi, 2011).

7. Bacteria associated with marine algae

The marine ambience is a complex ecosystem with an enormous plurality of forms of life that are associated between themselves, the most common associations found are between eukaryotic cells and microorganisms (Eganet al., 2008). The surface of all the marine eukaryotic organism are covered by microbes that live adherent to diverse communities often immersed in a matrix or forming a bilayer (Pérez–Matos et al., 2007). Also, the specificity of the guest organism also has been demonstrated in studies that show the presence of the only adherent stable communities to organisms of the same species through that they live on geographically distant regions (Webster & Bourne, 2007).

In the recent years, the bioactive properties of marine algae and marine microorganisms have been analyzed, and in both cases positive results have been obtained. Many of the marine algae species often come accompanied by several bacterial strains which have been taken of the sea together with the algae cells, or have been the result of a contamination in the algae culture. These mixed populations that are present in the culture and in the sea, show that the bacteria use organic substances secreted by living or dead algae cell. It has been observed that many types of seaweeds present a major growth in the presence of bacteria than in their absence. Some seaweed species need vitamins for their growth and possibly the bacteria are partially responsible for the production of these substances; some of them produce antibiotics (Jasti et al., 2005; Penesyan et al., 2010).

The number of natural products, discovered from several organism that include plants, animals and microorganisms, overcomes millions of compounds. Forty to sixty percent derives from terrestrial plants, from which twenty to twenty five % possesses bioactive properties such as antibacterial, anticancer, antifungal, antiviral and anti-inflammatory activity (Berdy, 2005).

Bacteria exist only in some seaweed species, as it is the case of *Leucobacter sp.*, collected in the Todos los Santos bay, BC. Mexico; which only was present in one out of six seaweeds analyzed (*Egregia menziessi*), (Villarreal-Gómez et al., 2010). This member of the *Actinobacteria* family, has also been associated to the nematode *Caenorhabditis elegans*(Muir & Tan, 2008). *Micrococcus* is another actinobacteria strain that has been associated only to *Egregia menziessi*. This strain is usually found in soil and water. It is catabolically versatile, with the skill of using unusual substrates like pyridine, herbicides, polychloric biphenyl's and oil. It can also biodegrade many environmental pollutants (Zhuang et al., 2003). The bacterial strain*Kocuria palustris* (Sm32), it is exclusively present in the brown seaweed *Sargassum muticum* that is considered to be an invasive species in many countries. This strain has industrial applications in the degradation of organic matter (Kovacs et al., 1999). The *Alcaligenes* found exclusively in the seaweed *Endarachne binghamiae*, are used for the industrial production of not standard amino acids (Madigan et al., 2005). Finally, the bacterial strain member of the genus *Alteromona*, was associated only with the seaweed *Laurencia pacifica*, generally isolated in sea water; this *Proteobacteria* has industrial use, since they produce polysaccharides of high molecular weight. Several of the bacterial strains phylogenetically related, have industrial application; therefore, it is necessary to study the chemical interactions seaweed - bacteria for a better understanding of the process of production of the different secondary metabolites, which produce these species.

8. Marine bacteria as producers of secondary metabolites

The marine microscopic communities are responsible of the change in the distribution of certain chemical elements in the sea. The autonomous aptitude of the marine organisms to produce substances biologically active that possibly accumulate, modify, kidnap and use toxins of other organisms, is a test of it (Mebs, 2000). For example, the lomaiviticins *a* and *b*, substances with antitumor potential were isolated for the first time from squids, and they contain the bacteria *Micromonospora lomaivitiensis*. In later experiments, this bacterium was isolated and cultivated in fermentation reactors, to finally determine that the bacteria were the real producers of lomaiviticin (He, 2001). Marine bacteria have often been considered to produce antibacterial and anticancer substances, allowing the ecological stability of the multiple marine ecosystems, the interrelations between epiphytic microorganism's ambiences, inhibiting the rival organisms and pathogenic microbes. The sharing or competition mechanisms that are known between these microorganisms are diverse, including antibiotic production, bacteriocines, siderofores, lysosomes, proteases and even the pH alteration through the production of organic acids (Avendaño-Herrera et al., 2005).

In recent studies done in Todos Santos Bay, B.C. to bacteria associated to the seaweed surface it was found that bacteria of the family *Firmicutes, Proteobacteria* and *Actinobacteria* produce compounds capable of inhibiting the growth of HCT-116 colorectal cancer cells (Villarreal-Gómez et al., 2010). Also, it was found that the bacteria *Microbulbifer thermotolerans*, and *Pseudoalteromonas sp*, are capable of producing biofilms and produce chemical compounds that protect them from the other protozoans. An example of these compounds is violacein, an alkaloid that it is synthesized predominantly in biofilm, it has been found that in nanomolar concentrations violacein inhibits protozoan cells and induces programmed cellular death in eukaryotic cells. This bacterial producing biofilm secretes specific chemical substances for defense purposes and contribute to the persistence of these bacterial strains in different environments and provide an ecological and evolutionary context for the discovery of bacterial metabolites against eukaryotic cells (Matz et al., 2008).

Bacillus sp species have been found to possess chemical compounds with anticancer activity. Although this type of bacteria can grow in almost any substrate, it is possible to suggest that this species seems to have acquired the skill to synthesize compounds capable of inhibiting HCT-116 colorectal cancer cells (Villarreal-Gómez et al., 2010).

Selective response mechanisms exist against certain organisms, as shown in marine biofilms of *Bacteriodetes, Planctomycetes, a,c* – and *d Proteobacteria*, where the production of chemical substances as violacein has been observed. This compound works as a defense mechanism against certain specific predators like the protozoa consuming bacteria. This allows the successful persistence of the bacterial biofilm in several marine environments (Matz et al., 2008). Studies done in seaweed collected in the same coastal area, share similar defense mechanisms and inhibit the growth of Gram negative bacteria *Proteus mirabilis* and *Klebsiella pneumoniae* (Villarreal-Gómez et al., 2010), creating an interesting ecological and biotechnological role, and becoming a great subject for the search of marine natural products.

The ethanolic extracts *of Grateloupia doryphora, Ahnfeltiopsis durvillaei, Prionitis decipiens, Petalonia fascia* and *feathery* Bryopsis of the central coast of Peru, presented antibacterial effect against the clinical strains *Staphylococcus aureus* ATCC 25923 and *Enterococcus faecalis* ATCC 29212 and not clinical strain *Staphylococcus aureus* ATCC 6633. The ethanolic extract of *B. plumosa* presented the biggest antibacterial effect against two strains of *S. aureus*, being evident in their inhibition halos, while the extract of *P. fascia* showed major antibacterial activity, acting on 3 mentioned above strains (Magallanes et al., 2003)

Studies done on bacteria associated with the marine worms of the *Polychaetes* species show a strong antimicrobial activity that can be used as a potential resource for the development of new medicines (Sunjaiy-Shankar et al., 2010).

The following table shows some examples of bacterial strains with bioactivity and the sources where they were obtained. It is possible to appreciate the diseases that can be fought utilizing the secondary metabolites from different types of bacteria. This emphasizes the importance of microorganisms as an ideal source of bioactive compounds.

Bacteria	Gram (+ or -)	Activity	Target organism	Disease	Source	Bibliography
Pseudomonas bromoutilis	-	Anticancer	Staphylococcus aureus, Streptococcus pneumoniae, Streptococcus pyogenes	Pneumonia, osteitis, arthritis, endocarditis, localized abscesses	Puerto Rico	Burkholder et al., 1966
Chromobacteria marinum	-	Antibacterial	Escherichia coli, Pseudomona aureginosa, Staphylococcus aureus	Pneumonia, osteitis, arthritis, endocarditis, localized abscesses	Seawater Of North Pacific	Anderson et al., 1974
Flavobacteria uliginosum	-	Anticancer	Sarcoma-180 cells	Viral tumor	Macroalgae (Sagami Bay Japan)	Okami, 1986
Bacillus sp.	+	Anticancer	HCT-116 cells	Colorectal Cancer	Mud near the Arctic Pole	Zhang et al., 2004
Lactococcus lactis	+	Anticancer	Human papilloma virus type 16 (HPV-16)	Cervical Cancer	Designed	Bermúdez-humarán et al., 2005
Staphylococcus aureoverticillatus	+	Anticancer	Tumor cells	Tumors	Marine sediments	Blunt et al., 2005
Marinobacter ydrocarbonoclasticus	-	Antibacterial (siderofore)	Mycobacteria tuberculosis, Bacillus anthracis	Tuberculosis, carbuncle (anthrax like)	Seawater	Pfleger et al., 2008

Table 1. Microorganism's producers of bioactive substances.

9. Interactions between bacterial species

Studying the diversity of bacterial species and knowing their inter and intraspecific interactions, makes the search for secondary metabolites in bacteria easier. For many years, researchers have managed to use different culturing methods that allow them to create dense bacterial populations that yield a great amount of extracts that can be used to investigate their bioactive properties.

Not only the seaweed - bacteria interactions can influence the secretion of bioactive substances, but also the interactions that exist between bacterial species that inhabit the same ecosystem. There are different types of interactions between bacterial species and other organisms; these can be positive (metabiosis and symbiosis) or negative (parasitism, predation and competition). For their high population density, nitrogen content and their relative incapability to escape from predators, the bacteria have served as food for diverse groups of organisms (Schlegel & Jannasch, 2006).

As a rule, bacteria have been considered to be the prey of many organisms, including other bacteria species or other types of microbes. This is the case of the genes involved in producing Pilli type IV which were identified in the periplasmic strain B of *Bdellovibrio bacteriovorus* 109J and in the epibiotic strain *Bdellovibrio sp.* JSS, both Gram negative proteobacteria, which are forced predators of other Gram negative bacteria (extracellular). Using immune fluorescence microscopy the presence of the pilli was observed in the phase of cellular attack, confirming that the Pilli type IV plays a role in the invasion of other Gram negative bacteria on the part of *Bdellovibrio* (Qin et al., 2010).

Another type of interaction between the bacteria is the competition. Some bacteria are eliminated by different species when the environmental resources are limited; therefore they produce compounds that impress negatively in their competitors. Generally, these antimicrobial compounds that are produced in the environment are difficult to detect, for example the bacteria incapable of producing antibiotic compounds must reproduce very fast compared to those that do. It has been found that in actinomycetes, bacterial species that have the slowest reproduction rates are the greatest producers of antibiotic compounds (Mahmoud & Koval, 2010).

In some cases, the seaweeds dissolve organic substances that are used as nutrients by the bacteria; nevertheless some of the bacteria do not obtain nutrients this way or from any other animals. They adhere to them only for physical support and obtain their nutrients directly from the surrounding environment (Rheinheimer, 1992).

Bacteria can have interactions among themselves in order to find mutual benefits. Such is the case of *Escherichia coli* and *Proteus vulgaris*; these two species can coexist in a rich in lactose and urea medium. In this case, *Escherichia coli* degrade lactose while *Proteus vulgaris* degrades urea. The final products of these degradations can be used by another organism as a carbon and nitrogen source (Rheinheimer, 1992).

10. Techniques used for bacteria identification

To this date, several bacterial identification methodologies are known to have contributed a great deal of information for the study of their diversity. Traditionally, the studies to characterize the bacterial diversity in environments are based on the assumption that the cultivation techniques allow the recovery of most of the microorganisms in a sample [39]. Nevertheless, in the database analysis of DNA sequences, it has been observed that the bacteria obtained by standard microbiological techniques represent just a limited proportion of samples from natural sources (Escalante-Lozada et al., 2004).

It has been proposed that non-cultivable bacteria are microorganisms phylogeneticaly related to the cultivable ones, but in a physiological state that makes them recalcitrant. The explanation is that some cultivable bacteria can turn in viable, but not cultivable when exposed to adverse environmental conditions and reverse to the cultivable state after the favorable conditions for their growth are restored (McDougald et al., 1998).

To study bacterial diversity, it is important to use molecular techniques that include the amplification of the 16S ribosomal gene using the polymerase chain reaction technique (PCR) in order to isolate and characterize their genetic material (Prieto-Davó et al., 2008).

11. Microbial diversity cultivation dependent techniques

To go as far as knowing all the chemical compounds produced by bacteria extracted from different environmental samples, their isolation and cultivation, is necessary to prepare organic extracts that could be evaluated chemically and biologically. The techniques used are generally known as microbial diversity cultivation dependent techniques or traditional methods of cultivation (Joint et al., 2010). These techniques are based on the need that the microorganisms have of taking from the environment a series of compounds that are used as energy source and synthesizing the cellular constituents necessary for their survival, like C, N, S, P, Ca, K, Na, Mg, Mo, Cu and Zn. Some microorganisms need to take the light energy or to oxidize chemical compounds; others need to obtain their carbon source from CO_2 or through carbonated organic compounds. Some microorganisms need a nitrogen source, for which they fix the atmospheric nitrogen, or take it from $NH_4{}^+$. Others are capable of reducing $NO_3{}^-$ or $NO_2{}^-$ to use it as an ammonium source, or, to use free amino acids. All these characteristics, as well as temperature, pH and salinity, have demonstrated to create a propitious ambient, for the isolation and purification of microorganisms originated from environmental samples (Rheinheimer, 1992).

Therefore, the collected environmental samples are exposed to different conditions and forms of cultivation that help to obtaining microorganisms capable of adapting to the established conditions. This allows to study the microorganism's diversity, as well as to cultivate them from different environmental samples.

12. Molecular techniques for bacteria identification

The molecular characterization of bacteria has had an enormous impact on the safety of microbiological industrial processes like the bio pharmacology and food industry as well as public health, since it establishes the source of contamination in a much more precise form and therefore to identify the strains involved in the process and to establish their trajectory.

The most used gene for bacterial molecular identification is the 16S ribosomal RNA gene, nevertheless, there have also been used other genes that can codify for 5S ribosomal proteins, and 23 S ribosomal RNA.

The ribosomes and ribonucleic acids (ribosomal RNA) are proteic complexes with the only function of synthesizing proteins. The ribosomes structure is preserved between the different life kingdoms (Plantae, fungi, animaleae, etc.). In the prokaryotes, the ribosomes are composed by two subunits: a big 50S, and small 30S subunits. The subunit 50S contains a 23S, one 5S rRNA and more than 30 different proteins. The subunit 30S contains one 16S rRNA and 20 additional proteins. Since the rRNA gene function is limited by the structure, certain regions in the rRNA genes that are in contact with other components in the ribosome must be preserved;

the sequences between the preserved regions have major mutation valuations. The preserved regions are useful to determine distant relations (genus), while the regions with higher mutation rate or variable regions are useful to distinguish organism closely related (species) (Eickbush & Eickbush, 2007). These characteristics of rRNA genes are excellent molecular chronometers for phylogenetic analysis and taxonomic classification of cellular organisms.

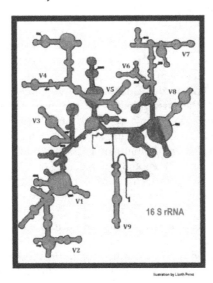

Fig. 2. Scheme of Bacterial 16S ribosomal RNA that shows the variable secuences (V1 to V9). The variable regions area used to characterized bacteria by species level, and the constant region is used to correlated the genus.

In general, the phylogenetic trees based on the 16S and 23S rRNA genes can be used in parallel (De Rijk et al., 1995), while the 5S rRNA gene it is considered not to contain sufficiently long sequences to do significant statistical comparisons. The detailed phylogenetic studies based on the genetic sequences 16S and 23S provide comparisons in three primary domains Bacteria, Archaea and Eukarya (Woese et al., 1990). Nevertheless, one of the disadvantages of the 23S rRNA gene on 16S rRNA in the phylogenetic studies is the rare use of the taxonomic classification for the absence of primers that amplify a considerable length sequence and the difficulty of genes sequences too long for the current technologies.

Previously the 16S rRNA and 23S rRNA genes have been compared and the results show that the 16S rRNA gene has more closely related typical sequences, with major length, insertions and/or deletions and possibly a better phylogenetic resolution for their high genetic change (Blunt et al., 2005). Nevertheless, recent studies indicated that the 23S rRNA gene also contains conserved regions for the design of primers capable of amplifying a wider length with one grade similar to the primers used in the of 16S rRNA genes (Hunt et al., 2006; Pei et al., 2009).

Molecular methods of characterization based on the amplification of fragments belonging to the 16S ribosomal RNA gene through automated techniques constitute a rapid, trustworthy and simple method of molecular genotyped bacterial and fungoid strains. The aptitude to identify microorganisms at species level and at the same time to establish a comparative analysis of the different analyzed strains constitutes a time saving way to identify the food pollutant potential and therefore to eradicate the pollutant focus or to facilitate the recognition of the producing strains of some bioactive metabolite capable of inhibiting the growth of other bacterial strains or carcinogenic cells.

The 16 S ribosomal gene (16S rRNA) is constituted by a region preserved through evolution, the mutations in this gene can usually be tolerated, since these mutations would only affect ribosomal RNA, nevertheless, the number of mutations are not completely well-known, the regions that are affected by them are met like "hot commercials" which present a considerable number of mutations, these areas are not the same in all species. The 16S rRNA gene mutations can affect the susceptibility to antimicrobial agents which can be an indicator to distinguish the phenotypic resistance to these agents. Nevertheless these characteristics do not affect the gene use for taxonomical identification at the genus and species level. 16S rRNA possesses a length of 1, 550 bp and it is composed by variable regions and conserved regions, with enough interspecific polymorphisms to provide a valid statistical characterization. The primers that are usually designed for the amplification of this gene, are based on the complementary chain of the conserved regions in the beginning of the gene at about 540 bp or at the end of the sequence   around 1,550 bp and the sequence of the variable region is used for taxonomic comparison intentions, making 1500bp the minimum length that should be used to compare DNA sequences. The 16S rRNA gene sequence has been determined for a big number of strains; these sequences are included in the biggest database of nucleotides known as GenBank. This database contains more than 20 million sequences of which more than 90, 000 correspond to the 16S rRNA gene. . In general, the comparison of the 16S rRNA gene sequences allows the differentiation between organisms at genus level in most bacteria phyla, also classifies the strains at multiple levels, including species and sub species (Clarridge III, 2004).

13. Extraction of bioactive compounds from bacteria associated with marine algae

To do a search of bioactive compounds from marine algae the following protocol is proposed:

a.   Isolation of the bacterial strains from the seaweed surface

After being collected, the seaweed is placed in an Erlenmeyer flask and is rinsed with distilled water; a small sample from the flask is taken with a sterile swab and is inoculated in a general media that allows the growth of most of the bacteria present in the seaweed surface. Then the media with inoculate is incubated for periods of 24 to 48 hours at 25 °C, until the developing colonies start to emerge. Later the bacteria will be purified up to the third generation to assure the integrity of pure colonies and finally it is important to take one of the colonies of every purified strain and cryopreserved it in 15 % glycerol at -70 °C.

b.   Macroscopic morphology and Gram Stain

The pure colonies are examined macroscopically evaluating their characteristics such as size, form, elevation, margin, color, type of surface, thickness, consistency, smell and pigments production. Consecutively microscopic classification of the bacterial strains, such as the Gram stain, is necessary (Gram, 1884). The Gram stain is a technique of bacterial characterization based on the chemical composition of the cellular wall of the bacteria. The Gram positive bacteria present a cytoplasmic membrane, have a thick peptidoglycan layer, contain teicoics acids and lipoteicoics that serve as chelating agents and certain type of adhesions; the bacteria representative of this group are *Firmicutes* and *Actinobacteria*, which includes many   well-known genus like *Bacillus, Listeria, Staphylococcus, Streptococcus, Enterococcus*, and *Clostridium.*

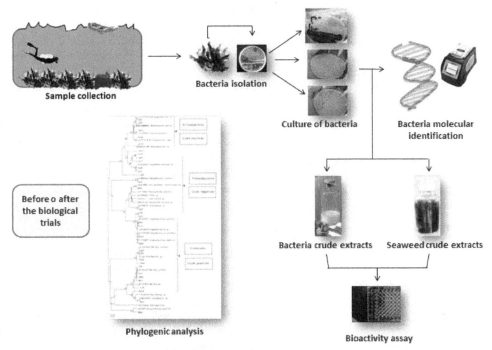

Fig. 3. Proposed methodology for the search of new bioactive compounds in bacteria associated with the surface of marine algae.

On the other hand, the Gram negative bacteria contain a thin cellular wall of peptidoglycan and an external membrane that covers their cellular wall. The external membrane contains diverse proteins, like purines or channels protein that allow the path of certain substances. Also they present a structure of lipopolysaccharides (LPS). The bacteria that predominate in this group are Proteobacteria, including to Escherichia coli, Salmonella and other enteric bacteria like Pseudomonas, Moraxella, Helicobacter, Stenotrophomonas, Bdellovibrio, acetic acid bacteria, Legionnaire's disease and the proteobacteria alpha like Wolbachia among others (Madigan et al., 2005).

Previous studies done with bacterioplacton demonstrated that most of the marine bacteria are Gram negative; but in recent studies with marine sediments evidence has shown that most of the bacteria that conform the marine sediments seem to be Gram positive (Gontang et al., 2007).

In studies made to the surface of the marine alga Monostroma undulatum (Gallardo et al., 2004) Gram negative bacteria belonging to the genus: Vibrio (20 %), inactive E. coli (18 %), Flavobacteria (11 %), Flexibacter (9 %), Moraxella (9 %), Pseudomona (9 %), Aeromonas (2 %), Acinetobacter (2 %), Cotophaga (2 %), Photobacteria (2 %) and Alteromonas (2 %); predominated and only one Gram positive was found, Staphylococcus.

c. Phylogenetic analysis

Since bacterial identification was mentioned previously and bacterial characterization with molecular techniques is crucial for the achievement of phylogenetic studies that allow to have an account of the cultivable bacterial species that are present in a similar community, there are several investigators who agree with the idea, that bacteria with closely related DNA sequences produce very similar compounds. If this is true, just by knowing the bacterial ecology of a certain area it will be possible to predict the types of compounds that could be isolated from the bacteria present in that particular area.

The DNA sequences can be analyzed using the BLAST database (Basic Local Alignment Search Tool) that is a GeneBank database integral function (Altschul et al., 1999). To align the existing sequences between themselves it is possible to use the Clustal X (Staley & Ta, 1985) and Bioedit programs (Hall & Brown, 2001) for manual alignments. For the phylogenetic tree construction it is possible to use the following indications: Bootstrap test of phylogeny (1000 repetitions), p-distance joining neighbor, using the MEGA4 program (Tamura et al., 2007) using segment sequences of up to 1500 bp.

d. Biological assays and mean lethal dose DL_{50}

A bioassay can be defined as any test that involves living organisms. With them, it is possible to evaluate the effects of any substances or material in terms of the biological answer they produce. The main target in this type of analysis is to evaluate the level of stimulus that is necessary to obtain a response in a group of individuals of a population. The level of stimulus that causes response in 50 % of the individuals under study is an important parameter of characterization denoted like average lethal dose (DL_{50}). The amount of time during which the stimulus is exhibited, must be specified, for example, 24 hours DL_{50}, this in order to compare and to estimate the relative potency of the stimulus. For the DL_{50}determination of the first step is securing the % of survival cell for every analyte and every target, the mortalities is corrected by Abbott's formula (Abbott, 1925).

$$M = \frac{me - mb}{1 - mb}$$

Where:
M = Mortality.
me = optical density in the extract.
mb = optical density in the target.

With the mortalities corrected, the obtained information is introduced in the software "BioStat 2008" (http://www.analystsoft.com) using Probit analysis to obtain the DL_{50}values (STATPLUS, 2008).

14. Conclusion

Natural products are a very important resource for the elaboration of medicines. Although a big number of plants, microbes and marine resources have been evaluated in the search of new bioactive compounds, it turns out to be insufficient and it is necessary and important to continue with the search of new secondary metabolites, especially those that are endophytes microorganisms of seaweed. The bad use given to antibiotics has resulted in the development of bacteria strains that are resistant to many of the known drugs. This situation has lead to a forced search for new antibiotic compounds, being the seabed a propitious site for exploration and future drug development. Also, the treatment with chemotherapy for the diverse causes of different types of cancer that at present today, appears effective, so the investigation becomes necessary in the chemistry of the natural products. The methods of bacterial culture and identification have become very promising especially, those done through molecular techniques, by which is possible to identify a strain up to species and sometimes at subspecies level. The diverse relationships that exist between microorganisms and their guests provoke that bacterial compounds can eventually be used as a source of new drugs for human well-being.

a. Future work in Drug Discovery

The strategies for drug discovery has been evolving constantly, today researchers do not conform with the finding of new and potent metabolites, now and for future days it has become important to do phylogenetic studies, structure elucidation   of chemical compounds, bioinformatics approaches, genomics, proteomics, reverse pharmacology and so on., One fundamental field that has to be developed is the improvement of more efficient culture media, because we need to culture all the strains of bacteria to be able to evaluate and separate its compounds, in the meanwhile, we can identify the genes of cultivable and non-cultivable bacteria by molecular techniques to compare and try to demonstrate that the strains of bacteria with very similar DNA sequences have equally similar metabolites and culture requirements, if this is true, before we start screening for drug discovery purposes, researchers must do phylogenetic studies of bacterial population of a certain determinate area and decide which strains cultivate and which not, these strategies will make the drug discovery process less expensive and faster.

Methodological improvements studies based on the characterization of the extracellular polymeric substance produced by marine microorganisms and a better understanding of host-microbe interactions, should be us to provide further insight into the adaptive strategies against microbial pathogens and establish the extent to which secondary metabolites regulate microbial interactions.

The new soft ionization methods: Matrix-Assisted Laser Desorption Ionization (MALDI) and Electrospray Ionization (EI) are the recent approach used in a variety of new and innovative Mass Spectrometric (MS) applications. With them, is easier to analyze surfaces, they are tolerant to impurities and do not require extensive sample preparations. A sensitive and precise Mass Spectrometric approach like Desorption Electrospray Ionization (DESI) should be used to measure the physical location and quantities of natural products on biological tissue surfaces, cells or even complex mixed-species assemblages. These MS imaging techniques known as "molecular eyes" are very precise and represent the last technological advance used to locate natural products in biological tissues (Esquenazi et al., 2009) allowing the study of the interface between the confluence of natural products chemistry, biology and ecology.

Bioinformatics is the part of molecular biology that involves working with biological data, typically using computers, with the goal of enabling and accelerating biological research. Bioinformatics comprises a wide range of activities: data capture, automated recording of experimental results; data storage and access, using a multitude of databases and query tools; data analysis; and visualization of raw data and analytical results (Pollock & Safer, 2001).

Today, many recently developed or discovered drugs with antibacterial and anticancer activity fail in clinical trials because of inefficiency for the anticipated indication or unexpected toxicity (Kola & Landis, 2004). Apparently, it remains hard to establish a clear link between antagonism or organism of a specific target and its influence in human illness and its target associated toxicity.

A significant cause for these high attrition rates is the often misjudged complexity of protein function in higher order organisms, in which, abundant protein-protein interactions, feedback loops and redundancies play a role. The collection of recognized pathways that can be found in public databases and commercial tools do not effectively address these issues because they are mainly a reflection of experimental data that are obtained from isolated cell lines and tissues. They address typically, the signaling events that lead to binding of transcription factors to the DNA, but do not detail the pleiotropic effects that arise downstream from the induced transcriptional program, which are most important in provoking the systems response to the signaling events and may determine, the capacity and toxicity of a drug (Pollock & Safer, 2001).

Most comparative genomics tools are intended at studying conservation of single genes or gene families, whereas computationally tools address orthologous biology, i.e. conservation of the entire pathways in which the target is involved, are unusual. This truly obstructs the output and success of translational investigation from pre-clinical to clinical studies (Pollock & Safer, 2001).

The developing of bioinformatics tools that addresses the above problems will allow for quicker and better experiments aimed at evaluating multiple targets and drugs for further clinical development. This will be a first step to reduce the high attrition rates associated with drug development (Pollock & Safer, 2001).

Clinical events or phenomena not reported previously following the administration of a known or new drug can offer valuable perceptions for drug development. Natural products have provided many such unexpected bedside interpretations. Researches in genomics, proteomics and metabolomics have stimulated the discovery of many new molecules, which are yet to be tracked for their drug-like activities. A new discipline called Reverse Pharmacology (RP) has been designed to decrease costs, time and toxicity.

The scope of reverse pharmacology is to understand the mechanisms of action at multiple levels of biology and to optimize safety, efficacy and acceptability of the leads in natural products based on relevant science in this approach, as the candidate travels a reverse path from 'clinics to laboratory' rather than classical "laboratory to clinics". Actual humans are used as the ultimate model and in-depth investigation of the effects of drugs and the nature of disease progression is becoming ever more feasible because of advances in clinical biomarkers and systems biology. This articulates both structure of the system and components to play indispensable role forming symbiotic state of the whole system (Patwardhan et al., 2008).

15. Acknowledgment

The authors thank A. Prieto-Davo and P.R Jensen (CMBB,U.C.S.D.) for their collaboration, A. Licea for his help in carrying out the bioassays (CICESE),M. Ritchie for facilitating the bacterial pathogens used, R. Aguilar (†) for assistance in the seaweed identification , A. Moreno and L. Pérez for photography and Illustration, respectively and A.M. Iñiguez-Martínez for reviewing the manuscript .

16. References

Abbott, W.S. (1925). A method for computing the effectiveness of an insecticide. *Journal of Economic Entomology*.18:265-7

Altschul, S.F.; Gish, W.; Miller, W.; Myers, E.W. & Lipman, D.J. (1990). Basic local alignment search tool. *Journal of Molecular Biology*. 215: 403-10

Anderson, R.J.; Wolfe, M.S. & Faulkner, D.J. (1974). Autotoxic antibiotic production by a marine Chromobacterium. *Marine Biology*. 27 (4): 281-285

Avendaño-Herrera, R.; Lody, M. & Riquelme, C.E. (2005). Producción de substancias inhibitorias entre bacterias de biopelículas en substratos marinos. *Revista Biología Marina y Oceanografía*. 40 (2)

Berdy, J. (2005). Bioactive microbial metabolites. A personal view. *Journal of Antibiotics*. 45: 581-26, and 43: 57-90

Bermúdez-Humarán, L.G.; Cortes-Pérez, N.G.; Lefevre, F.; Guimarães, V.; Rabot, S.; Alcocer-Gonzalez, J.M.; Gratadoux, J.J.; Rodriguez-Padilla, C.; Tamez-Guerra, R.S.; Corthier, G.; Gruss, A. & Langella, P. (2005). A Novel Mucosal Vaccine Based on Live *Lactococci* Expressing E7 Antigen and IL-12 Induces Systemic and Mucosal Immune Responses and Protects Mice against Human *Papillomavirus* Type 16-Induced Tumors. *The Journal of Immunology*. 175: 7297–7302

Blunt, J.W.; Copp B.R.; Munro, M.H.G.; Northcote, P.T. & Prinsep, M.R. (2006). Marine natural products. *Natural Products Report*. 23: 26–78

Blunt, J.W.; Copp B.R.; Munro, M.H.G.; Northcote, P.T. & Prinsep, M.R. (2005). Marine natural products. *Natural Products Report*. 22: 15-61

Burkholder, P.R.; Pfister, R.M. & Leitz, F.H. (1966). Production of a Pyrrole Antibiotic by a marine bacteria. *Applied Microbiology*. 14:649-653

Clarridge III, J.E. (2004). Impact of 16S rRNA Gene Sequence Analysis for Identification of Bacteria on Clinical Microbiology and Infectious Diseases. *Clinical Microbiology Reviews*. 17 (4). 840–862

Cragg, G.M.; Newman, D.J. (1999). Discovery & development of antineoplastic agents from natural sources. *Cancer Investigation*. 17: 153–163

De Rijk, P.; Van de Peer. Y.; Van den Broeck, I. & De Wachter, R. (1995). Evolution according to large ribosomal subunit RNA. *Journal of Molecular Evolution*. 41: 366–375

Egan, S.; Thomas, T. & Kjelleberg, S. (2008). Unlocking the diversity and biotechnological potential of marine surface associated microbial communities. *Current Opinion in Microbiology*. 11 : 219-225

Eickbush, T.H. & Eickbush, D.G. (2007). Finely orchestrated movements: evolution of the ribosomal RNA genes. *Genetics*. 175: 4777-485

Escalante-lozada, A.; Gosset-Lagarda, G.; Martínez-Jiménez A. y F.; Bolívar-Zapata, (2004). Diversidad bacteriana del suelo: Métodos de estudio no dependientes de cultivo microbiano e implicaciones biotecnológicas. Departamento de Ingeniería Celular y Biocatálisis. Instituto de Biotecnología. UNAM. *Agrociencia*. 38: 583-592

Esquenazi, E.; Dorrestein, P.C. & Gerwick, W.H. (2009). Probing marine natural product defenses with DESI-imaging mass spectrometry. *Proceedings of the National Academy of Sciences*. 6: 7269-7270

Faulkner, D.J. (2002). Marine natural products. *Natural Products Reports*. 19: 1–48

Gallardo, A.; Risso, S.; Fajardo, M. & Estevao, B. (2004). Characterization of microbial population present in the edible seaweed, Monostroma undulatum, Wittrock. Departamento de Bioquímica, Facultad de Ciencias Naturales, Universidad Nacional de Patagonia San Juan Bosco, Comodoro Rivadavia, Chubut, Argentina. *Archivos Latinoamericanos de Nutrición*. 54(3):337-45

Gontang, E.A.; Fenical, W. & Jensen, P.R. (2007). Phylogenetic diversity of gram-positive bacteria cultured from marine sediments. *Applied and environmental microbiology*. 73: 3273-82

Gram, C. (1884). The differential staining of Schizomycetes in tissue sections and in dried preparations. *Fortschitte der medicine*. 2:185-189

Hall, T.A. & Brown, J.W. (2001). The ribonuclease P family. North Carolina State University, Department of Microbiology. *Methods Enzymology*. 341: 56-77

He, H. (2001). Lomaiviticins A and B, potent antitumor antibiotics from *Micromonospora lomaivitiensis.Journal of the American Chemical Society*. 123: 5362-5363

Hunt, D.E.; Klepac-Ceraj, V.; Acinas, S.G.; Gautier, C. & Bertilsson, S. (2006). Evaluation of 23S rRNA PCR primers for use in phylogenetic studies of bacterial diversity. *Applied Enviromental Microbiology*. 72: 2221–2225

Isnard-Bagnis, C.; Moulin, B.; Launay-Vacher, V.; Izzedine, H.; Tostivint, I. & Deray G. (2005). Anticancer drug-induced nephrotoxicity. *Nephrology Therapy*. 1: 101–114

Jasti, S.; Sieracki, M.E.; Poulton, N.; Giewat, M.W. & J.N.; Rooney-Varga. (2005). Phylogenetic diversity and specificity of bacteria closely associated with *Alexandrium spp.* and other phytoplankton. *Applied Environmental Microbiology*. 71 (7): 3483-3494

John Davis, G.D. & Vasanthi, A.H.R. (2011). Seaweed metabolite database (SWMD): A database of natural compounds from marine algae. *Bioinformation*. 5(8). www.bioinformation.net

Joint, I.; Mühling, M. & Querellou, J. (2010). Culturing marine bacteria: an essential prerequisite for biodiscovery. *Microbial Biotechnology*. 3(5): 564-575

Kandhasamy, M. & Arunachalam, K.D. (2008). Evaluation of in vitro antibacterial property of seaweeds of southeast coast of India. *African Journal of Biotechnology*. 7 (12): 1958-1961

Kola, I. & Landis, J. (2004). Can the pharmaceutical industry reduce attrition rates?.*Nature Reviews Drug Discovery*. 3: 711-715

Kovacs, G.; Burghardt, J.; Pradella, S.; Schumann, P.; Stackebrandt, E. & Marialigeti, K. (1999). *Kocuria palustris sp. nov.* and *Kocuriarhizophila sp. nov.*; isolated from the rhizoplane of the narrow-leaved cattail (*Typhaangustifolia*). *International journal of systematic bacteriology*. 49: 167-173

Madigan, M.; Martinko, J. & Parker, J. (2005). *Brock Biology of Microorganisms*, Decimoprimeraedición, Prentice Hall. Southern Illinois University Carbondale. 1: 102-167

Magallanes, C.; Córdova, C. & Orozco, R. (2003). Antibacterial activity of ethanolic extracts of marine algae from central coast of Peru. *Revista peruana de biologia*. 10(2): 125- 132

Mahmoud, K.K. & Koval, S.F. (2010). Characterization of type IV pili in the life cycle of the predator bacteria *Bdellovibrio*. *Microbiology*. 156:1040

Manilal, S.; Sujith, S.; SeghalKiran, G.; Selvin, J.; Shakir, C.; Gandhimathi, R. & NatarajaPanikkar, M.V. (2009). Biopotentials of seaweeds collected from southwest coast of India. *Journal of Marine Science and Technology*. 17 (1): 67-73

Matz, C.; Webb, J.S.; Schupp, P.J.; Phang, S.Y.; Penesyan, A.; Egan, S.; Steinberg, P. & Kjelleberg, S. (2008). Marine Biofilm Bacteria Evade Eukaryotic Predation by Targeted Chemical Defense. School of Biological, Earth and Environmental Sciences and Centre for Marine Bio-Innovation, University of New South Wales, Sydney, Australia. *PLoSONE*. 3(7)

McDougald, D.; Rice, S.; Weichart, D. & Kjelleberg, S. (1998). Nonculturability: adaptation or debilitation. *FEMS Microbiology Ecology*. 25: 1-9

Mebs, D. (2000). Toxicity in animals- Trends in evolution? Toxicon. Elsevier Science, Ireland 39: 87-96

Muir, R.E.; & Tan, M.W. (2008). Virulence of *Leucobacterchromiireducens* subsp. solipictus to *Caenorhabditiselegans*: Characterization of a Novel Host-Pathogen Interaction. *Applied Environmental Microbiology*. 74(13): 4185–4198

Okami, Y. (1986). Marine microorganisms as a source of bioactive agents. *MicrobialEcology*. 12:65–78

Organización Mundial de la Salud (OMS). (2011). Centro de prensa. Nota descriptiva 194. http://www.who.int/mediacentre/factsheets/fs194/es/index.html

Patwardhan, B.; Vaidya, A.D.B.; Chorghade, M. & Joshi, S.P. (2008). Reverse Pharmacology and Systems Approaches for Drug Discovery and Development. *Current Bioactive Compounds*. 4, 000-000

Pei, A.; Nossa, C.W.; Chokshi, P.; Blaser, M.J. & Yang, L. (2009). Diversity of 23S rRNA Genes within Individual Prokaryotic Genomes. *PLoS ONE 4*. (5): e5437.doi:10.1371/journal.pone.0005437

Penesyan, A.; Kjelleberg, S. & Egan, S. (2010). Development of novel drugs from marine surface associated microorganisms, *Marine Drugs*. 8: 438-459

Pérez–Matos, A. E.; Rosado, W.; Govind, N.S. (2007). Bacterial diversity associated with the Caribbean *Tunicate Ecteinascidia* turbinate. *Antonie Van Leeuwenhoek*. 92: 155-164

Pfleger, B.; Sherman, D.; Nusca, T.; Scaglione, J. & Lee, Y. (2008). Petrobactin Biosynthesis: A Target for Antibiotics and a Platform for Producing Specialty Chemicals. *Chemical and Biological Engineering*, University of Wisconsin Madison, Madison, WI. Life Sciences Institute, University of Michigan, Ann Arbor

Prieto-Davó, A.; Fenical, W. & Jensen, P.R. (2008). Comparative actinomycete diversity in marine sediments. *Aquatic Microbial Ecology.* 52: 1–11

Pollock, S. & Safer, H. M. (2001). Bioinformatics in the Drug Discovery Process. *Annual Reports in Medicinal Chemistry*

Qin, S.; Xing, K.; Jiang, J.; Xu, L. & Li, W. (2010). Biodiversity, bioactive natural products and biotechnological potential of plant-associated endophytic actinobacteria. *Applied Microbiology and Biotechnology.* 89 (3): 457-473

Rheinheimer, G. (1992). *Aquatic microbiology.* 4th ed. John Wiley & Sons. West Sussex. United Kingdom

Salvador, N.; Gómez-Garreta, A.; Lavelli, L. & Ribera, L. (2007). Antimicrobial activity of Iberian macroalgae. *Science Marine.* 71: 101-113

Schlegel, H.G. & Jannasch, H.W. (2006). Prokaryotes and Their Habitats. *Prokaryotes.* Chapter 1.6. 1:137–184

Staley, J. & Ta, A.K. (1985). Measurement of in situ activities of non photosynthetic microorganisms in aquatic and terrestrial habitats. *Annual Reviews in Microbiology.* 39: 321-46

STATPLUS. (2008). *Probit analysis.* Biostat 2008. http://www.analystsoft.com

Sunjaiy-Shankar, C.V. Jeba- Malar, A.H. & Punitha, S.M.J. (2010).Antimicrobial activity of marine bacteria associated with *Polychaetes.* *Bioresearch Bulletin.* 1: 24-28

Tamura, K.; Dudley, J.; Nei, M. & Kumar, S. (2007). MEGA4: Molecular Evolutionary Genetics Analysis (MEGA) software version 4.0. *Molecular Biology and Evolution*

Teas, J. (2007). Seaweed and Soy: Companion Foods in Asian Cuisine and Their Effects on Thyroid Function in American Women. *Journal Medicinal Food.* 10: 90-100

Villarreal-Gómez, L.J.; Soria-Mercado, I.E.; Guerra-Rivas, G. & Ayala-Sánchez, N.E. (2010). Antibacterial and anticancer activity of seaweeds and bacteria associated with their surface. *Revista de Biología Marina y Oceanografía.* 45 (2): 267-275

Webster, N.S. & Bourne, D. (2007). Bacterial community structure associated with the Antarctic soft coral, *Alcyoniumantarcticum.* *FEMS Microbiology Ecology.* 59: 81-94

Woese, C.R.; Kandler, O. & Wheelis, M.L. (1990). Towards a natural system of organisms: proposal for the domains Archaea, Bacteria, and Eucarya. *Proceedings of the National Academy of Sciences of the United States of America.* 87: 4576–4579

Zhang, H. L.; Hua, H. M.; Pei, Y. H. & Yao, X.S. (2004). Anticancer activity and mechanism of *Scutellaria barbata* extract on human lung cancer cell line A549. *Chemistry Pharmacology Bulletin.* 52: 1029

Zhuang, W.; Tay, J.; Maszenan, A.; Krumholz, L. & Tay, S. (2003). Importance of Gram-positive naphthalene-degrading bacteria in oil-contaminated tropical marine sediments. *Lett Applied Microbiology.* 36 (4): 251-7

3

The Effect of CLA on Obesity of Rats: Meta-Analysis

Sejeong Kook and Kiheon Choi[1]
Duksung Women's University, Seoul,
South Korea

1. Introduction

Obesity is the noticeably fat mass, and an excessive nutrient intake is a cause of that. When the energy in body exceeds healthy limit, the much energy is stored as fat or in fatty tissue. Obesity was accelerated by eating fatty foods, insufficient exercise, and accumulation of stress.

Obesity has come to be recognized as a critical global health issue. Rates of obesity in North America and in most European countries have more than doubled in the last 20 years and over half the adult population are now either overweight or obese (1). According to the report which was released by Korea National Health Insurance Corporation (NHIC) in 2004, the obesity rate for subjects of the survey in 2003 was 56%, and increased in all age groups from 10s to 60s. In addition, the study by Oh *et al.* in 2005 reported that obesity was strongly associated with risk of some cancers, such as skin, liver, large intestine, thyroid gland, biliary tract, and uterus. It is well known that obesity is one of the causes that generate heart problems, diabetes, and other disorders (2).

Conjugated linoleic acid (CLA) is one of trans fatty acids and found mainly in the meat and dairy products of ruminants (3). CLA has received attention for its antioxidant and anti-cancer properties (4). Studies of CLA show that CLA supplementation tends to reduce body fat, improve serum lipid profiles, and decrease whole-body glucose uptake. However, the results of some studies on rats suggest that CLA supplement was not effective in reducing the fat accumulation (5). Though studies done on CLA have increased, they do not all show uniform results, showing it is necessary to summarize and analyze them.

Meta-analysis is a tool for summarizing the results of studies with related research hypotheses. It has three steps; deciding the association measure for detecting difference between groups, summarizing the association measure in assumed model, and identifying publication bias. The association measures in meta-analysis are the measure of related effects from the primary research such as the standardized mean difference, the mean difference, the risk difference, and an odds ratio. The decided measure is combined using only within variation of studies on the assumption that the results are homogeneity and

---

[1]Corresponding Author, Professor, Department of Statistics

uniformly distributed, and then it is tested on whether the homogeneity assumption is plausible. If the results are heterogeneity, the measure is recombined using within variation with the estimated the between variations of the studies. The former model is called the fixed effect model and the latter is the random effect model. Because significant results are more likely to be submitted or accepted than insignificant ones and the combined estimates are calculated by only published researches, the bias easily occurs in a small number of studies and is called the publication bias. If the publication bias is doubtable, the combined estimate is adjusted or the additional studies are required (6, 7).

The purpose of this study was to summarize the studies about effect of CLA on factors related to the anti-obesity in experimental rats by meta-analysis.

2. Method

2.1 Preparation of dataset for meta-analysis

The studies used in this meta-analysis were searched on the ScienceDirect in English, the DBpia database, and the KISS (Koreanstudies Information Service System) in Korean. The search keywords used were CLA, weight, health, or fat and the research was limited to the experimental rat studies. About 50 studies were collected and 12 studies were finally selected after omitting some studies with insufficient information, such as no sample variance or no sample size (5, 8-18).

The factors used to investigate the effect of CLA were collected from target studies, if it was studied in at least 2 studies. The selected factors were body fat (%), epididymal fat (g/100g), fat cell size (μm), final body weight (g), food intake (g/d), leptin (pg/ml), liver-TC (mg/g), liver-TG (mg/g), plasma-TC (mg/dl), plasma-TG (mg/dl), and weight gain (g/d). The unit of each factor was uniformly changed.

 In each study, there were two groups. One group of rats was treated with fat source such as beef tallow, coconut oil, corn oil, fish oil, safflower oil or soybean oil (FR), and the other group of rats was treated with fat source supplemented with CLA (FRC). Because the size of studies after grouping by treated fat source was not enough large, the each fat source has similar effect on obesity.

2.2 Measure and models for combining

Since the mean difference between FR and FRC was used for test on anti-obesity effect of CLA, the association measure was decided as the mean difference (MD) of each factor; mean of FRC minus mean of FR. The combined MD of each factor was calculated by the inverse variance method and the mean weighted by inverse variance in the primary studies. If the homogeneity was accepted by Cochran's Q test, it is assumed that the effect measured in the study population has a single value. The association measure is estimated by using a variation within the studies in the fixed effect model. To reduce the bias caused by heterogeneity, meta-regression models are used to analyze association between treatment effects and study characteristics. The period and amount of CLA supplementation were considered as covariates, and the estimated coefficients of covariates were investigated for possible sources of heterogeneity. If the homogeneity was rejected, the combined MDs of studies were calculated in random effect model in which total variation defined the variation within studies with estimated variation between studies.

2.3 Identification the publication bias

The existence of publication bias was checked by using a funnel plot or Egger's linear regression test. The funnel plot was a scatter plot of the association measure against a inverse of standard error of measure. If the shape of funnel plot about any factor is not funnel or cone around the combined MD, the publication bias was doubtable. Egger's linear regression test was used to test the null hypothesis that the funnel plot was not asymmetry. Egger's linear regression is a linear regression of standard normal deviate (defined as association measure over SE) against the inverse of SE, and there may be publication bias if the estimated intercept is significantly different from 0. A positive intercept indicates that more studies are associated with a bigger effect (19).

2.4 Software used for the meta-analysis

Version 1.5 of MIX program was used for combining MDs, plotting the funnel plots and checking the publication bias. The STATA program was used for meta-regression.

3. Result

3.1 Combined mean difference and homogeneity test of studies (Fat/Weight/Plasma/Liver)

The combined MDs between FR and FRC were presented with p-value to test heterogeneity of results in Table 1. In the fixed effect model, the MDs were significantly different from 0 for body fat, final body weight, liver-TC, plasma-TC, plasma-TG and weight gain (<0.05). Especially the p-values tested for body fat, final body weight, and weight gain were less than 0.01, showing CLA significantly decreased the level of each of them.

	Fixed effect model [a)]		Heterogeneity [b)]	Random effect model [a)]	
	estimate	p-value [c)]	p-value	estimate	p-value [c)]
Body fat(%)	-1.2504	<0.0001	<0.0001	-1.1239	0.1020
Epididymal fat (g/100g)	-0.0131	0.6988	<0.0001	-0.0470	0.4512
Fat cell size(μm)	-0.6088	0.4707	0.4213	-0.6088	0.4770
Final body weight(g)	-3.5922	0.0004	<0.0001	-3.3114	0.3941
Food intake(g/d)	-0.1639	0.1190	0.1829	-0.2235	0.2634
Leptine(pg/ml)	-6.6161	0.7563	0.0696	33.5415	0.4738
Liver-TC(mg/g)	-0.1859	0.0352	0.0146	-0.5187	0.2483
Liver-TG(mg/g)	-0.2007	0.2689	<0.0001	-0.8575	0.2539
Plasma-TC(mg/dl)	-11.5280	<0.0001	0.0087	-15.0920	0.0009
Plasma-TG(mg/dl)	-3.2202	0.0446	0.0730	-4.0647	0.1287
Weight gain(g/d)	-0.5377	<0.0001	<0.0001	-0.1561	0.5062

Table 1. Combined MDs and Homogeneity test in fixed effect model and random effect model. [a)]

The fixed effect model was inadequate in terms of body fat, epididymal fat, final body weight, liver-TC, liver-TG, plasma-TC, plasma-TG and weight gain, because of the heterogeneity of studies for them (p-value<0.05). The MD about heterogeneity factor was regressed against two covariates, amount and period of CLA supplementation, by meta-regression method in Table 2. Only the MD of the final body weight was significantly correlated with the period of CLA supplementation (p-value <0.05), and the coefficient of period was estimated -0.0698. That meant the MD of the final body weight was estimated to decrease by 0.0698 per unit increase in the period.

The random effect model was used to combine MD on heterogeneity factors; body fat, epididymal fat, final body weight, liver-TC, liver-TG, plasma-TC, plasma-TG, and weight gain. The value of the MD of plasma-TC was significantly negative, but the value of the MD of other factors was not negative.

Factor(unit)	Intercept	Coefficient of amount	Coefficient of period
Body fat(%)	-0.1278	-1.9423	0.0038
Epididymal fat(g/100g)	0.0203	-0.3859	-0.0003
Final body weight(g)	1.7030	4.5770	-0.0698 b)
Liver-TC(mg/g)	NA		
Liver-TG(mg/g)	-2.6548	-13.8573	0.0947
Plasma-TC(mg/dl)	-1.2486	13.5180	0.0032
Plasma-TG(mg/dl)	-2.1294	-10.2354	0.0681
Weight gain(g/d)	-0.0165	-0.0914	0.0004

a) If it is found to be hetero in the effect of treatment between studies, then meta-regression can be used to analyze associations between treatment effect and covariates in study.

b) The estimated coefficient of period was significantly efficient to combined mean difference about final body weight between the FR and the FRC(p-value <0.05), thus the MD of final body weight is estimated to decrease by 0.0698 per unit increase in the period.

Table 2. Estimated intercept and coefficient of covariates in meta-regression.

Factor(unit)	Egger's linear regression	
	Intercept a)	p-value b)
Body fat(%)	14.1719	0.3649
Epididymal fat(g/100g)	-1.5864	0.0756
Fat cell size(μm)	1.5284	0.5107
Final body weight(g)	-0.0471	0.9677
Food intake(g/d)	-0.1382	0.7718
Leptine(pg/ml)	1.8242	0.2223
Liver-TC(mg/g)	NA	
Liver-TG(mg/g)	-2.4299	0.3152
Plasma-TC(mg/dl)	-2.8889	0.0807
Plasma-TG(mg/dl)	-0.7732	0.2357
Weight gain(g/d)	3.1444	0.5980

a) The data were expressed the intercept of Egger's linear regression about factors.

b) The expressed p-values were for testing that the estimated intercept is not significantly different with 0. If the p-value is less than 0.05, the combined MD in Table 1 possibly exist the publication bias, and the additional analysis is required.

Table 3. Estimated intercepts and p-values of Egger's linear regression test

3.2 Publication bias

To check the publication of the funnel plot expressed in Fig. 1. The funnel plots of fat cell size, food intake, leptine and plasma-TG were checked in the fixed effect model due to the homogeneity. The dots of fat cell size and the dots of leptine were not symmetric by combined

MD (solid line), so the publication biases were doubtable. The dots of factors in the random effect model were symmetric by solid line, thus the publication bias were ignorable. Since the intercepts of Egger's linear regression by factors in Table 3 were not significant (p-value >0.05), the funnel plot on each factor was symmetric and the publication bias was ignorable.

Factor	Fixed effect model	Random effect model

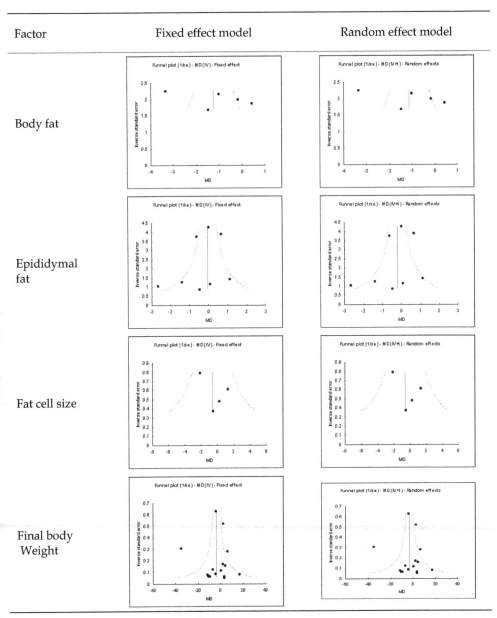

Fig. 1. Funnel plot: In fixed effect model and random effect model

Factor	Fixed effect model	Random effect model

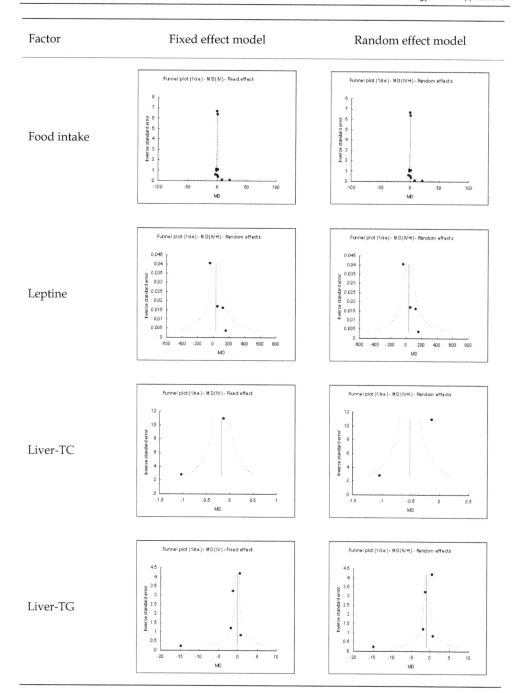

Fig. 1. Funnel plot: In fixed effect model and random effect model (Continuation)

Factor	Fixed effect model	Random effect model

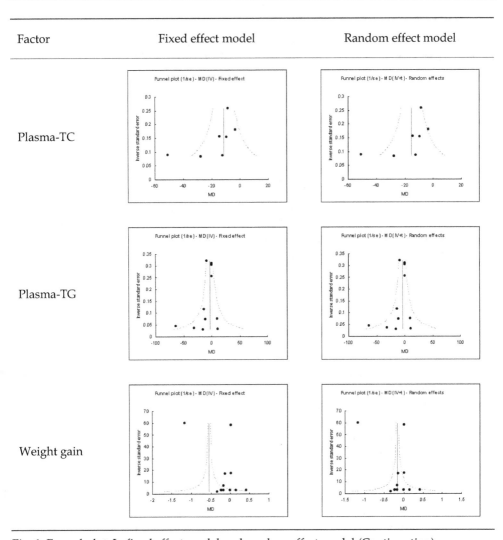

Fig. 1. Funnel plot: In fixed effect model and random effect model (Continuation)

4. Discussion

4.1 The effects of CLA on the fat depositions

The factors related with fat deposition were body fat and epididymal fat, and our study showed that CLA supplement did not affect the fat deposition. The results of studies on body fat and epididymal fat were significantly heterogeneous. In fixed effect model, body fat was significantly affected by CLA. Under the heterogeneity of studies on body fat, the effect of CLA was not significant in random effect model. We could not check the condition of energy expenditure, and that was one of reasons for explaining about heterogeneity.

4.2 The effects of CLA on the weights of body

The results of studies done on the final body weight and weight gain were very heterogeneous (p-value < 0.001), and the result of efficiency test of CLA in fixed effect model differed from one in random effect model. This situation happened due to extreme one of combined results. The extremely negative MD had a small standard error, so the high influence of them on combination made significantly negative value in fixed effect model but not in random effect model. After omitting the extreme value in analysis, the homogeneities results on final body weight and weight gain were accepted (p-value >0.05), and the effects of CLA on them were not significant.

4.3 The effects of CLA on the plasma lipid

The level of plasma-TC was significantly decreased in FR than in FRC regardless of the homogeneity of the factors. Only in fixed effect model, the level of plasma-TG was significantly decreased in FR. According to Hwang and Kang, the reduction effect of CLA on plasma-TG was significant in the case rats that ware fed with beef tallow for 1 and 4 weeks and in the case rats that was fed with fish oil only for 4 weeks (20). The anti-obesity effect of CLA was assured like Hwang and Kang's study.

5. Conclusion

In this study, the researchers studied about effect of CLA on factors, such as body fat (%), epididymal fat (g/100g), fat cell size (µm), final body weight (g), food intake (g/d), leptin (pg/ml), liver-TC (mg/g), liver-TG (mg/g), plasma-TC (mg/dl), plasma-TG (mg/dl), and weight gain (g/d). The CLA supplement was significantly effective on reduction of body fat (%), final body weight (g), liver-TC (mg/g), plasma-TC (mg/dl), plasma-TG (mg/dl), and weight gain (g/d) with homogeneity assumption between studies. However, only plasma-TG was significantly decreased with the random effect model on heterogeneity factors. Further study should focus on what the effect of CLA depend on fat source.

6. Acknowledgment

This work was supported by Priority Research Centers Program through the National Research Foundation of Korea (NRF) funded by the Ministry of Education, Science and Technology(2010-0029692).

7. References

Akahoshi A. et al. (2003). Metabolic effects of dietary conjugated linoleic acid (CLA) isomers in rats. *Nutritional Research*, 23, pp. 1691-1701

Andreoli M, Scalerandi M, Borel I, Bernal C. (2007). Effects of CLA at different dietary fat levels on the nutritional status of rats during protein repletion. *Nutritional*, 23, pp. 827-835

Belury MA. (2002). Inhibition of carcinogenesis by conjugated linoleic acid: Potential mechanisms of action. *Journal of Nutrition*. 132, pp. 2995–2998

Chin SF, Storkson JM, W. Liu, Albright KJ, and Pariza MW. (1994). Conjugated linoleic acid(9,11- and 10, 12-octadecadienoic acid) is produced in conventional but germ-free rats fed linoleic. *J Nutr*, 124, pp. 694-701

Choi K, Kook S. (2007). *Meta-analysis using program MIX*. Free Academy, Seoul

Chol N, Won D, Yun S, Jung M, Shin H. (2004). Selectively hydrogenated soybean oil with conjugated linoleic acid modifies body composition and plasma lipids in rats. *Journal of Nutritional biochemistry*, 15, pp. 411-417

Hwang TH, Kang KJ. (2005). The time course effects of conjugated linoleic acids on body weight, adipose depots and lipid profiles in the male ICR mice fed different fat sources. *The Korean Nutrition Society*, 8, pp. 205-211

Jason C.G Halford. (2006). Pharmacotherapy for obesity. *Appetite*, 46, pp. 6-10

John Wiley & SonsSutton A, Abrams K, Jones D, Sheldon T and Song F. (2000). *Methods for Meta-Analysis in Medical Research*. Wiley

Kang KJ, Kim KH, Park HS. (2002). Dietary Conjugated Linoleic Acid did not Affect on Body Fatness, Fat Cell Sizes and Leptin Levels in Male Sprague Dawley Rats. *Nutritional Sciences*, 5, pp. 117-122

Kang KJ, Park HS. (2001). Effects of Conjugated Linoleic Acid Supplementation on Fat Accumulation and Degradation in Rats. *The Korean Nutrition Society*, 34(4), pp. 367-374

Kloss R, Linscheid K, Mark K. (2005). Effects of conjugated linoleic acid supplementation on blood lipids and adiposity of rats fed diets rich in saturated versus unsaturated fat. *Pharmacological Research*, 51, pp. 503-507

Leonhart M, Munch S, Westerterp-Plantenga M, Langhans W. (2004). Effects of hydroxycitrate, conjugated linoleic acid, and guar gum on food intake, body weight regain, and metabolism after body weight loss in male rats. *Nutrition Research*, 24, pp. 659-669

Oh SW, Yoon WS, Shin SA. (2005). Effect of excess weight on cancer incidences depending on cancer sites and histologic findings among men: Korea national health insurance corporation study. *Journal of Clinical Oncology*, 23, 00. 4742-4754

Park Y, Albright K, Pariza M. (2005). Effects of conjugated linoleic acid on long term feeding in Fischer 344 rats. *Food and Chemical Toxicology*, 43, pp. 1273-1279

Purushotham A, Shrode G, Wendel A, Liu L, Belury M. (2007). Conjugated linoleic acid does not reduce body fat but decreases hepatic steatosis in adult Wistar rats. *Journal of Nutritional Biochemistry*, 18, pp. 676-684

Rahman S, Huda M, Uddin M. (2002). Akhteruzzaman, S. Short-term Administration of Conjugated Linoleic Acid Reduces Liver Triglyceride Concentration and Phosphatidate Phosphohydrolase Activity in OLETF Rats. *Journal of Biochemistry and Molecular Biology*, 35, pp. 494-497

Rothstein H, Sutton A, Borenstein, M. (2005). *Publication bias in meta-analysis*.

Tsuzuki T, Kawakami Y, Nakagawa K, Miyazawa T. (2006). Conjugated docosahexaenoic acid inhibits lipid accumulation in rats. *Journal of Nutritional Biochemistry*, 17, pp. 518-524

West DB, Delany JP, Camet PM, Blohm F, Truett AA, Scimeca J. (1998). Effects of conjugated
    linoleic acid on body fat and energy metabolism in the mouse. *Am J Physiol*, 44, pp.
    667-672

Yamasaki M, Ikeda A, Oji M, Tanaka O, Hirao A, Kasai M, Iwata T, Tachibana H, Yamada K.
    (2003). Modulation of Body Fat and Serum Leptin Levels by Dietary Conjugated
    Linoleic Acid in Sprague-Dawley Rats Fed Various Fat-Level Diets. *Nutrition*, 19,
    pp. 30-35

4

The Welfare of Transgenic Farm Animals

Michael Greger

The Humane Society of the United States,
USA

1. Introduction

As part of a burst of deregulatory activity in the dwindling days of the Bush administration, the U.S. Food and Drug Administration proposed guidelines for the approval of genetically engineered farm animals for the American food supply. Imagine "double muscled" beef cattle born so enormous they can be extracted only via Caesarian section, a dairy cow capable of generating ten times more milk than a calf could suckle (if she were allowed, that is), a hen laying so many eggs she risks a prolapse (laying her own uterus), turkeys so top-heavy they are physically incapable of mating, and chickens with such explosive growth they have to be starved lest they risk aortic rupture (Renema, 2004). Imagine a world in which farm animals have been so genetically modified for rapid muscling that billions suffer in chronic pain from skeletal disorders that impair their ability to even walk.

Unfortunately, this is the world we already live in. All of these abominations exist today, products of conventional techniques of genetic manipulation, such as artificial selection and insemination, hormone-induced superovulation, and embryo splitting and transfer. Genetic engineering, the creation of transgenic farm animals whose genes have been modified through biotechnology, goes a step further, giving agribusiness an additional tool to stress animals towards their biological limits at the expense of their health and welfare—and, potentially, ours as well.

2. Production-related disease related to extant breeding technologies

2.1 Mighty mice

Ever since the early Eighties when it was demonstrated that one could nearly double the size of mice by engineering them to produce rat or human growth hormones, the livestock industries have been clamoring to make use of this technology (Palmiter et al., 1982, 1983; Westhusin, 1997). Double-muscling is a genetic defect maintained in certain breeds of beef cattle caused by a mutation of a gene which regulates muscle growth. Not only do births of such calves require surgical intervention, their tongue muscles may be too enlarged to suckle, leading to death (Lips et al., 2001; Uystepruyst et al., 2002). Unable to move sufficiently in the womb due to their unnatural size they may be born with their joints locked in place and be unable to stand (Lips et al., 2001). The inherent welfare problems have led some European countries to consider banning the intentional breeding of such cattle on animal welfare grounds (Anonymous, 2010a). Reads one editorial in the *British*

Veterinary Journal: "I wonder how many [farmers] could truly claim to be proud of breeding animals which they know are unfit to survive without recourse to elective surgery" (Webster, 2002).

The creation of the "mighty mouse," however, with up to *triple* the muscle mass has reignited hopes within the agribusiness community that this mutation could successfully be transferred to sheep, pigs, chickens, turkeys, and fish (McPherron et al., 1997; McPherron & Lee, 1997; Rodgers & Garikipati, 2008).

2.2 Milk machines

Today's dairy cows endure annual cycles of artificial insemination, mechanized milking for 10 out of 12 months (including 7 months of their 9-month pregnancies), and giving birth. Over the past century, selective breeding has tripled the annual milk yield per cow to about 20,000 pounds. It took the first half of the century to force the first ton increase, but since the 1980s, the industry has managed to milk an extra ton of production per cow every eight or nine years. According to Bristol University Emeritus Professor John Webster in the Department of Clinical Veterinary Science, "The amount of work done by the [dairy] cow in peak lactation is immense. To achieve a comparable high work rate a human would have to jog for about 6 hours a day, every day." This excessive metabolic drain overburdens the cows, who are considered "productive" for only two years and are slaughtered for hamburger around their fourth birthday when their profitability drops, a small fraction of their natural lifespan (Dewey, 2001).

Turning dairy cows into milk machines has led to epidemics of so-called "production-related diseases," such as lameness and mastitis, the two leading causes of dairy cow mortality in the United States. We all remember the sick and crippled dairy cows being dragged and beaten at that California dairy cow slaughter plant in 2008. That loss of body condition is in part a direct result of this extreme selection for unnaturally high milk yields.

That slaughter plant investigation triggered the largest meat recall in history for fear the cows might be infected with mad cow disease. A cow's natural diet—grass—can no longer sustain such abnormally high levels of milk production. They must be fed feed concentrates such as grains or slaughter waste. Today's dairy cows may be forced to eat a pound a day of meat, blood, and bone meal, euphemisms, as described by a leading feedstuffs textbook, for "trimmings that originate on the killing floor, inedible parts and organs, cleaned entrails, fetuses...."

Ever increasing rates of mastitis, (United States Department of Agriculture [USDA], 2008a) udder infections, have led to the extensive use of antibiotics in the dairy industry, including classes of drugs important to human medicine such as penicillin, erythromycin, and tetracycline (USDA, 2008b). A 2005 survey of Pennsylvania dairy herds even uncovered that about 1 in 5 operations were injecting cows with a third generation cephalosporin, a class of antibiotics critical for the treatment of serious infections in children (Call et al., 2008). The concern is that by selective breeding for an overstressed caricature of an animal, the dairy industry's reliance on pharmacological crutches may in turn breed antibiotic resistance to drugs necessary for human medicine (Alcaine et al., 2005).

The mastitis epidemic in the national dairy herd also affects milk quality. American milk has the highest allowable pus cell concentration in the world, legally allowing over 300 million "somatic" cells per glass, 90% of which are neutrophils (pus cells) when there is an udder infection of that severity (United States Public Health Service, 2003; Ruegg, 2001). The industry, however, has always argued that it doesn't matter how inflamed and infected the udders of our factory farmed dairy cows are, because of pasteurization—it's essentially cooked pus, so there's no food safety risk. But just as parents may not want to feed their children fecal matter in meat even if it's irradiated fecal matter, they might not want to feed their children pasteurized pus.

And you can taste the difference. A 2008 study published in the *Journal of Dairy Science* found that cheese made from high somatic cell count milk had both texture and flavor defects as well as increased clotting time compared to milk conforming to the much more stringent European standards. Treating animals as mere commodities is not only bad for the animals and risky for people, but may result in an inferior product.

2.3 Sweating like a pig

The intensive breeding of pigs for increased muscle mass has led to a susceptibility to porcine stress syndrome, in which electric prodding and other stressors can trigger muscle rigidity, high fever, and acute death from what's called malignant hyperthermia, also known as "hot death" (Casau, 2003; Wendt et al., 2000; Anonymous, 2010b). It costs the industry hundreds of millions of dollars a year, but the reason the genetic defect hasn't been eliminated is that the mutation that puts pigs at risk for this disease is the same mutation that adds 2-3% muscle mass to the dressed carcass weight (MacLennan & Phillips, 1992).

Postmortem, their muscles can become pale, soft, and sweaty, which many consumers find objectionable. As the director of the Muscle Biology Laboratory at the University of Wisconsin told the *New York Times*, "You don't want to sell your deli meat with a spoon" (Casau, 2003).

2.4 Shell game

Whereas ancestors to the modern-day chicken laid about 25 eggs a year, today's laying hens produce more than ten times that number. After about a year they are considered "spent." By the end their bones are so brittle because of the excessive draw of calcium from their skeletons for egg shell formation that up to a third of hens have freshly broken bones at slaughter.

The loss of muscle tone from excessive egg-laying, along with consumer demand for "jumbo" eggs, places hens at risk for the prolapse of part of their reproductive tracts during egg-laying (Keshavarz, 1990; Zuidhof, 2002). This can lead to bleeding, infection, and death from cloacal cannibalism, as stressed and overcrowded cage-mates peck at the exposed organ (Newberry, 2004; Zuidhof, 2002). The steroidal sex hormone activity associated with heavy egg production is thought to be why both benign and cancerous tumors are so common in commercial birds and also why hens are predisposed to salpingitis. This pelvic infection can lead to the buildup of masses of caseous exudate (oozing material with a cheese-like consistency), which can expand and end up fatally filling the body cavity.

2.5 Butterballs

Ben Franklin's tree-perching "Bird of Courage" has been transformed into a flightless mammoth bred to grow so fast, a group of veterinary researchers concluded, "that they are on the verge of structural collapse" (Wise & Jennings, 1972). Wild turkeys grow to be 8 pounds (Healy, 1992). The average turkey grown today is more than 28 pounds (National Agricultural Statistics Service, 2007). Their skeletons cannot adequately support such weight, leading to degenerative hip disease, spontaneous fractures, and up to 20% mortality due to lameness in problem flocks (Julian, 1984). An editor at *Feedstuffs*, the leading U.S. agribusiness weekly, wrote that "turkeys have been bred to grow faster and heavier but their skeletons haven't kept pace, which causes 'cowboy legs.' Commonly, the turkeys have problems standing...and fall and are trampled on or seek refuge under feeders, leading to bruises and downgradings as well as culled or killed birds" (Smith, 1991).

Commercial strains may not only outgrow their skeletons, but their cardiovascular systems as well. Modern day turkeys have been bred to grow so fast that up to 6% of modern-day turkey flocks simply drop dead from acute heart failure at just a few months of age (Mutalib & Hanson, 1990). It still may make good economic sense in the end. The sudden deaths of turkeys has in fact been regarded by some in the industry as a sign of "good flock health and fast growth rate as in the case of sudden death syndrome (flip-over) in broiler chickens" (Mutalib & Hanson, 1990). As one producer wrote, "Aside from the stupendous rate of growth...the sign of a good meat flock is the number of birds dying from heart attacks" (Baskin, 1978).

2.6 Why they can't cross the road

The commercial breeds of chickens raised for meat, so-called "broilers," probably suffer the most. Compared to 1920, broilers now grow twice as large in half the time, reaching slaughter weight in around 6 weeks. To put their growth rate into perspective, the University of Arkansas Division of Agriculture calculates, "If you grew as fast as a chicken, you'd weigh 349 pounds at age 2" (Boersma, 2001).

Their hearts and lungs can't keep up. Due to this breeding for rapid growth, a hundred million chickens in the United States every year succumb to sudden death or "flip-over" syndrome, since the birds are often found on their backs after dying in a fit of convulsions and wing-beating (Julian, 2004). These are baby birds, only a few weeks old, dying of heart attacks. One poultry specialist mused in the trade publication *World Poultry*, "Mathematically, it is evident that the present rate of improvement in growth cannot be continued for more than a couple of decades, or the industry will be faced with a bird that virtually explodes upon hatching" (Urrutia, 1997).

Today's broiler chickens are crippled with inbred physical disabilities, from "twisted-leg" deformities to avulsed and ruptured tendons. At six weeks, broiler chickens have such difficulty supporting their grossly overweight bodies that they are forced to spend most of their time lying in their own waste, leading to an increased incidence of painful contact burns such as "breast blisters" from the ammonia released from the decomposing excrement (Estevez, 2002; Weeks et al., 2000). Those unable to hobble using their wing tips as crutches or crawl on their shanks to food and water won't make it to slaughter.

A review published in 2003 in an industry text *Poultry Genetics, Breeding and Biotechnology* concluded, "There is no doubt that the rapid growth rate of birds used for meat production is the fundamental cause of skeletal disorders, nor that this situation has been brought about by the commercial selection programmes used over a period of 40-50 generations" (Whitehead et al., 2003). There is no doubt that the industry is to blame. More than a quarter of broilers have been found to have difficulty walking in chronic pain. With 9 billion chickens produced annually in the United States, that means billions of animals are made to suffer every year because of this genetic selection for extreme productivity, a pedigree of pain.

According to Professor Emeritus Webster, "Broilers are the only livestock that are in chronic pain for the last 20% of their lives. They don't move around, not because they are overstocked, but because it hurts their joints so much" (Erlichman, 1991). This chronic pain experienced by our freakishly heavy modern day chickens and turkeys "must constitute," Webster concludes, "in both magnitude and severity, the single most severe, systematic example of man's inhumanity to another sentient animal" (Webster, 1995).

Two prominent poultry researchers, however, offer the following economic analysis:

"Two decades ago the goal of every grower was to ensure that the flock grew as rapidly as possible. However, the industry has developed a broiler that, if grown as rapidly as possible, will achieve a body mass that cannot be supported by the bird's heart, respiratory system or skeleton.

"The situation has forced growers to make a choice. Is it more profitable to grow the biggest bird possible and have increased mortality due to heart attacks, ascites, and leg problems, or should birds be grown slower so that birds are smaller, but have fewer heart, lung and skeletal problems?...A large portion of growers' pay is based on the pound of saleable meat produced, so simple calculations suggest that it is better to get the weight and ignore the mortality" (Tabler & Mendenhall, 2003).

2.7 Look what's coming to dinner

Chickens and turkeys aren't the only lame birds. In its final days, the Bush Administration's promise to "sprint to the finish" involved rolling back restrictions on smokestack emissions, commercial ocean fishing, and mountaintop removal coal mining (Smith, 2008). Overshadowed by election coverage, in September 2008 the FDA released draft guidelines to move the approval of genetically engineered farm animals in the U.S. food supply one step closer to reality (Food and Drug Administration, 2009).

The Biotechnology Industry Organization claims that the genetic engineering of farm animals offers "tremendous benefit to the animal by enhancing health, wellbeing, and animal welfare" (Gottlieb, 2002). Theoretically this technology could be used by industry giants to ameliorate some of the inbred animal "illfare" they have created, but if past performance is any predictor of future behaviour, genetic engineering will just be used to further industry goals of production efficiency at nearly any cost. The meat, egg, and dairy industries recognize that enhanced productivity generally comes at the expense of animal health and well-being. The reason given for not using existing breeding programs to relieve suffering is presumably the same reason biotech resources won't be diverted to improve welfare: doing so, in the words of livestock geneticists in the *Journal of Animal Science*, may "result in less than maximal progress in economic traits" (Kanis et al., 2005).

The primary goal set out for transgenic food animals has explicitly been to improve productivity, so-called "quantitative genetic engineering" concerned with increasing "economic fitness" (Dickerson & Willham, 1983; Pinkert & Murray, 1999). Consider the most widespread current use of biotechnology in animal agriculture, recombinant bovine somatotropin. The injection of this genetically engineered growth hormone increases milk yield in dairy cattle, but also increases the rates of mastitis, lameness, and poor body condition. Yet millions of U.S. dairy cows are repeatedly injected with this genetically engineered hormone throughout their short lives, demonstrating dairy industry priorities — profits at the expense of animal health. More is not always better.

Agribusiness claims in its public relations materials that biotechnology will be used to improve animal welfare, but to date gene constructs designed to express growth factors constitute the largest class of transgenes so far experimentally transferred into livestock (Murray, 1999). It is instructive that the first report of the successful creation of transgenic livestock was the "Beltsville pigs," engineered at a USDA lab in Beltsville, Maryland to express human growth hormone (Hammer et al., 1985). Yes, their feed efficiency modestly improved, but many became lame, lethargic, and uncoordinated with thickened skin and bulging eyes. These pigs also suffered with ulcerated stomachs, inflamed hearts and kidneys and severe joint degeneration. Several of them died during or immediately after confinement in a restraint device, suggesting an increased susceptibility to stress (National Research Council, 2002). Seventeen of the nineteen transgenic pigs created didn't last a year.

Though animal scientists have cited the Beltsville pigs research "as an excellent example of the value of this technology," the results of the now infamous experiment have been used by critics for over 20 years to condemn the genetic engineering of farmed animals on animal welfare grounds (Wheeler et al., 2003). Reliance on laboratory freaks as the centerpiece of one's argument, though, ignores the fact that the technology has improved since those early experiments and, more importantly, overlooks the much larger concern. Given the inefficiency and unpredictability of this still emerging science, attention has been drawn to the unintended consequences, but considering the power of the technology and the sheer number of animals raised for food every year — 50 billion land animals alone — what of the secondary effects of the *intended* consequences (Mak, 2008)? Due to agribusiness prioritizing economic fitness over physical fitness, billions of farm animals are already in pain. Now they want to stitch in growth hormones to strain animals even further past their breaking point? DNA technologies have been considered by corporate breeders as the fast lane on what they call the "road to the biological maximum" (Buddiger & Albers, 2007).

Grahame Bulfield, the head of the Roslin Institute, the creator of the cloned sheep Dolly, is quoted as saying:

"The view I take on animal welfare is that the technology itself is a red herring. If an animal is lame because of genetic modification or selective breeding or poor nutrition, or because I kick it, it is wrong that it's lame. So you have to pay attention to the phenotype — that is, to the animal itself — rather than the technique that produces the problem" (Klotzko, 1998).

The speed, power, and ecologically disruptive potential, however, of genetic engineering is unique. Selective breeding is a powerful tool; it is, after all, what enabled humankind to turn a wolf into a poodle, but that was over a period of 14,000 years (Pennisi, 2002). Dramatic changes can be induced by gene manipulation in a single generation, and few of the usual checks and balances imposed by natural selection may apply. In natural or artificial selection, the trait that is chosen comes coupled to a constellation of linked attributes that may help the animal maintain homeostatic balance, as teetering as it may be. Due to the single-gene nature of transgenic change, however, engineered animals may suffer a greater loss of fitness than their selectively bred counterparts in conforming form to function. This has been clearly demonstrated in transgenic fish.

3. Ecological concerns raised by farm animal transgenesis

3.1 Bigger fish to fry

Several species of genetically engineered fish stand ready to be marketed worldwide, transgenic tilapia in Cuba, transgenic salmon in the United States and Canada, and transgenic carp in the People's Republic of China (Kaiser, 2005). The North American AquAdvantage™ salmon is positioned to become the first transgenic animal available for human consumption. Like all farmed animals, farmed fish undergo genetic manipulation through selective breeding to enhance economically favored traits such as rapid growth rate. Genetic selection for salmon size over a period of ten years has been shown to increase average weights by about 60% (Hershberger et al., 1990). Salmon engineered with transgenic growth hormones can be 1100% heavier on average, though, with one fish weighing out at 37 times normal (Devlin et al., 1994).

Yes, some of these transgenic salmon suffer from severe and sometimes fatal cranial disfigurements, but the larger concern surrounds their overall average fitness. The critical swimming speed of salmon genetically engineered to grow twice as fast is twice as slow as the speed of same-sized normal salmon, impairing their ability to forage and avoid predators. Similarly, normal catfish exhibit better predator-avoidance skills compared to transgenic catfish. The concern is that should these transgenic fish escape into the wild – a common occurrence in aquaculture – they could lead to species population extinction (Muir & Howard, 1999).

Male medaka fish genetically engineered with a salmon growth hormone, for example, possess an overwhelming mating advantage compared to wild-type medaka males due to their large body size. While they preferentially attract mates, in the end bigger is not necessarily better. Should their offspring bear a viability disadvantage, mathematical modeling suggests a "Trojan gene effect," where a combination of mating advantage with survivability disadvantage could ultimately lead to the rapid collapse and extinction of both the transgenic and wild fish population in as few as approximately 50 generations should a transgenic male escape into the wild (Howard et al., 2004).

A report from the National Academy of Sciences on animal biotechnology concluded that these potential environment impacts present the greatest science-based safety concern (National Research Council, 2002). The one class of species – fish – considered to pose the highest risk is the one closest to commercialization. The risk is so great that biologic, rather than physical, containment of these animals may be necessary, such as induced sterility. The incorporation of so-called "suicide genes" is under consideration to prevent the genetic pollution of environment (as well as to protect corporate intellectual property rights and investment) (Wheeler et al., 2003). On the other hand, some Purdue University scientists have expressed hope that the Trojan gene effect could itself be harnessed as a tool for biological control: transgenic males could be created and released intentionally to drive wild populations of unwanted species to extinction (Muir & Howard, 1999).

3.2 Narrowing the genetic base

There is also a concern that biotechnology will lead to the loss of genetic biodiversity within farm animal species. An international analysis of commercial poultry breeds published in the *Proceedings of the National Academy of Sciences* found that about half of the genetic diversity of chickens has already been lost. According to the Food and Agriculture Organization of the United Nations, over the last century 1,000 farm animal breeds—about one-sixth of the world's cattle and poultry varieties—have disappeared, with breeds continuing to go extinct at a rate of one or two every week. Transgenesis will be subject to the same pressures that have already led to such a narrowing of the genetic base. Should an engineered line of animals gain a clear competitive advantage, competitors may replace varieties viewed as obsolete. Genetic uniformity increases the vulnerability of monocultures of animals to diseases that could spill over into human populations.

4. Anthropocentric concerns raised by farm animal transgenesis

4.1 Bred to be contagious

The biotech industry touts human benefits as well. Under consideration are cows that produce milk with fewer disease-causing components — fewer allergens, less lactose, or less saturated butterfat, though the industry fears the latter could deleteriously impair the whipping of cream (Gibson, 1991; Jost et al., 1999; Karatzas & Turner, 1997; Reh et al., 2004). Adding human breast milk genes to dairy cow udders has been suggested to improve baby formula (Wall et al., 1997). Incorporation of a humanized version of a roundworm gene into pigs could potentially make pork a source of omega-3 fatty acids and take pressure off diminishing global fish stocks. Pigs have even been implanted with spinach genes (Lai et al., 2006; Young, 2002). Dr. Seuss's signature dish may soon be realized.

Human health concerns, as expressed for example in a February 2009 review in *Critical Reviews in Food Science and Nutrition*, have largely been limited to the potential for growth hormones from genetically enhanced animals to promote human colon, breast, and prostate cancer (Dona & Arvanitoyannis, 2009). The physiological trade-off between productivity and immune function may pose a broader risk, however.

Genetic manipulation for accelerated muscle, milk, and egg production carries an inverse relationship with immune function, a trade-off that has been empirically demonstrated in chickens, pigs, and both beef and dairy cattle. This has been explained by the "resource-allocation hypothesis," which suggests that protein and energy diversion from host defense to breast muscle mass production, for example, explains why chickens with accelerated growth are at risk for increased immune dysfunction, disease morbidity, and disease mortality. As breast mass enlarges, the lymphoid tissue, the immune system organs themselves, shrink.

Before domestication, natural selection chose strong immune systems for survival. After domestication, though, artificial selection concentrated on improvement of production traits with less attention to resistance to disease, resulting in survival of the fattest rather than the fittest. The reason this may pose a human health hazard is that three quarters of emerging human infections diseases have come from animals. Whether it's mad cow, bird flu, porcine Nipah virus, *Strep suis*, or poultry and aquaculture-related foodborne disease, how we breed and raise animals can have global public health implications.

As a crutch to compensate for the imposed immunodeficiency (as well as the often overcrowded, stressful, unhygienic conditions on factory farms) agribusiness pours millions of pounds of antibiotics straight into chicken, pig, cattle, and fish feed to promote further growth and stave off disease, a practice banned in the European Union and condemned by the American Academy of Pediatrics, the American Medical Association, the World Health Organization, and hundreds of other medical and public health organization. Antibiotic resistant bacteria, including the "superbug" MRSA found recently in 70% of pigs tested in Iowa and Illinois, may then transfer to people via contaminated air, water, soil, or food. We may be sacrificing a future where antibiotics will continue to work for treating sick people by squandering them today on animals that are not yet sick at all.

4.2 Chicken surprise

A 1997 scientific expedition to Alaska further underscored the threat of weakened farm animal immunity. Digging up victims of the 1918 flu pandemic discovered frozen in the permafrost for tissue samples, scientists allied with the U.S. Armed Forces Institute of Pathology were able to decipher the genetic code of the killer virus, solving perhaps the greatest medical detective story of all time. The 1918 pandemic was the worst plague in human history, killing more people in 25 weeks than AIDS has killed in 25 years, an estimated 50 million people dead. In 2005, with the entire genome of the 1918 virus finally decoded, the mystery was solved. Humanity's greatest mass murderer turned out to be a bird flu virus. This finding, combined with the unprecedented recent emergence of highly pathogenic bird flu viruses around the world such as H5N1, means that disease losses from selecting or engineering fast growing breeds of chickens with essentially built-in immune dysfunction can no longer just be factored in to the corporate bottom line. Millions of human lives may be at stake.

There has been interest in trying to genetically engineer our way out of these problems. Instead of stopping the cannibalistic feeding of slaughterhouse waste, blood, and manure to cows, for example, researchers are trying to create mad cow disease resistant cattle (Cyranoski, 2003). Instead of removing the strain on overproducing dairy cattle, researchers are working on creating cows that secrete an antibiotic substance directly into their milk to prevent udder infections (Wall et al., 2005). Production-related diseases have become preferred technofix targets presumably because they represent barriers to even greater productivity. The industry may be able to squeeze extra tons of milk from cows secreting antibiotics without rampant mastitis, but the metabolic, musculoskeletal, and painful hoof problems associated with overproduction would be further aggravated. Issues surrounding the Enviropig™ offer a parallel.

4.3 Trojan pig

Trumpeted by the pork industry as the "biggest breakthrough in pig farming since the invention of the trough," a new line of transgenic pigs incorporating a composite of mouse and bacterial genes has been patented to produce manure with less phosphorus: the Enviropig™ (Vestel, 2001). This may allow for the further expansion of swine CAFOs, confined animal feeding operations. Already some CAFOs store hog waste in massive open-air manure pits the size of several football fields, which can burst, spilling millions of gallons of excrement into local watersheds. In one year, 1991, an estimated one billion fish were killed from farm animal manure run-off in North Carolina alone (Zakin, 1999). Enviropigs may produce less phosphorus, but what about the other pollutants in manure — the nitrates which end up in the groundwater leading to miscarriages, birth defects, and "blue-baby syndrome," the hydrogen sulfide emissions that have killed CAFO workers, the ammonia contributing to acid rain, potent greenhouse gases such as methane and nitrous oxide, and the increased asthma rates in adjoining school districts and elevated infant mortality? The pigs aren't the problem; CAFOs are the problem.

In the United States, farm animals produce an estimated 2 billion tons of manure each year, the weight of 20,000 Nimitz-class aircraft carriers. Manure has been found to be the source of more than 100 pathogens and parasites that can infect people, as well as antibiotics, hormones, pesticides, and toxic heavy metals. Enviropigs won't rid CAFOs of the odor, disease, pollution and occupational hazards inherent to intensive confinement. They will, however, be trumpeted as exemplars by the biotechnology industry of the golden age that transgenic farm animals are to herald, as golden rice was used to tout genetically modified crops.

Golden Rice was hyped as the salvation for millions of children threatened with blindness, but cynics argued that Golden Rice was more about the salvation of the beleaguered biotech industry (Anonymous, 2008). The cynics may have been right. In the eight years since its development not a single grain has been sown for consumption, whereas during that same period hundreds of millions of tons of Roundup Ready® crops have been planted worldwide, increasing the global ecological burden of herbicides and herbicide resistance. Similarly, the industry may publicly peddle concepts like the Enviropig™ as a ploy to dampen criticism while slipping past the more lucrative and damaging applications of transgenic livestock.

5. Using biotechnology to improve the welfare of farm animals

5.1 Cui bono?

Animal agriculture has undergone a mass consolidation in recent decades. For example, a handful of corporations now supply most of the breeding stock for all the world's poultry. Soon, the industry predicts, there may essentially be only three poultry breeders in the world. Today, a single pedigree cockerel can potentially give rise to two million broiler chickens. This means that selected or engineered traits can be propagated around the world at an unprecedented rate. The industry can now replace practically the entire global chicken flock in a space of three or four years, affecting the welfare of 50 billion animals for better or for worse.

The genetic engineering of farmed animals is not *necessarily* harmful. Like nearly all tools and technologies, the consequences depend on how it's used. Theoretically, there are numerous applications of biotechnology that could indeed improve the lives of farm animals by undoing the harm of selective breeding, but one has to consider who owns and stands to profit from the technology? Based on the livestock industry's track record one can be certain that in nearly any conflict that arises between production efficiency and animal suffering, profitability will win the day, but there are rare circumstances in which producers and animals may both benefit.

Today's laying hens produce more than ten times the number of eggs than their ancestors, leading to uterine prolapses and critically weakened, broken bones as their skeletal calcium is disproportionately mobilized for shell formation. Egg-laying breeds have been so genetically manipulated—through conventional selection—that it's not profitable to raise male offspring for meat. So hundreds of millions of male chicks every year in the United States are gassed, ground up alive, or just thrown in dumpsters to suffocate or dehydrate to death after hatching. Economically it doesn't make sense to even waste feed on male chicks because they haven't been bred for excessive muscle mass. Engineering hens that lay only female chicks would double the yield for the breeding industry while sparing hundreds of millions of animals a tragic death. Similarly, constructing dairy cows to preferentially deliver females could save a half million male calves from their doomed fate in the veal industry.

Tens of millions of piglets are castrated without anesthesia or postoperative painkillers every year in the United States to prevent "boar taint" of carcasses, a quality considered amenable to genetic manipulation. No federal regulations protect animals on the farm and "standard agricultural practices" such as castration and dehorning are typically exempt from state anti-cruelty statutes.

Dehorning of beef cattle is another painful surgical procedure performed without anesthesia primarily to protect carcass quality, but could be obviated by knocking out the single gene responsible for horn production (Rollin, 1995). Polled (congenitally hornless) breeds already exist, a fact that may make cattle genetically engineered without horns more palatable to the public. Of course if beef cattle weren't crammed so tightly into feedlots there wouldn't be the level of bruising from horns that leads to so much carcass wastage. This raises the question: is it preferable to engineer animals to fit industrial systems, or rather to engineer systems that fit the animals in the first place?

5.2 Carving square pegs into round holes

More than 95 percent of egg-laying hens in the United States are crammed five to seven together into file-cabinet sized wire "battery cages," affording each hen less than a sheet of paper of space on which to live for over a year before she is killed. Nobel Laureate and noted father of modern ethology Konrad Lorenz wrote: "The worst torture to which a battery hen is exposed is the inability to retire somewhere for the laying act. For the person who knows something about animals it is truly heart-rending to watch how a chicken tries again and again to crawl beneath her fellow cagemates to search there in vain for cover." What if this nesting urge could be removed through genetic tinkering, though? This brings to mind the ill-famed blind chicken experiments.

In 1985 poultry scientists published a series of experiments showing that under conditions of intensive confinement congenitally blind hens are more efficient at laying eggs than hens that can see. Under the stressful, barren, overcrowded battery cage conditions, hens can peck each other to death, so the ends of their sensitive beaks are burned off as chicks to minimize the damage they can do. They still peck at one another, though, which can increase feed requirements because body heat is lost from exposed skin due to feather loss. But blind hens don't seem to peck at each other as much, not do they seem to move as much either, another big cost saver in terms of feed efficiency. Feed "wasted" on movement means less energy directed to egg production. The researchers concluded that "genetically blind birds were more efficient in converting feed into products. It is therefore worthwhile to explore further the potential of this mutation in egg-laying strains under cage systems" (Ali & Cheng, 1985).

The general public reacts negatively to the notion of the industry deliberately breeding hens to be blind in order to save on feed costs, but the larger issue remains unaddressed (Lassen, 2006). What has the system come to when animals have to be literally mutilated — whether via debeaking, dehorning, detoeing, desnooding, disbudding, mulesing, comb removal, teat removal, teeth cutting, or tail docking — to fit the industrial model? Rather than creating blind chickens better adapted to confinement, an informed public would likely reject stuffing birds in tiny cages in the first place, as California voters did in 2008, passing a ballot initiative that phases out battery cages by a landslide 27 point spread victory, making it the most popular citizens' initiative in California history. A 2007 American Farm Bureau poll found that a majority of Americans are in agreement that farm animals shouldn't be raised in cages and crates.

5.3 Mike the headless chicken

If demand for the cheapest possible meat continues to grow unabated, some animal welfare scientists have suggested going beyond the design of sightless birds, and moving to brainless. Mike the headless rooster (1945-1947) became a circus sideshow phenomenon after an incomplete decapitation left him with his brainstem intact. He was able to walk, balance on a perch, and, fed with an eyedropper, lived 18 months with no head. In this vein one could theoretically engineer headless chickens, stick tubes down their neck, and have all the meat with none of the misery. Though aesthetically abhorrent, which is worse: raising brainless chickens or a system in which animals might be better off braindead than fully alive?

CIP, Congenital Insensitivity to Pain, is a rare neurological disorder in which children are born unable to feel pain. Due to their susceptibility to injury, they don't live very long, but the industry doesn't need farm animals to live very long. The moral outrage such a breeding program would engender might be tempered should the public become aware of the current paradigm, in which billions of animals are raised to suffer in chronic pain.

These scenarios speak to how far we've strayed from tradition concepts of animal husbandry, how far out of step animal agribusiness is now from mainstream American values — and the industry knows it. Professor Emeritus of Animal Science Peter Cheeke wrote in his collegiate textbook Contemporary Issues in Animal Agriculture:

"One of the best things modern animal agriculture has going for it is that most people…haven't a clue how animals are raised and 'processed.' In my opinion if most urban meateaters were to visit an industrial broiler house to see how the birds were raised…some, perhaps many, of them would swear off eating chicken and perhaps all meat. For modern animal agriculture, the less the consumer knows about what's happening before the meat hits the plate the better" (Cheeke, 2004)

5.4 Meat without feet

The answer may lie in producing meat "ex vivo," outside of a living animal. In 1932, Winston Churchill predicted: "Fifty years hence we shall escape the absurdity of growing a whole chicken in order to eat the breast or wing by growing these parts separately under a suitable medium." He was a few years off, but in 2000 NASA scientists showed that one could start to grow fish flesh in a petri dish.

The first In Vitro Meat Consortium Symposium took place in 2008 at the Norwegian Food Research Institute, bringing together an international cadre of research scientists working on the issue. With the right mixture of nutrients and growth factors, muscle cells may be able to be coaxed to multiply enough times to produce processed meat products such as sausage, hamburger, or chicken nuggets. Meat scientists at Utrecht University in conjunction with a Sara Lee sausage manufacturer subsidiary are currently working off a grant from the Dutch government to produce cultured meat as part of a national initiative to reduce the environmental impact of food production. Theoretically, the entire world's meat supply could be produced from a single cell taken painlessly from a single animal.

Reasoned one animal scientist at the Portuguese Institute for Molecular and Cell Biology:

"Frankly, if the end product is to be the white meat of a month-old broiler chicken or the minced meat of a hamburger, prepared without care and eaten absent-mindedly, why make the detour through a sentient vertebrate which needs kilos of grain just to keep upright and has a brain that may feel fear and frustration?"

Imagine victimless meat, minus manure and methane, fished out oceans, and jungles deforested for fodder. Meat could be grown hygienically, eliminating million of cases of foodborne illness, and more efficiently, since the vast majority of corn, soy, and grain we feed animals now is lost to metabolism—just keeping the animals alive—and making inedible structures like the skeleton. Unnatural, yes, but so is most of what we eat, from bread to yogurt to hydroponic vegetables. There is arguably very little natural about the way our meat is produced today. Biotechnology has the potential to dramatically affect the welfare of farm animals on a massive scale, but whether this effect is positive or negative depends on how it's used and how it's regulated.

6. Meeting consumer expectations

In a dismissal of the charge that biotechnology leads to the treatment of animals as mere commodities, bioethicists at the Danish Centre For Bioethics and Risk Assessment respond: "There is already a tendency to treat animals as mere things in industrial farming" (Sandøe & Holtug, 1993). This doesn't justify further erosion of consideration for farm animals, but rather constitutes a call for critical reflection on contemporary practices. As the complete genomic sequences of all farm animals become available, there will be an increasing need for guidelines and guidance as to what is and is not ethically permissible.

Colorado State University Distinguished Professor Bernard Rollin, professor of animal sciences, biomedical sciences, and philosophy, has introduced as a guiding principle the concept of "conservation of welfare": when genetically engineering animals, the transgenic animals should be no worse off afterwards than their parents were (Pew, 2005). Given the volume of current suffering imposed by conventional techniques, though, rather than arguing for the status quo, perhaps a "remediation principle" would be more appropriate. Society could mandate that transgenesis for increased production require the resulting farm animals be better off than their parents. Equipped with such powerful new tools, animal agriculture could use biotechnology to bring itself more in line with rising societal expectations for farm animal care.

In order for biotech companies to recoup their R&D investments and for agribusiness corporations to sell products of this technology, a broad public acceptance is necessary. The most extensive international study of public perceptions was a survey of more than 34,000 residents of 34 countries in Africa, Asia, the Americas, Europe and Oceania in 2000. Only 35% of global consumers were in favor of using biotechnology to increase farm animal productivity (Environics International, 2000). In the United States the percentage of those who found it acceptable to use biotechnology to create faster-growing fish dropped from 32% in 1992, to 28% in 1994, to 23% in 2000 (Hoban, 2004). According to a nationwide survey conducted in 2003 by the Pew Initiative on Food and Biotechnology, the majority of Americans (58%) even oppose scientific research into the genetic engineering of animals (PEW, 2005).

At the same time there has been a groundswell in public awareness and scrutiny over the treatment of animals raised for food. According to a 2007 American Farm Bureau poll executed by Oklahoma State University, 95% of consumers agreed with the statement that "[i]t is important to me that animals on farms are well cared for" and furthermore, 76% disagreed that "[l]ow meat prices are more important than the well-being of farm animals" (Lusk et al., 2007). An Ohio State University survey found that 81% felt farmed animal well-being is as important as the well-being of companion animals, such as dogs and cats (Rauch & Sharp, 2005). The Farm Bureau found that the majority of surveyed Americans oppose the way hundreds of millions of farm animals are raised every year in the United States—the intensive confinement of animals in cages and crates. Three quarters of Americans would vote for a law that would require farmers to treat their animals more humanely, a sentiment reflected in a 2008 Gallup poll recognizing widespread support for the passage of "strict laws" concerning the treatment of farm animals. "It was a little surprising the extent to which the issue of humane treatment of animals is ingrained and widespread in our society," the director of public relations for the Farm Bureau told *Meat & Poultry* magazine. "There's a lot of interest in this" (Newport, 2008).

This emerging social ethic for the welfare of farm animals could be an opportunity for the biotech industry rather than an impediment. A consumer backlash against biotechnology resulting from an application perceived to worsen the plight of billions of farm animals could undermine confidence not only in the food system but adversely affect the public's view regarding medical applications of biotechnology as well as the science of genomics as a whole (Pew, 2005). According to an extensive national survey and focus group discussions published in 1993 by the North Carolina Cooperative Extension Service, the least acceptable applications of biotechnology reportedly appeared to include genetically engineering food animals for accelerated growth (Hoban & Kendall, 1993). By instead redressing the pain and suffering caused by conventional breeding, the biotech industry could improve its public image and reduce the stigma hindering the technology, and agribusiness could address societal concerns while potentially expanding its market share. Either way, the debate over transgenic farm animals may bring to light the excesses of the current breeding paradigm and force the meat egg, and dairy industries to revisit practices they have so far taken for granted.

7. Conclusion

The Pew Commission on Industrial Farm Animal Production was formed to conduct a comprehensive, fact-based, and balanced examination of key aspects of the farm animal industry. This prestigious independent panel was chaired by former Kansas Governor John Carlin and included former U.S. Secretary of Agriculture Dan Glickman, former Assistant Surgeon General Michael Blackwell, and James Merchant, then Dean of the University of Iowa College of Public Health. They released their report in 2008. It concluded: "The present system of producing food animals in the United States is not sustainable and presents an unacceptable level of risk to public health and damage to the environment, as well as unnecessary harm to the animals we raise for food." Animals have already in effect been manufactured to be damaged and diseased (Ott, 1996).

In their report, the National Academy of Science and National Research Council's Committee on Defining Science-Based Concerns Associated with Products of Animal Biotechnology expressed concern that certain farmed animals have already been pushed to the edge: "Indeed," they concluded, "it is possible that we already have pushed some farm animals to the limits of productivity that are possible by using selective breeding, and that further increases only will exacerbate the welfare problems that have arisen during selection" (National Research Council, 2002). Biotechnology could be used to reverse some of the damage, but given animal agriculture's track record of willful neglect, the incorporation of genetic engineering will likely just reinforce current practices and worsen an already broken system.

8. References

Alcaine, S.; Sukhnanand, S.; Warnick, L.; Su, W.; McGann, P.; McDonough, P. & Wiedmann, M. (2005). Ceftiofur-resistant salmonella strains isolated from dairy farms represent multiple widely distributed subtypes that evolved by independent horizontal gene transfer. *Antimicrobial Agents and Chemotherapy*, Vol. 49, No. 10, (October 2005), pp. 4061-4067, ISSN 0066-4804

Ali, A. & Cheng, K. (1985). Early egg production in genetically blind (*rc/rc*) chickens in comparison with sighted (*rc+/rc*) controls. *Poultry Science*, Vol. 64, No. 5, (May 1985), pp. 789-794, ISSN 0032-5791

Anonymous. (2008). Vitamin A-rich Golden Rice, touted as salvation for millions of children threatened with blindness or premature death. *Food Chemical News*, (4 February 2008), ISSN: 0015-6337

Anonymous. (2010a) Sweden faces challenge over Belgian Blue cattle ban. *Agra Europe*, Issue AE2413 (21 May 2010), ISSN: 0002-1024

Anonymous. (2010b). Malignant hyperthermia, In: *Merck Manual*, C. Kahn, (Ed.), Merck, ISBN 9780911910933, Whitehouse Station, NJ, USA.

Baskin, C. (1978). Confessions of a Chicken Farmer. *Country Journal*, (April 1978), pp. 38, ISSN 0898-6355

Boersma, S. (2001). Managing rapid growth rate in broilers. *World Poultry*, Vol. 17, No. 8, (2001), pp. 20-21, ISSN 1388-3119

Buddiger, N. & Albers, G. (2007). Future trends in turkey breeding, In: *Hybrid Turkeys*, 10.07.2008, Available from: <http://www.hybridturkeys.com/Media/PDF_files/Management/Mng_future>

Call, D.; Davis, M. & Sawant, A. (2008). Antimicrobial resistance in beef and dairy cattle production. *Animal Health Research Reviews*, Vol. 9, (September 2008), pp. 1-9, ISSN 1466-2523

Casau, A. (October 7, 2003). When pigs stress out. In: *The New York Times*, 17.03.2008,

Cheeke, P. (2004). *Contemporary Issues in Animal Agriculture*, Pearson Prentice Hall, ISBN 0131125869, Upper Saddle River, NJ, USA

Cyranoski, D. (2003). Koreans rustle up madness-resistant cows. *Nature*, Vol. 426, No. 6968 (December 2003), pp. 739-911, ISSN 0028-0836

Devlin, R.; Yesaki, T.; Biagi, C.; Donaldson, E.; Swanson, P. & Chan, W. (1994). Extraordinary salmon growth. *Nature*, Vol. 371, No. 6494, (1994), pp. 209-210, ISSN 0028-0836

Dewey, T. (2001). Bos Taurus, In: *University of Michigan Museum of Zoology Animal Diversity Web*, 17.03.2008, Available from: <http://animaldiversity.ummz.umich.edu/site/accounts/information/Bos_tauru s.html.

Dickerson, G. & Willham, R. (1983). Quantitative genetic engineering of more efficient animal production. *Journal of Animal Science*, Vol. 57, pp. 248-264, ISSN 0021-8812

Dona, A. & Arvanitoyannis, I. (2009). Health risks of genetically modified foods. *Critical Reviews in Food Science and Nutrition*, Vol. 49, (2009), pp. 164-175, ISSN 1040-8398

Environics International. (2000). *Proceedings of International Environmental Monitor 2000*, Toronto, Canada, 2000

Erlichman, J. (1991). The meat factory. *The Guardian*, (14 October 1991)

Estevez, I. (2002). Poultry welfare issues. *Poultry Digest Online*, Vol. 3, No. 2, (2002), pp. 1-12, ISSN 0032-5724

Food and Drug Administration. (2009). Regulation of Genetically Engineered Animals Containing Heritable Recombinant DNA Constructs

Gibson, J. (1991). The potential for genetic change in milk fat composition. *Journal of Dairy Science*, Vol. 74, (1991), pp. 3258, ISSN 0022-0302

Gottlieb, S. (2002). Genetically Engineered Animals and Public Health. *New York Sun*, (May 2002), pp. 3-5

Hammer, R.; Pursel, V.; Rexroad Jr, C.; Wall, R.; Bolt, D.; Ebert, K; Palmiter, R. & Brinster, R. (1985). Production of transgenic rabbits, sheep and pigs by microinjection. *Nature*, Vol. 315, (June 1985), pp. 680-683, ISSN 0028-0836

Healy, W. (1992). Behavior, In: *The Wild Turkey: Biology and Management*, J.G. Dickson, (Ed.), Stackpole Books, ISBN 9780811718592, Harrisburg, Pennsylvania

Hershberger, W.; Myers, J.; Iwamoto, R.; McAuley, W. & Saxton, A. (1990). Genetic changes in the growth of coho salmon (*Oncorhynchus kisutch*) in marine net-pens, produced by ten years of selection. *Aquaculture*, Vol. 85, (1990), pp. 187-197, ISSN 0044-8486

Hoban, T. & Kendall, P. (1993). *Consumer attitudes about food biotechnology*. North Carolina Cooperative Extension Service, Raleigh, North Carolina

Hoban, T. (2004). Public attitudes towards agricultural biotechnology. *Food and Agriculture Organization of the United Nations*, ESA Working Paper No. 04-09, (May 2004)

Howard, R.; DeWoody, J. & Muir, W. (2004). Transgenic male mating advantage provides opportunity for Trojan gene effect in a fish. *Proceedings of the National Academy of Sciences*, Vol. 101, No. 9, (March 2004), pp. 2934-2938, ISSN 0027-8424

Jost, B.; Vilotte, J.; Duluc, I.; Rodeau, J. & Freund, J. (1999). Production of low-lactose milk by ectopic expression of intestinal lactase in the mouse mammary gland. *Nature Biotechnology*, Vol. 17, (1999), pp. 160-164, ISSN 1546-1696

Julian, R. (1984). Tendon avulsion as a cause of lameness in turkeys. *Avian Diseases*, Vol. 28, No. 1, (January-March 1984), pp. 244-249, ISSN 0005-2086

Julian, R. (2004). Evaluating the impact of metabolic disorders on the welfare of broilers, In: *Measuring and Auditing Broiler Welfare*, C. Weeks & A. Butterworth, (Eds.), CABI Publishing, ISBN 9780851998053, Wallingford, United Kingdom

Kaiser, M. (2005). Assessing ethics and animal welfare in animal biotechnology for farm production. *Rev. sci. tech. Off. int. Epiz.*, Vol. 24, No. 1, (2005), pp. 75-87

Kanis, E.; De Greef, K.; Hiemstra, A. & van Arendonk, J. (2005). Breeding for societally important traits in pigs. *Journal of Animal Science*, Vol. 83, (2005), pp. 948-957, ISSN 0021-8812

Karatzas, C. & Turner, J. (1997) Toward altering milk composition by genetic manipulation: current status and challenges. *Journal of Dairy Science*, Vol. 80, (1997), pp. 2225-2232, ISSN 0022-0302

Keshavarz, K. (1990). Causes of prolapsed in laying hens. *Poultry Digest*, (September 1990), pp. 42, ISSN 1444-8041

Klotzko, A. (1998). Voices from Roslin: the creators of Dolly discuss science, ethics, and social responsibility. *Cambridge Quarterly of Healthcare Ethics*, Vol. 7, (1998), pp. 121-140, ISSN 0963-1801

Lai, L.; Kang, J.; Li, R.; Wang, J.; Witt, W. & Yong, H. (2006). Generation of cloned transgenic pigs rich in omega-3 fatty acids. *Nature Biotechnology*, Vol. 24, (2006), pp. 435-436, ISSN 1546-1696

Lassen, J., Gjerris, M., & Sandøe, P. (2006). After Dolly. *Theriogenology*, Vol. 65, No. 5, pp. 992-1004, ISSN 0093-691X

Lips, D.; De Tavernier, J.; Decuypere, E. & Van Outryve, J. (2001). Ethical objections to caesareans: implications on the future of the Belgian White Blue, *Preprints of EurSafe 2001: Food Safety, Food Quality, Food Ethics*, Florence, Italy, October 2001

Lusk, J.; Norwood, F. & Prickett, R. (2004). Consumer preferences for farm animal welfare, Oklahoma State University Department of Agricultural Economics

MacLennan, D. & Phillips, M. (1992). Malignant Hyperthermia. *Science*, Vol. 256, (May 1992), pp. 789-794, ISSN 0036-8075

Mak, N. (2008). Animal Welfare for Sale: Genetic Engineering, Animal Welfare, Ethics, and Regulation. In: *American Anti-Vivisection Society*, (November 2008)

McPherron, A.; Lawler, A. & Lee, S. (1997). Regulation of skeletal muscle mass in mice by a new TGF-β superfamily member. *Nature,* Vol. 387, pp. 83-90, ISSN 0028-0836

McPherron, A. & Lee, S. (1997). Double muscling in cattle due to mutations in the myostatin gene. *Proceedings of the National Academy of Sciences USA,* Vol. 94, No. 23, (November 1997), pp. 12457-12461, ISSN 0027-8424

Muir, W. & Howard, R. (1999). Possible ecological risks of transgenic organism release when transgenes affect mating success, *Proceedings of the National Academy of Sciences,* Vol. 96, No. 24, (November 1999), pp. 13853-13856, ISSN 0027-8424

Murray, J. (1999). Genetic modification of animals in the next century. *Theriogenology,* Vol. 51, (1999), pp. 149-159, ISSN 0093-691X

Mutalib, A. & Hanson, J. (1990). Sudden death in turkeys with perirenal hemorrhage: field and laboratory findings. *Canadian Veterinary Journal,* Vol. 31, (1990), pp. 637-642, ISSN 0008-5286

National Agricultural Statistics Service. (2007). Overview of the U.S. turkey industry, In: *The United States Department of Agriculture,* 25.11.2008

National Research Council. (2002). *Animal Biotechnology: Science-Based Concerns,* National Academies Press, ISBN 0309084393, Washington, District of Columbia

Newberry, R. (2004). Cannibalism, In: *Welfare of the Laying Hen. Poultry Science Symposium Series 27,* G.C. Perry, (Ed.), CABI Publishing, ISBN 0851998135, Wallingford, UK

Newport, F. (2008). Post-Derby tragedy, 38% support banning animal racing, In: *Gallup,* 15.05.2008

Ott, R. (1996). Animal selection and breeding techniques that create diseased populations and compromise welfare. *Journal of the American Veterinary Medical Association,* Vol. 208, No. 12, (June 1996), pp. 1969-1974, ISSN 0003-1488

Palmiter, R.; Brinster, R.; Hammer, R.; Trumbauer, M.; Rosenfeld, M.; Birnberg, N. & Evans, R. (1982). Dramatic growth of mice that develop from eggs microinjected with metallothionein-growth hormone fusion genes. *Nature,* Vol. 300, (December 1982), pp. 611-615, ISSN 0028-0836

Palmiter, R.; Norstedt, G.; Gelinas, R.; Hammer, R. & Brinster, R. (1983). Metallothionein-human GH fusion genes stimulate growth of mice. *Science,* Vol.222, No. 4625, (November 1983), pp. 809-814, ISSN 0036-8075

Pennisi, E. (2002). A shaggy dog history. *Science,* Vol. 298, pp. 1540-1542, ISSN 0036-8075

Pew Initiative on Food and Biotechnology. (2005). *Proceedings of Exploring the Moral and Ethical Aspects of Genetically Engineered and Cloned Animals,* Rockville, MD

Pinkert, C. & Murray, J. (1999). Transgenic Farm Animals, In: *Transgenic Animals in Agriculture,* J.D. Murray, G.B. Anderson, A.M. Oberbauer, M.M. McGloughlin (Ed.), 1-18, CABI Publishing, ISBN 0 85199 293 5, New York, NY

Rauch, A. & Sharp, J. (2005). *Ohioans' attitudes about animal welfare,* Ohio State University Department of Human and Community Resource Development, (2005)

Reh, W.; Maga, E.; Collette, N.; Moyer, A.; Conrad-Brink J. & Taylor, S. Hot topic: using a stearoyl-CoA desaturase transgene to alter milk fatty acid composition. *Journal of Dairy Science,* Vol. 87, (2004), pp. 3510-3514, ISSN 0022-0302

Renema, R.A. & Robinson, F.E. (2004). Defining normal, *World's Poultry Science Journal,* Vol. 60, (December 2004), pp. 508-522, ISSN 0043-9339

Rodgers, B. & Garikipati, D. (2008). Clinical, agricultural, and evolutionary biology of myostatin: a comparative review. *Endocrine Reviews*, Vol. 29, No. 5, (August 2008), pp. 513-534, ISSN 2008-0003

Rollin, B. (1995). Research issues in farm animal welfare, In: *Farm Animal Welfare*, Iowa State University Press, ISBN 0813801915, Ames, IA, USA.

Ruegg, P. (2001). Milk Secretion and Quality Standards, Cooperative Extension of the University of Wisconsin

SandØe, P. & Holtug, N. (1993). Transgenic animals — which worries are ethically significant? *Livestock Production Science*, Vol. 36, (1993), pp. 113-116, ISSN 0301-6226

Smith, J. (2008). White House to Ease Many Rules, *Washington Post*, (31 October 2008)

Smith, R. (1991). Cutting edge poultry researchers doing what birds tell them to do. *Feedstuffs*, (September 1991), pp. 22, ISSN 0014-9624

Tabler, G. & Mendenhall, A. (2003). Broiler nutrition, feed intake and grower economics. *Avian Advice*, Vol. 5, No. 4, (2003), pp. 8-10

United States Department of Agriculture. (2008a) Changes in the U.S. Dairy Cattle Industry, 1991-2007, IN: *Dairy 2007 Part II*

United States Department of Agriculture. (2008b) Antibiotic Use on U.S. Dairy Operations, 2002 and 2007

United States Public Health Service. (2003). Grade "A" Pasteurized Milk Ordinance, In: *U.S. Food and Drug Administration*, 17.03.2008

Urrutia, S. (1997). Broilers for next decade: what hurdles must commercial broiler breeders overcome? *World Poultry*, Vol. 13, No. 7, (1997), pp. 28-30, ISSN 1388-3119

Uystepruyst, C.; Coghe, J.; Dorts, T.; Harmegnies, N.; Delsemme, M.; Art, T. & Lekeux, P. (2002). Optimal timing of elective caesarean section in Belgian white and blue breed of cattle, *The Veterinary Journal*, Vol. 163, No. 3, pp. 267-282, ISSN 1090-0233

Vestel, L. (2001). The next pig thing, In: *Mother Jones*, 17.03.2008

Wall, R.; Kerr, D. & Bondioli, K. (1997). Transgenic dairy cattle: genetic engineering on a large scale. *Journal of Dairy Science*, Vol. 80, (1997), pp. 2213-2224, ISSN 0022-0302

Wall, R.; Powell, A.; Paape, M.; Kerr, D.; Bannerman, D. & Pursel, V. (2005). Genetically enhanced cows resist intramammary Staphylococcus aureus infection. *Nature Biotechnology*, Vol. 23, (2005), pp. 445-451, ISSN 1546-1696

Webster, A. (2002). Rendering unto Caesar, *The Veterinary Journal*, Vol. 163, (2002), pp. 228-229, ISSN 1090-0233

Weeks, C.; Danbury, T.; Davies, H.; Hunt, P. & Kestin, S. (2000). The behaviour of broiler chickens and its modification by lameness. *Applied Animal Behaviour Science*, Vol. 67, (2000), pp. 111-125, ISSN 0168-1591

Wendt, M.; Bickhardt, K.; Herzog, A.; Fischer, A.; Martens, H. & Richter, T. (2000). Porcine stress syndrome and PSE meat, *Berl Munch Tierarztl Wochenschr*, Vol. 113, No. 5, (May 2000), pp. 173-190, ISSN 0005-9366

Westhusin, M. (1997). From mighty mice to mighty cows. *Nature Genetics*, Vol. 17, (September 1997), pp. 4-5, ISSN 1061-4036

Wheeler, M.; Walters, E. & Clark, S. (2003). Transgenic animals in biomedicine and agriculture: outlook for the future. *Animal Reproduction Science*, Vol. 79, (April 2003), pp. 265-289, ISSN 0378-4320

Whitehead, C.; Fleming, R.; Julian, R. & Sorenson, P. (2003). Skeletal problems associated with selection for increased production, In: *Poultry Genetics, Breeding, and*

Biotechnology, W.M. Muir & S.E. Aggrey, (Eds.), CABI Publishing, ISBN 9780851996608, Wallingford, United Kingdom

Wise, D. & Jennings, A. (1972). Dyschondroplasia in domestic poultry. *Veterinary Record,* Vol. 91, (1972), pp. 285-286, ISSN 0042-4900

Young, E. (2002). GM Pigs are both meat and veg. *New Scientist: The World's No. 1 Science & Technology News Service,* Vol. 12, No. 30, (January 2002), ISSN 0262-4079

Zakin, S. (1999). Nonpoint pollution: the quiet killer. *Field & Stream,* (August 1999), pp. 84-88, ISSN 8755-8599

Zuidhof, M. (2002). Common laying hen disorders Prolapse in laying hens, In: *Alberta Agriculture and Rural Development,* 24.07.2008

5

The Bumpy Path Towards Knowledge Convergence for Pro-Poor Agro-Biotechnology Regulation and Development: Exploring Kenya's Regulatory Process

Ann Njoki Kingiri

African Centre for Technology Studies (ACTS),
Off United Nations Crescent, Gigiri Nairobi,
Kenya

1. Introduction

Contestations over the regulatory trajectory that developing countries should take to embrace the benefits of biotechnology[1] have been debated widely. Predominant debates in the global arena have focussed on two competing regulatory approaches namely; the more permissive approach of the United States (US) that presents biotechnology as posing no risk to the environment or human health unless proven otherwise through scientific risk evaluation and the more restrictive approach of European Union (EU) that imposes precautionary restrictions on use of products of biotechnology even when scientific knowledge about risks is uncertain. There is now a wide body of literature looking at trade conflicts brought about by these divergent regulatory policies leading to regulatory polarization (Paarlberg, 2001, 2008; Bernauer and Aerni, 2007; Bernauer, 2005; Falkner, 2007; Murphy and Levidow, 2006). Arguably, these polarised debates only expose the political dynamics of biotechnology from a very narrow view, primarily trade imperatives (Clapp, 2006). Some analysts departing from what they perceive to be a narrow approach to this subject have attempted to explain the bumpy path to biotechnology deployment and regulation in developing economies that exhibit different characteristics. Millstone and van Zwanenberg (2003) for instance looking at GM policies in the South have shown that the scientific conflicts embedded in GM safety compel countries to pursue divergent regulatory choices. Research has also shown that local context dictates technological dynamics and

---

[1] Here I use the term biotechnology to mean the manipulation of living organisms to produce goods and services useful to human. I make distinction between traditional (or conventional) and modern biotechnologies. The traditional approach allows the development of new products (such as seed varieties) by the process of selection from genetic material already present within a species, while the modern (transgenic) approach develops products (such as seed varieties) through insertion of genetic material from different species into a host plant. These products are popularly known as Genetically Modified Organisms (GMOs).

should be given a place in biotechnology development and regulation (van Zwanenberg et al., 2008). Falkner and Gupta (2009) have also noted that despite the EU-US international regulatory conflicts, developing countries are responding to related pressures in different and unique ways.

Important for this paper is the dynamics brought about by regulatory pressures and what this means for knowledge use towards productive debates that could lead to pro-poor biotechnology development. In view of this, there is need to re-orient the discussions around how actors in the respective value chains ought to respond to regulatory demands brought about by biotechnology, and how this impacts knowledge production dynamics. The paper argues that biotechnology development will only contribute to economic development if knowledge (regulatory, social and scientific knowledge), emanating from different knowledge nodes is allowed to converge to a point where it can consequently inform productive innovation policy processes. This argument is based on the understanding that requisite innovation capacities need to be built in order for actors to use the resources at their disposal towards behavioural change for biotechnology regulation (Hall, 2005). Knowledge is one of the resources and how it is applied is crucial for biotechnology innovation process or trajectory.

This paper relooks at knowledge production dynamics through an empirical account that documents the biotechnology regulatory trajectory in Kenya. The analytical context for the paper is backed by the political nature under which biotechnology development and biosafety regulation have co-evolved for almost two decades (Harsh, 2005; Sander, 2007; Kingiri and Ayele, 2009, Kingiri, 2010, 2011a,b). Analysis drawn from these papers suggests that scientific knowledge predominantly directs biotechnology development and regulation (cf Kingiri, 2010). In addition, the fragmented nature of actors' infrastructure and their belief systems derail the knowledge convergence efforts (Kingiri, 2011a). Although this process lacked legal direction in terms of policy (Biotechnology policy and Biosafety Act were only approved in 2006 and 2009 respectively), the paper suggests that lessons learnt in the Kenya's regulatory process should move the country biotechnology sector to a higher level towards putting the research products which have been in the pipeline to use. The objective of the paper therefore is to explore and understand how knowledge convergence can be attained towards moving biotechnology science forward towards innovation.

The paper is structured as follows. Political scenario under which biotechnology regulation occurs is discussed first. This is followed by an analytical context under which this paper is situated. Next, the paper illuminates the dynamics associated with biotechnology regulation using Kenya as an example. Lastly, the paper discusses the emerging dynamics associated with regulation and role of knowledge actors. It then concludes by drawing lessons that might inform a productive regulatory process towards knowledge convergence.

2. Political nature of biotechnology regulation: An overview

It is now understood that agricultural production constraints like pests and diseases have been perpetuating the cycle of food insecurity and poverty in sub-Saharan Africa. The questions that many have been asking relate to whether biotechnology can be exploited as a possible solution to these and other production constraints. Proponents are optimistic about this while opponents are pessimistic citing safety concerns around human health and environment. Answers to these and related questions provide a more complex and charged debate about biotechnology development and regulation.

Despite the undisputed consensus about biotechnology as a tool for agricultural development in poor economies (FAO, 2004), political polarization on GMOs has been increasing. The participation of many interested stakeholders in charting a supposedly sustainable regulatory pathway has confounded the process due to the value based nature of divergent perspectives (Paarlberg and Pray, 2007; Leach et al., 2007). It has been noted that policy processes embraced in advanced economies particularly EU and USA has significantly shaped public opinion and regulation in most African countries (Newell, 2003). EU for instance has been associated with advocacy groups opposing biotechnology introduction even in regions where food security challenges persist (Herring, 2010; Paarlberg, 2001, 2008). Other analysts have further explored the problem more broadly and have added that a more holistic approach to biotechnology debate is needed to embrace the context that varies with regions, localities and social preferences (van Zwanenberg et al., 2008; Glover, 2010).

This paper is in line with holistic view of biotechnology regulation that take cognisance of the context specific nature of domestic knowledge dynamics prompted by biotechnology. This includes environmental and social economic context, political and cultural context in relation to how decision processes are pursued to promote legitimacy and transparency among others (Glover, 2010). Arguably, these factors shape knowledge production dynamics giving a context to the analysis.

3. Setting the scene: Conceptual and analytical context

Building on some of the studies on biotechnology governance in Kenya alluded to elsewhere; this section calls attention to the importance of the various aspects of technological, regulatory and social local contexts in which the knowledge actors (including the organisations involved) and regulatory process are embedded. It seeks to provide the analytical context for the paper as well as situate the multiple actors engaged in biotechnology research and development (R & D) for the last two decades within the process of regulation implementation. By doing this, the paper exposes the motivations and opportunities for actors in their engagement with biosafety regulatory process and formulation of regulatory instruments, and the institutional challenges and strengths related to modern biotechnology governance.

The paper seeks to analyse how biotechnology regulation (which includes instituting a biosafety regulatory system for management of biotechnology) may have affected efforts to bring about a knowledge convergence in biotechnology regulation. Kenya was selected backed by the rich political context that prevailed during the establishment of a regulatory system for management of biotechnology Research & Development (R & D).

3.1 Research context and methodology

Kenya presents an excellent case to investigate knowledge management associated with modern biotechnology in terms of regulatory policy environment and context. This is because the initiation of biotechnology R & D activities that commenced in 1990's paralleled the establishment of the requisite regulatory process providing an exemplary context to investigate the dynamics around knowledge production with both technological and regulatory orientations. This parallel process engaged communities in research, policy and public arenas in an iterative manner bringing about interesting biotechnology and institutional innovations. Secondly, policy initiatives like the strategy for revitalising agriculture (RoK, 2005) and the Vision 2030 embrace an integrated approach to innovation towards economic development.

The context described above created a conducive environment to undertake qualitative in-depth semi-structured interviews with over 50 individual knowledge actors who had (or claimed to have) a stake in decisions pertaining to biotechnology as researchers, policy makers, employees of nongovernmental organisations (NGOs) and members of the public (mainly consumers and farmers). The research period was between 2006 and 2011. This was complemented by observations carried out during different scientific and public workshops in biosafety and biotechnology held during this period, and analysis of relevant secondary documents. Interviewees' points of engagement in the regulatory activities and decision processes are seen in the context of effort to provide knowledge (e.g. information, expertise and other resources) to influence policy outcomes. Consequently, the data analysis captured the different ways knowledge is used in the regulatory processes and what factors come into play.

Unless otherwise stated, codes are used to report all information cited in this paper in order to guarantee anonymity of some of the interviewees as requested. For instance, NGOco-NS4 refers to a non scientist interviewee from a civil society organisation.

3.2 An overview of Kenya's biotechnology development and regulation

3.2.1 Milestones in Kenya's biotechnology sector

Modern biotechnology has revolutionised many sectors including agriculture and embraces a wide range of applications including tissue culture, markers assisted selection and genetic engineering (GE) also referred elsewhere in this paper as modern biotechnology. All these are being applied in Kenya, but the latter is the focus of this paper. Just like many African countries, GE is relatively new, but GE products have been handled indirectly through trade in form of food aid (Kagundu, 2008).

Agricultural R&D has long been recognised as central to knowledge creation, technology development and innovation. During the pre-independence period, the R & D agenda was set by the British colonial government, which recognised the importance of Science and Technology (S & T) in agricultural production (Ochieng, 2007). It is not until early 1990's that biotechnology innovations in form of tissue culture received considerable attention (Wambugu, 2001). Actual work involving advanced GE commenced in 1991 when Kenyan scientists went to USA and in collaboration with scientists there, engineered a virus resistant sweet potato (Odame et al., 2003). Thereafter in 1998, the transformed plants required regulatory approval for this research to continue in Kenya. However, actual process of regulatory process and implementation had commenced prior to 1998.

To date, over six GE R & D initiatives have been evaluated in public institutions in conjunction with local and international partners (see Kingiri, 2011a for details). These

The Bumpy Path Towards Knowledge Convergence for Pro-Poor Agro-Biotechnology Regulation and Development:
Exploring Kenya's Regulatory Process

69

activities include *Bt* maize and *Bt* cotton engineered for resistance to insect pests, cassava for resistance to viruses and sorghum for resistance to striga weed. The recombinant rinderpest vaccine initiative targeted control of rinderpest disease in cattle and other viruses in small ruminants. Other initiatives are in the pipeline for example the sorghum fortified with nutrients funded by the Bills and Melinda gates foundation through the Africa Harvest Biotechnology Foundation International (see www.africaharvest.org). Since the approval of the first transgenic crop- the sweet potato in 1998, no product has reached the farmers and the furthest the biotechnology activities have gone towards a product is the (CFTs)[2]. It is hoped that with the establishment of a functional biosafety framework, the situation will change. In addition, the food insecurity related issues have prompted the government to take drastic policy measures approving importation of GM maize to avert a food crisis in the country (Daily Nation, 2011).

3.2.2 Biosafety regulatory mechanism

Biosafety encompasses the regulatory mechanisms that the government has put in place for the governance of GE activities. Article (8g) of the Convention on Biological Biodiversity (CBD, 2000) and Article (16) of the Cartagena Protocol provide for establishment of appropriate mechanisms to regulate, manage and control risks associated with Living Modified Organisms (LMOs). The protocol emphasises on risk assessment (RA) and risk management, and provides guidelines to achieve this (Annex III). There are several ways in which risk identified during RA can be managed, e.g. confinement, restricted use, provision of guidance, technical advice and record keeping (Halsey, 2006).

At the early stages of biotechnology research activities, Kenya opted to use the existing infrastructure, the Science & Technology Act (RoK, 1980) to institute regulatory mechanisms through the drafting and adoption of the *Regulations and Guidelines for Biosafety in Biotechnology in Kenya* (RoK, 1998). There were concerns that these regulations came long before the biotechnology policy and have not been legally binding as required by the law. In an effort to legalise the regulations as well as the biotechnology activities, *the National Biotechnology Development Policy* was drafted and later approved in 2006 (RoK, 2006). This was followed by Biosafety Act, 2009 (RoK, 2009). Kenya signed and ratified the Cartagena Protocol in May 2000 and January 2002 respectively. This further obligated the government to put up regulatory structures to operationalise it. This Biosafety Act therefore primarily seeks to operationalise the Protocol. The controversial developments surrounding its formulation over the years are at the centre of this paper. Kingiri (2011a) captures some of the main developments, revealing the dynamics that include the engaged different actors and the nature of engagement between 2002 and 2009. Within this period, various versions of the biosafety bill were drafted and discussed before the final version (RoK, 2009) was approved to become an Act in Feb. 2009. Meanwhile, regulations to be appended to the Act were drafted under the Program for Biosafety Systems (PBS) support and recently became operational from July 2011 after signing by the Minister for Higher Education, Science and Technology.

---

[2] This is a field trial of GM plants not approved for general release in which measures for reproductive isolation and material confinement are enforced in order to confine the experimental plant material and genes to the trial site (Halsey, 2006:4).

Previously, all the involved government actors and other nongovernmental players involved in biotechnology governance were brought together under the National Biosafety Committee (NBC) coordinated by the National Council for Science and Technology (NCST). NBC acted as a boundary organisation overseeing the management of biotechnology research through regulation. This role has since been taken over by the National Biosafety Authority (NBA) formed under the provision of the Biosafety Act.

3.2.3 Theoretical framework

To analytically situate the discussion in sound theoretical debates, this paper draws upon insights from the integrated knowledge management literature. Scholars try to explain the changing role of science in policy deliberations and the changing integrated knowledge production architecture prompted by new technological developments (Gibbons et al., 1994; World Bank, 2006). In the case of biotechnologies, this brings about governance challenges linked to biosafety regulation imposed to promote technological competitiveness and encourage public acceptance of these new technologies (Lyall, 2007).

4. Dynamics associated with biotechnology regulation: An empirical exploration of Kenya's case

In this section, practical reasons why and how stakeholders got entangled in Kenya's regulatory process is explored and the kind of reactions this generated. This helps us understand the challenges that may hamper a productive regulation towards knowledge convergence.

4.1 Challenges confronting the evolving agricultural R & D and modern biotechnology governance terrain

4.1.1 Contract research

It is widely argued that biotechnology is a key tool for 21st century sustainable development. However, most people agree that this may remain a dream unless certain challenges are addressed that include political support through provision of incentives for research and regulation (cf Bananuka, 2007). In Kenya, government support for S & T including biotechnology R & D has been minimal (Beintema et al., 2003; Odame et al., 2003). The dwindling research funds and other policy reforms have encouraged collaborative research, technology development and deployment (RoK, 2005; KARI, 2005; RoK, 2007). Although the government continues to fund public agricultural research, a significant support comes from donor organisations (Beintema, et al., 2003:5). Kenya Agricultural Research Institute (KARI) being the major research institute involved in modern biotechnology research (and the only one undertaking biotechnology CFTs) has undergone significant restructuring in response to these reforms and challenges. These changes have contributed to a rise in contract research characterised by increased donor funding (Beintema, et al., 2003:5). For instance all the agricultural GM trials are being undertaken through Public Private Partnerships (PPPs) arrangement (Ayele et al., 2006).

A lot has also been documented regarding the tissue culture bananas contract research (cf Smith, 2004). However in the case of modern biotechnology, the nature of contractual research is still under-researched and is undergoing changes at unprecedented rate due to the evolving institutional and regulatory contextual issues. What seems to lack is information on how actors in the value chain have been responding to the institutional changes associated with regulation of biotechnology science. This is crucial for related knowledge convergence efforts.

4.1.2 Biotechnology and biosafety capacity

According to Bananuka (2007), the need for regulatory capacity evolves alongside an operational biotechnology sector, and this has been the case in Kenya. Since the biotechnology programme was initiated in early 1990's, capacity in both modern biotechnology techniques and biosafety (human, infrastructural and institutional) has been built over the years. For instance, a modern biosafety Level II greenhouse has been put up at the KARI biotechnology centre. According to a report prepared for policy makers (Handbook for Policy Makers, 2007), the number of scientists trained in biotechnology countrywide has gone up, with 45% of those trained being actively engaged in GE work. In addition, capacity in regulatory institutions like Kenya Plant Health Inspectorate Service (KEPHIS), Kenya Bureau of Standards (KEBS), Department of Veterinary Services (DVS) and Department of Public Health (DPH) has been strengthened and as argued in this report, these institutions are in a state to oversee the implementation of biosafety regulations. Arguably, these rhetorical claims were advanced by GMOs proponents, in their endeavour to lobby for the enactment of the bill (Kingiri, 2011a, b). However, these capacity building and biopolicy developmental efforts have been collaborative. Despite these milestones, both infrastructural and human capacity remains far from being adequate. This is attributed to several factors among them inadequate government support for research discussed above and lack of regulatory policy environment to spur development (Wafula et al., 2007), that would further encourage and favour capacity building efforts. The increased cross-over of trained scientists from public institutes to international organisations locally and abroad has also contributed to the unsustainable capacity building efforts, a trend which is prevalent in the African region as a whole (Hastings, 2009).

4.2 Stakeholders' proactive role in regulatory process

From the foregoing analysis, it is emerging that various challenges have hampered the evolution of the twin processes - biotechnology innovation and regulatory regime. These relate to partly the technical and institutional capacities, but this analysis does not address an important question related to how the actors (individuals, organisations and related links) deal with the analysed challenges. The implications in terms of how challenges are dealt with are important in informing the dynamics around knowledge use and regulatory decision making processes.

This section tracks empirically the Kenya's regulatory trajectory paying attention to the involvement of knowledge actors in this process, exposing the tensions that this generated. It is important to note that many interviewees desired a regulatory environment that would enhance deployment of products of GE science. Biosafety bill was a gateway towards achieving that goal. Media reports analysed during field work confirm some activism by the scientific and non scientific communities in support or against the biosafety bill. Biosafety formulation process as a pertinent step in legalising the regulatory regime engaged the scientific community intensely. Scientists collectively educated policy makers and journalists, sensitizing them on GE thus making "*a case for biotechnology*" as well as persuading them to support it (RSIn-GP2, Dec. 2007). This was however viewed with suspicion by some interviewees, who were concerned with what they viewed as biotechnology promotional agenda and associated politics. Several documents obtained during field work and numerous media reportage by both proponents and opponents seemed to confirm this pro-activeness (see Kingiri, 2011a for a detailed empirical account of dynamics involved).

4.2.1 Motivations, opportunities, interests and implications associated with biotechnology

When biotechnology research was initiated in early 1990's it provided an incentive for researchers with many seizing this opportunity to pursue their knowledge and technology transfer endeavours. This triggered public reaction with this behaviour being interpreted by non scientific communities from the civil societies as unwarranted excitement and hype. To moderate the different interests and concerns, amongst stakeholders, the government imposed biosafety regulations (RoK, 1998) to guide in subsequent knowledge generation endeavours and decision making processes. Regulations were interpreted in different ways by different stakeholders (see Kingiri, 2011b for details). Perhaps because of conflicts between different motivations and opportunities presented by GE research, and the different challenges associated with biosafety regulations and implementation, the stakeholders exposed certain varying behavioural practices. This generated varying reactions as reported in the subsequent sections.

4.2.2 Scientists credibility and transparency questioned

The conduct of GE trials was perceived to require a substantial level of credibility on the part of scientists due to sensitivity of GE technology. However, research scientists were perceived by a number of interviewees to be untrustworthy and dishonest. Perhaps credibility is one aspect that regulations should be promoting, prompting regulators to emphasise appropriate monitoring of research trials and scientists: "*scientist…will do things that you cannot believe it is possible.*" (RSPu-PS7, Nov. 2008). Credibility was however found to be constrained by institutional obligations and compromises that both scientists in policy and practice were forced to make. The interim regulations prior to the Biosafety Act were unclear about how credibility as an ethical practice is linked to compliance and monitoring. However, the Biosafety Act provides for intensive monitoring through designated biosafety experts (Articles 43 & 45).

4.2.3 Attitude towards regulations and regulators

Many interviewees in policy arena including regulators described scientists as having a negative attitude towards regulations. The reasons behind this negative attitude were attributed to conflicting motivations like research interests discussed previously, making scientists view regulations as "*hindrance to science*" (LABp-NS8, Jan. 2008). Others explained that scientists find it difficult to adjust from their normative basic research behaviour to a supposedly demanding research practice like the one demanded by GE research. The attitude of researchers towards regulators and vice versa promoted suspicions and misunderstandings amongst them, constraining effective regulatory practice.

The Bumpy Path Towards Knowledge Convergence for Pro-Poor Agro-Biotechnology Regulation and Development:
Exploring Kenya's Regulatory Process

73

4.2.4 Poor public communication and biased reporting

This section reports on practice of scientists related to how they disseminate research information emanating from biotechnology research trials. The regulatory instruments prior to the Biosafety Act and the Act itself are implicit about how this reporting process should be managed. They however emphasise on transparency that should promote public trust. RoK (1998) in particular recommends "openness" to "safeguard public interest" through transparent handling of information and adhering to regulations (executive summary). In RoK (2009), NBA is wholly responsible for information handling and management including consequent public awareness. A register will also be maintained as a repository for biosafety information. It is however unclear how interested parties should access it. Accounts of interviewees suggested that scientists have poor public communication skills on biotechnology matters. In addition, when they communicate (as demanded by the sensitive nature of this technology), there are weaknesses that are revealed through the reports and the communication strategies they adopt. However, many interviewees were in agreement that scientists have a very important role to play in communicating scientific and technical facts to the public about their GE work. Some perceived this as the only way of demystifying the prevailing negative publicity around GE technology. There were however perceived weaknesses and challenges in the articulation of this role which are explored next.

4.2.4.1 Communicating science versus public understanding of science

Some interviewees claimed that research scientists use "scientific jargon" that need to be "toned down" for lay people to understand. The use of technical and scientific language was perceived to be an indicator of poor communication skills that purportedly differentiates pro-GE scientists from anti-GE activists. This discussion seems to point out that scientists have not come to the level of the non scientists or the public when communicating technical aspects of GE research. This analysis does not however expose the reasons behind this seemingly uncomfortable behaviour and repercussions.

4.2.4.2 Public communication constrained by fear of misinterpretation

Scientists argued that they deliberately avoid communicating scientific findings to the public because of fear of misinterpretation, propaganda and potential negative impact this may have, for instance on their careers and research reputation. Fear of propaganda was associated with activists, who some claimed unjustifiably fight biotechnology impacting on scientists' reporting behaviour:

"So [research scientists] have avoided bringing negative stories and even when they see them they remove them and instead keep quiet. Experience has shown that, any negative you bring will be used against you. So we have to continue in the way I think we are at least less risky." (TAR-NSS1, researcher & technology advocacy, international NGO, Feb. 2008)

This has implications as many scientists asserted. The fear of reporting non-factual and unverified or unconfirmed findings constrain effective and timely reporting, leaving room for misinterpretation by counter groups. Arguably, scientists are held back from freely sharing their findings with the public for fear of repercussions associated with misinterpretations. This has implications for practice on the part of the scientists in respect of information and knowledge management, and how this is interpreted by others.

4.2.4.3 Communicating the positives and transparency

Majority of scientists admitted that when scientists communicate about GE science, it is basically the positive and promotional information that highlights benefits more than risks. Misinterpretation was affecting the way scientists communicate, compelling them to talk more of tangible benefits and less on unverified or "*unknown*" risks. Several non scientist interviewees corroborated the "biased reporting" linked to provision of information inclined more to successes:

"In Kenya, all we are hearing are the positive aspects. We know that no technology in this world is without risks. So why is the potential risk side [of GE technology] silent? That in itself sends alarm bells to us [civil society]." (NGOf-NS1, farmers' rights advocacy, civil society, Nov. 2007)

Defending this practice, some researchers argued that, the nature of biological science training encourages them to pursue only facts, compelling them to withhold information that cannot be validated. This was discussed in connection with confidence and easiness in reporting facts as opposed to unverifiable information like cases of uncertainty. They further argued that reporting on GE risks may cause panic among the public if negative non-validated aspects related to scientific "*process*" are highlighted. However other interviewees claimed that, scientists withholding of some information was linked to "*a normative rigid research practice*" that compels them to vet what they report (ATp-PS3, Nov. 2007). This analysis seems to portray scientists as self centred, and tends to put to doubt their previous claims of fear of misinterpretation. Questionably, there is a disconnect between constrained communication and the unbalanced information consequently disseminated.

4.2.4.4 Unreliable & biased information and multiple obligations

Exogenous pressures were perceived by a number of scientists and most civil society interviewees to be limiting the reporting freedom of researchers, prompting them to produce what was referred to as "*biased*" and "*unreliable*" information, presumably manipulated to suit certain interests. Many felt that, reports emanating from research trials were unreliable because of the partnerships environment under which the trials are undertaken. This was perceived to be prompting reporting that favoured multiple obligations commensurate with different interests. This created tension amongst the civil societies: "*it is difficult to say per se that in the current [donor] context the information from those researchers would be fully reliable*" (NGOco-NS4, consumers' network, Jan. 2008).

The preceding analysis suggests that certain technical and non technical factors largely influence the behavioural practice exhibited by scientists in knowledge production endeavours linked to biotechnology regulation. Some factors are associated with opportunities presented by GE science, while others are linked to challenges that confront actors including scientists as they engage in biotechnology research and regulatory process.

4.2.5 Technical experts and conflicts of interest

National efforts to establish a legally binding regulatory regime in compliance with Cartagena Protocol engaged stakeholders in various ways. One of the roles of the NBC according to RoK (1998) was to draw policies and procedures to govern biotechnology. In this regard, this gave NBC the legal powers to spearhead the policy-making process. However, NBC coordination role in the biosafety bill formulation process was perceived to

The Bumpy Path Towards Knowledge Convergence for Pro-Poor Agro-Biotechnology Regulation and Development:
Exploring Kenya's Regulatory Process

75

be blurred by the activism of other actors, a view shared by both scientists and non scientists. Arguably, the scientists and their allies became the main drivers of the bill formulation process:

"The main players were the biotechnology industry, and the scientists make much of the industry. The whole process was supposed to be an initiative of the government but the interest was with people from the biotechnology industry than what we would call the broader section of Kenyan society." (JO-NS6, journalist, local daily, Apr. 2008)

NBC was also largely made up of scientists representing different organisations with two representatives from the civil society. This being the case, it can be concluded that scientists and their affiliated institutions played vital roles as technical experts (see Kingiri, 2010). This role is however threatened by perceived motivations and interests likely to bring about conflicts of interest. It was a concern of non-scientists from the civil society that technical information used in risk assessments (RA) and consequent decision making pertaining to GE trials was solicited by scientists from technology developers who are interested parties.

The relationships established around the regulatory process in the Kenyan context were mutual in that the participating players expected to benefit. Scientists and the government were for instance receiving financial support from non state actors and donors. These relationships and partnerships were perceived by many interviewees to have positively enhanced the regulatory process. Further, some interviewees were in agreement that the government has inadequate capacity to support the regulatory process, so these other supporting parties were filling in that gap. From these accounts, resources and in particular financial support was a key incentive cementing these relationships.

4.3 The never ending controversy

The Biosafety Act (2009) approved in Feb 2009 may be seen as a victory for agro biotechnology development towards benefiting the poor. The formation of an administrative entity, the NBA to legally manage biosafety controversies under the provision of this Act may also be seen as a plus towards development endeavours. However, the broader food security issues as well as socio economic and political environment mask smooth biotechnology development and regulation. On 14th February 2011, the Kenyan cabinet made a political pronouncement that approved immediate importation of GM maize to avert a looming food crisis. This development received considerable media reportage which subsequently generated wide public protests led by civil society (see Opiyo, 2011; Omondi, 2011; Kinuthia, 2011). The proponents who include scientists did not see anything wrong with the importation citing scientific evidence that has shown GM products to be safe for human consumption. The opponents expressed scepticism citing unconfirmed risks posed by these products to human health. This controversy suggests that the debate surrounding biotechnology clearly continues to remain polarised making it harder for the public to endorse biotechnology products. This is a major drawback to science advancement as well as a threat to its longstanding authority in providing solutions to societal problems.

5. Discussion and conclusion

The foregoing empirical exploration of Kenya's regulatory process exposes a controversial engagement in knowledge production dynamics. This is in part linked to weak governance structures in terms of both biotechnology delivery and related regulatory mechanisms. This has implications for productive knowledge convergence efforts. Firstly, in relation to who should be involved as stakeholders in the regulatory process and how they should be engaged, the Kenya's case presents major participation and transparency challenges. The regulatory process although enlisting participation of both technical and non technical experts sidelined the public expertise (see also Kingiri, 2010). This may imply that expertise that may bring on board socio, economic and cultural contexts of Kenya's broader agricultural terrain could have been ignored in decision making processes. For instance, the public private partnerships that are currently evident in Kenyan biotechnology initiatives are largely triggered by technical and financial constraints (Intellectual Property Rights-IPRs, infrastructure, funding, individual scientists interests etc) as opposed to the needs and production constraints that can benefit farmers. Secondly, the resources (including knowledge and information shared and disseminated amongst players, regulatory instruments, legal and administrative structures, media as avenue for information dissemination and experts) that purportedly steered the regulatory dynamics were not devoid of conflicts of interests and influence. This has implications for productive knowledge convergence efforts as it generates suspicion, luck of trust and perhaps potential rejection of science.

In a complex science policy terrain like biotechnology regulation, multiple contexts may work for or against the intended innovation and public policy. Thus, the following question posed by Haas, (1992:1) is very valid. Can policy makers or scientists themselves *identify national interests and behave independently of pressures of social groups they nominally represent?*" He argues that, actors can learn new patterns of reasoning informed by a wider stakeholder needs and interests. The general argument advanced here is that technical experts that include scientists can genuinely play their part to influence positive change in policy-making through appropriate use of knowledge and information (Haas, 1992:3). The scientific community has a major role to play in this because they understand the complexities and uncertainties associated with biotechnology better than the non-scientists and policy makers (Bradshaw and Borchers, 2000). In addition, inclusion of a wide range of expertise that encompasses non-technical professionals is a positive way to democratise the regulatory process towards a socially robust knowledge production infrastructure (Nowotny, 2003; Nowotny et al., 2001, 2003).

The behavioural practice exhibited by Kenyan scientific experts in generation and handling of biosafety related information could be a concern for a productive knowledge convergence that is intended to promote biotechnology development and adoption. To promote credibility and transparency and consequently enhance trust associated with biotechnology; this paper further suggests a change of attitude of actors towards a socially responsible process. The scientific community and policy makers and those groups that claim to represent the farmers and public must be honest with no hidden agenda (Ammann and Ammann, 2004). In addition, reflexivity should be encouraged. As a value based practice, reflexivity is the process by which individuals involved in knowledge production try to operate from the standpoint of all experts involved (Gibbons et al., 1994). For the purpose of enhancing knowledge convergence in biotechnology development, expertise from different stakeholders should be considered in biosafety regulation and other decision making processes.

6. Lessons towards knowledge convergence

This section looks at insights that this paper can draw upon from almost two decades of Kenya's biotechnology and regulatory regime co-evolution in terms of practice. Three distinct aspects are key in putting the lessons discussed here into context:

Dynamism: Biotechnology innovation is advancing at an unprecedented pace, perhaps faster than the capacities of actors and institutions to adjust in order to accommodate the requisite changes needed to foster innovation and responsive engagement of stakeholders, including regulation (Tait et al., 2006:379). This has called for new styles of governance that urge for participative decision making processes that consider all stakeholders' interests and values (Lyall and Tait, 2005).

Multifaceted: Both biotechnology innovation and the embedded regulatory process involve many actors with each process being multifaceted. Consequently, the accompanying practices that actors chose to adopt or pursue are perceived to be problematic. According to Murphy and Chataway (2005), this may be attributed to influence of different policy cultures at the global level (e.g. EU versus US) and regional level (e.g. African Union). In the case of Sub Saharan Africa, this is also connected to influence of policy cultures at national levels (Mugwagwa, 2008).

Complexities related to shifting regulatory practice: The entire biotechnology and regulation revolution involves complex trade related and institutional dynamics (Fukuda-Parr, 2006) which inevitably impacts behaviour of actors like scientists. In Kenya, the behavioural shifts are sometimes encouraged by the inadequate and specialised biotechnology-biosafety knowledge capacities needed to move the regulatory process forward. This may not be construed to be a bad thing because within a dynamic and functional system like biotechnology, this may promote cumulative knowledge and learning. However, how learning and knowledge are managed is important for practice.

Considering these dynamics, a number of lessons can be drawn in relation to knowledge use and policy making as explored next.

6.1 Harnessing the positive aspects and dealing with the negative aspects

We cannot rule out the important learning that has taken place in the evolving Kenya's biotechnology regulatory system for almost two decades both at the institutional and individual levels, much of which constitute tacit knowledge. The government has to look for ways of using this accumulated knowledge. One way it can do this is to compile a list of experts who have been involved, and perhaps include and consider them as official experts. They would then be called upon from time to time in biotechnology and biosafety awareness campaigns and capacity building efforts targeting the wider stakeholder communities. In addition to sensitising people about specific technical subjects, they would also be requested to talk about their experiences in biosafety regulatory process, providing a platform for meaningful deliberations that can bring about knowledge convergence promoting pro-poor and pro-biotechnology innovation agenda.

It is possible that the regulatory dynamics discussed in Section 5above may have a negative impact on future biotechnology deployment and adoption. For instance, the scientific community's active participation in the regulatory process may have resulted into more of technical and scientific knowledge informing the policy deliberations. This may have ignored some other relevant knowledge which may enhance convergence efforts. These possible negative aspects cannot be ignored and have to be factored into future decision-making processes. How can this be done?

- The government has a major role to play by adopting a governance approach to public policy processes through weighing and analysing the types of knowledge that inform the process. The objective would be to ensure that socially desirable knowledge informs the final policy outcome (Nowotny et al., 2001).
- The government needs to build and sustain technical capacities thereby have a wide pool of experts in which to draw expertise from. It should also spread its wings to other academic and non academic institutions to solicit expertise not only for regulatory instruments, but also for overall risk assessment and environmental safety reviews.

6.2 Reconceptualising policies formulation process

In addition to the above lessons, the significant shift in behavioural practice associated with knowledge actors demonstrated empirically in the Kenyan case that accompany the biotechnology and biosafety revolution lead to a compelling urge to reconsider how policy and regulatory formulation processes are conceptualised and articulated. Regulatory practice, if it is to achieve greater effect in reconciling the governance agenda of modern biotechnology on the one hand, and role of actors in providing evidence-based expertise into the process; it must factor into the process this inevitable change in practice. This is not to denounce the economic and institutional factors in which this shift is embodied, but rather to suggest that this becomes an additional consideration in policy processes. Since this shift is exhibited by actors spread out in different institutions (academic, policy, NGOs, public), effective policy and regulatory processes must first acknowledge its potential to influence policy directions. Consequently, strategies should be devised that encourage a reflexive and responsive behaviour (Lyall, et al., 2009: 261). This may enrich how policies are implemented considering that cultural practices in biotechnology are linked to values and interests (Laurie et al., 2009).

In conclusion, the paper appeals to the policy, public and scientific communities to adopt a reflexive approach to biotechnology development and regulation in order to enhance convergence of knowledge for sustainable development.

7. Acknowledgements

This paper is based on research conducted in Kenya over the period 2006-2011 funded by the Open University, UK, the UK Economic and Social Research Council (ESRC), Innogen centre and partly by the United Kingdom, Department for International Development (UK-DFID) - Research Into Use (RIU) program, and African Centre for Technology Studies (ACTS). The author gratefully acknowledges this support. The views expressed in the paper do not necessarily reflect those of the Open University, ESRC Innogen centre, DFID and ACTS.

8. References

Ammann, K. & Ammann, P. (2004). Factors influencing public policy development in agricultural biotechnology. In Shantaram, S. (Ed.), *Risk Assessment of Transgenic*

Crops. Handbook of Plant Biotechnology, Vol. 9, pp. 1552. Wiley and Sons, Hoboken, NJ, USA.

Ayele, S.; Chataway, J. & Wield, D. (2006). Partnerships in African crop biotech. *Nature Biotechnology,* Vol. 24, pp. 619-621.

Bananuka, J.A. (2007). Biotechnology capacity building needs in Eastern Africa. In ICTSD and ATPS, *Biotechnology: Eastern African perspectives on sustainable development and trade policy,* pp. 1-17. ICTSD, Geneva, Switzerland and ATPS, Nairobi, Kenya.

Beintema, N. M.; Mureithi, F. M. & Mwangi, P. (2003). Agricultural Science and Technology Indicators (ASTI). Kenya. *ASTI Country Brief* (8).

Bernauer, T. (2005). The causes and consequences of international trade conflict over agricultural biotechnology. *International Journal of Biotechnology* Vol. 7, pp. 7-28.

Bernauer, T. & Aerni, P. (2007). Competition for public trust: Causes and consequences and consequences of extending the transatlantic biotech conflict to developing countries. In R. Falkner (Ed.), The international politics of genetically modified food: *Diplomacy, trade and law,* pp. 138-154. Basingstoke: Palgrave Macmillan.

Bradshaw, G. A. and Borchers, J. G. (2000). Uncertainty as information: Narrowing the science-policy gap. *Conservation Ecology,* 4 (1): Article 7. Accessed on March 16, 2009 at http://www.consecol.org/vol4/iss1/art

CBD, (2000). Secretariat of the Convention on Biological Diversity. Cartagena Protocol on Biosafety to the Convention on Biological Diversity: Text and Annexes. Montreal, Canada.

Clapp, J. (2006). Unplanned exposure to genetically modified organisms: Divergent responses in the global south. *Journal of Environment and Development,* Vol. 15, pp. 3-21.

Daily Nation, (2011). Kenya approves import of GMO maize. The Cabinet through the President issues a statement that improved importation of GMO maize following a purportedly food crisis. *Article by Nation reporters.* July 14, 2011.

Falkner, R. (2007). The political economy of 'normative power' Europe: EU environmental leadership ininternational biotechnology regulation. *Journal of European Public Policy,* Vol. 14, pp. 507-526.

Falkner, R. & Gupta, A. (2009). Limits of Regulatory Convergence: globalization and GMO politics in the South. *International Environmental Agreements,* Vol 9, pp. 113-133.

Food & Agriculture Organisation (FAO), (2004). The state of food & agriculture. Agricultural biotechnology: meeting the needs of the poor? FAO, Rome, Italy.

Fukuda-Parr, S. (2006). Introduction: Global actors, markets and rules driving the diffusion of genetically modified (GM) crops in developing countries. *Int. J. Technology and Globalisation,* Vol. 2 (1/2), pp. 1-11.

Gibbons, M.; Limoges, C.; Nowotny, H.; Schwartzman, S.; Scott, P. & Trow, M. (1994). The new production of knowledge: the dynamics of science and research in contemporary societies, London: Sage.

Glover, D. (2010). Is *Bt* Cotton a Pro-Poor Technology? A Review and Critique of the Empirical Record. *Journal of Agrarian Change,* Vol. 10 (4), pp. 482-509.

Haas, P. M. (1992). Introduction: epistemic communities and international policy coordination. *International Organization,* Vol. 46 (1), pp. 1-35.

Hall, A. (2005). Capacity development for agricultural biotechnology in developing countries: an innovation systems view of what is and how to develop it. *J. Int. Dev.*, *Vol.* 17, pp. 611–630.

Halsey, M. E. (2006). Integrated confinement system for genetically engineered plants. St.Louis, Missouri, USA, Donald Danforth Plant Science Center & PBS, and IFPRI-USAID. Accessed online at www.ifpri.org/pbs/pbs.asp, on March16, 2009.

Handbook for Policy Makers, (2007). Status of biotechnology in Kenya. International Service for Acquisition of Agri-biotechnology Applications (ISAAA), AfriCenter.

Harsh, M. (2005). Formal and informal governance of agricultural biotechnology in Kenya: participation and accountability in controversy surrounding the draft biosafety bill. *J. Int. Dev.*, Vol. 17, pp. 661–677.

Hastings, A. (2009). Science training: if governments lead, others will help. SciDev.Net, Opinion,March 11, 2009. Accessed at www.info@scidev.net, on March 16, 2009.

Herring, R. J. (2010). Epistemic brokerage in the bio-property narrative: contributions to explaining opposition to transgenic technologies in agriculture. *New Biotechnology*, Vol. 00, No. 00. June 2010.

Jasanoff, S. (2004). Ordering knowledge, ordering society. In Jasanoff, S. (Ed.), *States of Knowledge: the co-production of science and social order*, pp.13-45. Routledge, London & New York.

Kagundu, A.M. (2008). Risk assessment mechanisms for genetically engineered plant products at official entry points in Kenya. A paper presented in the 1ˢᵗ *all Africa congress on biotechnology*, September 22-26, 2008. Nairobi, Kenya.

Kenya Agricultural Research Institute (KARI), (2005). Agricultural innovations for sustainable development. *Strategic Plan 2005-2015*. June, 2005.

Kingiri A. & Ayele S. (2009). Towards a smart biosafety regulation: the case of Kenya. *Environ. Biosafety Res.* Vol. 8, pp. 133-139.

Kingiri, A. (2010). An analysis of the role of experts in biotechnology regulation in Kenya. *Journal of International Development, Vol.* 22, pp. 325–340.

Kingiri, A. (2011a). Underlying tensions of conflicting advocacy coalitions in an evolving modern biotechnology regulatory subsystem: Policy learning and influence of Kenya's regulatory policy process. *Science and Public Policy Vol.* 38, (3), pp. 199-211.

Kingiri, A. (2011b). The contested framing of Biosafety Regulation as a tool for enhancing public awareness: Insights from the Kenyan regulatory process and BioAWARE strategy. International Journal of Technology and Development Studies (IJTDS), Vol. 2 (1), pp. 64-86.

Kinuthia, S. (2011). Kenyans are faced with a serious crime against humanity – feeding GMOs? An open letter to the Kenyan government and copied to relevant key Ministries and individuals in biotechnology arena. July 21, 2011. Letter signed on behalf of two civil society groups.

Laurie, G.; Bruce, A. & Lyall, C. (2009). The roles of values and interests in the governance of the life sciences: learning lessons from the "ethics+" approach of UK biobank. In Lyall, C., Papaioannou, T. and Smith, J. (Eds.), *The limits to governance. The challenge of policy-making for the new life sciences*, pp. 51-77. Farnham, Ashgate.

Leach, M.; Scoones, I. & Stirling, A. (2007). Pathways to sustainability: an overview of the STEPS Centre approach. *STEPS Approach Paper*, Brighton, STEPS Centre.

The Bumpy Path Towards Knowledge Convergence for Pro-Poor Agro-Biotechnology Regulation and Development:
Exploring Kenya's Regulatory Process

81

Levidow, L. (2007). European public participation as risk governance: enhancing democratic accountability for agbiotech policy? *East Asian Science, Technology and Society: an International Journal*, Vol. 1, pp. 19-51.

Lyall, C. (2007). Governing genomics: new governance tools for new technologies. *Technology Analysis & Strategic Management*, Vol. 19 (3), pp. 369-386.

Lyall, C. & Tait, J. (2005). Shifting policy debates and the implications for governance. In Lyall, C. and Tait, J. (Eds.), *New modes of governance. Developing an integrated policy approach to science, technology, risk and the environment*, pp. 3-17. Aldershot, Ashgate.

Lyall, C.; Papaioannou, T. & Smith, J. (Eds.) (2009). Governance in action in the life sciences: some lessons for policy. In Lyall, C., Papaioannou, T. and Smith, J. (Eds.). *The limits to governance. The challenge of policy-making for the new life sciences*, pp. 261-273. Farnham, Ashgate.

Millstone, E. & van Zwanenberg, P. (2003). Food and agricultural biotechnology policy: How much autonomy can developing countries exercise? Development Policy Review, Vol.21, pp. 655–667.

Mugwagwa, J. T. (2008). Supranational organizations and cross-national policy convergence: the case of biosafety in Southern Africa. PhD Thesis, Development Policy and Practice, Faculty of Mathematics, Computing and Technology. The Open University.

Murphy, J. & Chataway, J. (2005). The challenges of policy integration from an international perspective: The case of GMOs. In Lyall, C. and Tait, J. (Eds.), *New modes of governance: Developing an integrated policy approach to science, technology, risk and the environment*, pp. 159-176. Aldershot, Ashgate.

Murphy, J. & Levidow, L. (2006). Governing the transatlantic conflict over agricultural biotechnology: Contending coalitions, trade liberalisation and standard setting. London: Routledge.

Newell, P. (2003). Globalization and the governance of biotechnology. Global Environmental Politics, Vol. 3, pp. 56–71.

Nowotny, H. (2003). Democratising expertise and socially robust knowledge. *Science and Public Policy*, Vol. 20 (3), pp. 151-156.

Nowotny, H.; Scott, P. & Gibbons, M. (2001). Re-thinking science: knowledge and the public in an age of uncertainty. Polity Press, Cambridge, UK.

Nowotny, H.; Scott, P.; & Gibbons, M. (2003). Mode 2 revisited: the new production of knowledge. *Minerva*, Vol. 41, pp. 179–194.

Ochieng, C.M. (2007). Development through positive deviance and its implications for economic policy-making and public administration in Africa: the case of Kenyan agricultural development, 1930-2005. *World Development*, Vol. 35 (3), pp. 454-479.

Odame, H.; Kameri-Mbote, P. & Wafula, D. (2003). Governing modern agricultural biotechnology in Kenya: implications for food security. *IDS Working Paper*, 199, Institute of Development Studies (IDS), University of Sussex, Brighton, UK.

Omondi, G. (2011). Kenya: State plans drive to popularise GMOs amid raging debate. *Business Daily*, July 20, 2011.

Opiyo, D. (2011). Kenya: The shocking reality about GMOs. accessed at www.allaafrica.com on July 11, 2011.

Paarlberg, R. (2008). Starved for science: how biotechnology is being kept out of Africa. Cambridge, MA: Harvard University press.

Paarlberg, R. (2001). The politics of precaution: genetically modified crops in developing countries. IFPRI. The Johns Hopkins University Press.

Paarlberg, R. & Pray, C. (2007). Actors on the Landscape. AgBioForum, Vol. 10(3), pp. 144-153.

Republic of Kenya (RoK), (1980). The Science and Technology Act. Government printer, Nairobi, Kenya.

RoK, (1998). Regulations and Guidelines for Biosafety in Biotechnology for Kenya. *National Council for Science and Technology (NCST)*, No. 41.

RoK, (2005). Strategy for Revitalising Agriculture (SRA): 2004-2014. (Short version), Feb 2005.

RoK, (2006). National Biotechnology Development Policy. Government Printer, Nairobi, Kenya.

RoK, (2007). Kenya Vision 2030. Government printer, Nairobi, Kenya.

RoK, (2009). The Biosafety Act, 2009. Kenya Gazette Supplement No. 10 *(Acts No. 2)*, Government Printer, Nairobi, Kenya, 13 February, 2009.

Sander, F. (2007). A construction of Kenya's Biosafety Regulations and Guidelines. How international donor agencies interact with regulatory innovation actor-network. Msc. Thesis. Science and Technology Studies, Faculty of Social and Behavioural Sciences, University of Amsterdam.

Smith, J. (2004). The anti-politics gene: biotechnology, ideology and innovation systems in Kenya. *Innogen Working Paper*, 31.

Tait, J.; Chataway, J.; Lyall, C. & Wield, D. (2006). Governance, policy, and industry strategies: pharmaceuticals and agro-biotechnology. In Mazzucato, M. and Dosi, G. (Eds), *Innovation, growth and market structure in high-tech industries: the case of biotech-pharmaceuticals*, pp. 378-401. Cambridge: Cambridge University press.

van Zwanenberg, P.; Ely, A. & Smith, A. (2008). Rethinking regulation: international harmonisation and local realities. *STEPS Working Paper*, 12, Brighton: STEPS Centre.

Wafula, D.; Persley, G.; Karembu, M. & Macharia, H. (2007). Applying biotechnology in a safe and responsible manner: justification for a biosafety law in Kenya. *Biosafety Policy Brief*, August, 2007. Washington, D.C. IFPRI.

Wambugu, F. (2001). Modifying Africa. How biotechnology can benefit the poor and hungry, a case paper from Kenya. (2nd Ed.).

World Bank. (2006). *Enhancing Agricultural Innovation: How to go beyond the Strengthening of Research Systems*. Economic Sector Work Report. The World Bank: Washington, DC, pp. 149.

6

Establishment of Functional Biotechnology Laboratories in Developing Countries

Marian D. Quain, James Y. Asibuo,
Ruth N. Prempeh and Elizabeth Y. Parkes

Council for Scientific and Industrial Research, Crops Research Institute, Kumasi, Ghana

1. Introduction

Traditionally, biotechnology is defined as making use of living organisms or genetic material from living organisms to provide new products for agricultural, industrial, and medical uses. This definition includes the use of fermentation in the leavening in the 10000 BC. This technology over the years has advanced into Modern Biotechnology. According to the Cartagena protocol (Secretariat of the Convention on Biological Diversity, 2000), Biotechnology is defined as any technological application that uses biological systems, living organisms, or derivatives thereof, to make or modify products or process for a specific use. According to the Convention on Biological Diversity, Art.3 (i), "Modern biotechnology" means the application of:

a. *In vitro* nucleic acid techniques, including recombinant deoxyribonucleic acid (DNA) and direct injection of nucleic acid into cells or organelles, or

b. Fusion of cells beyond the taxonomic family, that overcome natural physiological reproductive or recombination barriers and that are not techniques used in traditional breeding and selection.

Plant Biotechnology encompasses tools such as tissue culture and molecular biology which are used in crop improvement. Although these technological tools are applied in advanced countries, their use in agricultural research and development in developing countries is limited. However, these countries need to enhance the utilization of tissue culture and molecular biology to increase agriculture productivity. The prospects of biotechnology as a modern tool for addressing various productivity problems and challenges in agriculture in the face of present day changing climatic conditions and starvation are now well known. Agriculture accounts for about 40% of Ghana's GDP, contributes 35% of foreign exchange earnings, and provides employment for over 60% of the population. More than 80% of the rural populations depend on it for their livelihood.

In recognition of the need to use biotechnology tools in agriculture in sub-Sahara Africa, in June 2003 at a Worldwide Ministerial Conference, in Sacramento, USA 112 Ministers of Science and Technology from 117 countries recommended the facilitation of access of Developing countries to Science and Technology innovations as means to reach Millennium development goals (MDGs). As a follow-up to that, in June 2004, a West African Ministerial Conference on "the use of Science and Technology to improve agricultural productivity in Africa" was held in Ouagadougou (Burkina Faso). At this meeting, participants recognised the need to:

- Develop an information strategy on new technologies and especially on biotechnology
- Establish a strong scientific partnership between research institutions of Africa and developed countries
- Put in place a Regional Biotechnology Centre

Subsequently, in November 2004, there was the Economic Community of West African States (ECOWAS) Ministerial Conference on "Agriculture and Biotechnology" in Abuja (Nigeria) the meeting recommended the following:

- The need to establish Regional Biotechnology Centres of Excellence in countries having comparative advantages
- The promotion of *in situ* Research and Development activities on priority areas to support the emergence and growth of the biotechnology industry in West Africa
- The transfer of biotechnology product packages and their commercialisation in relevant areas
- The reinforcement of private-public sectors collaboration in order to boost the local Biotechnology industry
- The reinforcement of regional and national capacities in Biosafety

There was thus adoption of the Biotechnology and Biosafety Programme (BBP) Action Plan by the ECOWAS Ministerial Conference in Accra (Ghana), in May 2007. The BBP general objective is to use the development and exploitation of biotechnology products as means to increase agricultural productivity and competitiveness in West Africa

The Specific Objectives are to:

1. Promote the use of Biotechnology in agriculture
2. Develop a regional approach for Biosafety
3. Establish and make effective at the regional level, a mechanism for coordination of initiatives, fund raising and communication in the field of Biotechnology and Biosafety.

Thematic areas addressed in priority being:

- The application of Molecular Markers
- The application of Genetic Engineering
- The application of Molecular Diagnostics for animal and plant diseases
- Plant tissue/cell culture and micro-propagation techniques
- Vaccines for livestock production
- Animal Reproduction Technologies

This meeting was a significant landmark in the application of Science and Technology for improving performance in the agribusiness sector in the West African sub-region. This is because the meeting brought together biotechnology and biosafety experts, as well as the ECOWAS Ministers of Environment, Science and Technology, and Agriculture. At the end of the meeting, the communiqué issued indicated that the Ministers fully support the application of biotechnology in addressing some of the numerous problems facing agriculture in Africa, particularly towards improvement in production, competitiveness and sustainable management of natural resources. Although the Ministers pledged their support, they stressed the need to have in place safety measures to ensure effective and sustainable application of the technologies.

All these initiatives have harnessed the application of modern techniques, however, in Ghana, utilization of biotechnology tools in agricultural research and development is associated with several setbacks: which may lead to nonfunctional and unsustainable laboratories. This paper thus focuses on how functional laboratories have been established in Ghana, with specific reference to some research output by the Biotechnology Unit of the Council for Scientific and Industrial Research (CSIR) – Crops Research Institute (CRI).

2. Establishment of biotechnology laboratories and some research output

2.1 Tissue culture

Tissue culture is the most applied biotechnology tool and its establishment is vital for the application of various techniques. By means of definition, plant tissue culture is the growth or maintenance of plant cells, tissues or organs or whole plant on a nutrient culture medium under aseptic conditins *in vitro*. Tissue culture employs the principle of "Totipotency" which is the cell characterestics in which the potential for forming all the cell types in the adult organsim is retained, or the capacity of differentiated cells to retain their full genetic potentialities and express them under appropriate conditions, or potential of cells or tissues to form all cell types and/ or to regenerate plants (Murch & Saxena 2005).

Listed below are applications and advantages of plant tissue culture

- Production of clones
- Large-scale plant multiplication
- Mutation-assisted breeding
- Induction of genetic variability- somaclonal variation
- *In vitro* selection
- International germplasm exchange
- All year round availability of tissue culture derived plants
- High commercial prospects – floriculture and vegetative crops
- Plants as a bioreactor for producing vaccines, and chemicals
- Saves time and space
- Long-term storage of elite genetic material
- Establishment of germplasm bank
- Labour oriented- employment generation, socio-economic impact
- Secondary metabolite production- medicine

Plant tissue culture techniques include:

- Anther/pollen/microspore/ovary culture
- Protoplasts
- Embryo rescue
- *In vitro* fertilization

- *In vitro* micro-grafting
- Micropropagation
- Somatic embryogenesis
- Callus and cell suspension
- Cryopreservation
- Cold storage
- Encapsulation
- Bioreactor
- Gene transfer

In Ghana, one of the first agricultural based research organizations to set up a tissue culture laboratory was the Ghana Atomic Energy Commission (GAEC) in the mid 1980s. This was followed by the Department of Botany of the University of Ghana, Legon in 1988, to train students. Subsequently, the Council for Scientific and Industrial Research (CSIR) Crops Research Institute (CRI) also established a tissue culture laboratory in 1996. All these set ups had to cope with interrupted supply of water and electricity, however, putting in place efficient water storage systems as well as bore hole and also standby generator for electricity helped solve these problems. Other setbacks included lack of regular and reliable source of consumables, glassware, equipment, equipment maintenance, not to mention source of funds, since limited funds were received from central government. However, it is worth mentioning that the CGIAR centers have been instrumental in training human resources and assisting with supplies through projects. The Consultative Group on International Agricultural Research (CGIAR) centers include the International Institute for Tropical Agriculture (IITA), International Center for Tropical Agriculture (CIAT), International Crops Research Institute for the Semi-Arid-Tropics (ICRISAT), just to mention a few.

When setting up the tissue culture facility of the CSIR-CRI, the laboratory initially used to share laminar flow cabinet with the microbiology research group. This was very frustrating since it took us a while to establish clean cultures. Once we established our ability to produce results in tissue culture, the laboratory in 1998 collaborated in research activities sponsored by German Technical Cooperation (GTZ), West Africa seed Development Unit (WASDU) for the production of clean plating materials of yam and cassava in West Africa (Quain, 2001). Another project with IITA with funding from GATSBY UK, produced clean *Musa* planting material for field evaluation and selection of hybrids with tolerance to the Black Sigatoka Disease in Ghana which started in 1998 this project resulted in the selection an release of two hybrid Musa species for release and utilization in Ghana. Also, in 1999, as the institute's sweetpotato breeding group worked towards the selection of varieties to be released to farmers under the Root and Tuber Improvement Project (RTIP), the laboratory with assistance from IITA, produced clean planting materials for multilocational trial. The sweetpotato varieties cleaned through tissue culture techniques in the laboratory, when established in the field was highly accepted by Agriculture extension officers and farmers. The tuber yield resulted in a 30% increase when compared with the conventional planting material (Otoo & Quain 2001).

Subsequently, the CSIR-CRI tissue culture laboratory has optimized existing protocols for local crop varieties, some of these recent research outputs include publications on the following: Multiple Shoot Generation Media for *Musa sapientum* L. (False Horn, Intermediate French Plantain and Hybrid Tetraploid French Plantain) Cultivars in Ghana

(Quain et al., 2010a). This paper considered plantains (*Musa sapientum*), a major staple in Ghana, which encounter several production constraints including availability of adequate healthy planting materials at the time the crop needs to be planted. In attempts to improve production, tissue culture methods were employed, using one medium. It was however realized that optimization of *invitro* rapid propagation protocol for mass production of different accessions of *Musa* was paramount. Excised buds from cultures with proliferating buds were used as explants in this experiment. The cultures of proliferating buds had been generated from excised apical meristem of four *Musa* varieties (False Horn; local names – Osoboaso and Apantu, intermediate French plantain; local name – Oniaba and FHIA 21, which is a hybrid tetraploid French plantain) which were cultured on Murashige and Skoog (MS) medium (Murashige and Skoog, 1962) containing indole-3-acetic acid (IAA), citric acid, and 0-20 µM benzyl amino purine (BAP). The most popular local plantain variety, Apantu, only produced proliferating buds profusely when placed on routine medium (MS medium containing IAA, citric acid and 20 µM BAP). Reducing the concentration of BAP generated an average of more than 4 shoots/culture in 8 weeks. Medium not supplemented with any plant growth regulators also generated an average of less than 2 shoots/culture in 8 weeks. The other three *Musa* varieties generated 4-8 shoots/culture from proliferating buds, indicating that each cultivar has optimum concentrations at which rapid plantlet formation can be optimized to meet growing demands for planting material. This protocol has presently been adapted by the laboratory which produces about 3000 musa plants yearly through tissue culture for farmer, NGOs and interested organizations.

Other research activities have also established *in vitro* manipulation protocols for *Dioscorea* species (yam), which is a vital staple. In the yam tissue culture research, effect of various hormonal (growth regulators) combinations on *in vitro* sprouting of various species of *Dioscorea* spp under light and dark conditions (Ashun, 1991). *In vitro* studies on micropropagation of various yam species (*Dioscorea* species) (Ashun, 1996), indicated that where various concentrations of phytohormone Naphthalene Acetic Acid (NAA), 2,4-dichlorophenoxycetic acid (2,4-D), and BAP are used to culture *Dioscorea spp in vitro* using complete Murashige and Skoogs medium, the concentrations of 0.5µM and 5µM NAA, enhanced plantlet development. BAP concentrations of 5µM and above were lethal to explant development whereas 5µM and above NAA enhanced callus development in *D. alata* cv. 145 used. These studies also established that during yam explants initiation in culture with explants derived from vine, the age of the explants is critical. Explants aged two to 20 weeks were used in this study and for the different *Dioscorea* species used, the highest growth for *D. bulbifera* was in 2 week old explants, *D. dumentorum* were six weeks and four week old explants for *D. alata* (Quain & Achempong, 2001). *Dioscorea* species produce tubers and it is an important staple in Ghana and sub-saharan African countries. The above mentioned tissue culture studies therefore provide the basic tissue culture tools applicable in modern biotechnology, that can be used in the improvement of the crop in the sub-region.

Current tissue culture research activities are aiming at producing protocol for the *in vitro* manipulation of local bananas, plantains as well as various local root and tuber crops. The aim of these is to establish schematic mass production systems to benefit the commercial farmer. Protocols for long-term *in vitro* conservation of germplasm are also being optimized. The development of all these protocols will facilitate the adaptation of other modern biotechnology tools for the maintenance and improvement of local crop varieties to meet agricultural production constraints.

2.2 Molecular biology

Molecular biology is the aspect of biology that deals with the molecular basis of biological activity. This aspect of science is related with other areas of biology, chemistry, genetics and biochemistry. Mostly, molecular biology chiefly covers interactions between the various systems of a cell, namely between the different types of DNA, RNA and protein biosynthesis, and how these interactions are regulated. In 1961, William T. Astbury, made a statement that "molecular biology implies not so much a technique as an approach, an approach from the viewpoint of the so-called basic sciences with the leading idea of searching below the large-scale manifestations of classical biology for the corresponding molecular plan. It is concerned particularly with the *forms* of biological molecules and with evolution, exploitation, and ramification of those forms of the ascent to higher and higher levels of organisation. Molecular biology is predominantly three-dimensional and structural — which does not mean, however, that it is merely a refinement of morphology. It must at the same time inquire into genesis and function" (Astbury, 1961).

Molecular biologists have since the late 1950s and early 1960s learned to characterize, isolate, and manipulate the molecular components of cells and organisms. These components include firstly, DNA, which is the storehouse of genetic information. Secondly is RNA, which is a close relative of DNA with functions ranging from serving as a temporary working copy of DNA to actual structural and enzymatic functions, as well as a functional and structural part of the translational apparatus. Thirdly, are proteins; the major structural and enzymatic type of molecule in cells (Molecular Biology *Source*: http://en.wikipedia.org/w/index.php?oldid=417169395).

The aspects of biology listed below can all be found under Molecular Biology:

- Genomics
- Proteomics
- Molecular Microbiology
- Genetic transformation
- Molecular modelling
- Molecular breeding
- Molecular marker selection
- Mutation
- Bioinformatics
- Genetic fingerprinting

Following the successful establishment of tissue culture facility at the CSIR – CRI, it became apparent that a molecular biology laboratory be establish to complement the biotechnology activities. Project proposals developed under the then Agricultural Services Sub-Sector Investment Programme (AGSSIP) project secured the necessary basic equipment for molecular biology research. The project also contributed toward the training of two researchers; one in marker assisted breeding and another in genetic transformation in advanced laboratories. These scientists returned to Ghana, with the basic laboratory consumables to initiate the molecular biology laboratory. With assistance from a research assistant who had just completed using molecular techniques in disease diagnosis in a Ghanaian university, the molecular biology laboratory started operation in 2006 supported with inputs from the Generation Challenge Programme (GCP), CIAT and International Atomic Energy Agency (IAEA). In that same year, funds were also secured from the Government of Ghana to provide more equipment and consumables for the laboratory.

These gave the laboratory a sound basis to be really established. Since then the CSIR-CRI molecular biology laboratory has been locally adjudged the best biotechnology laboratory in Ghana and is yearly conducting training courses for researchers and students both locally and within the West African sub-region. Several research publications have been released and these include the following:

Assessing transferability of Sweetpotato EST SSR primers to cocoyam and micropropagation of nine elite cocoyam varieties in Ghana; Cocoyam, an important staple crop in Ghana, provides edible leafy vegetable and starchy cormels. Due to difficulty in getting primers for cocoyam, sweetpotato EST SSR primers were used to amplify genomic DNA of elite cocoyam lines. Genomic DNA was isolated from 10 sweetpotato and nine cocoyam cultivars. Ten sweetpotato accessions were screened alongside three cocoyam cultivars, using 22 EST Sweetpotato SSR primers, 13 of which could amplify cocoyam sequences and were subsequently used to screen nine cocoyam cultivars. Thirteen random primers were also used for diversity study. Cocoyam cultivars were established *in vitro*. Dendogram generated after screening cocoyams alongside sweetpotato, grouped sweetpotato varieties in two main clusters and cocoyam in one cluster. The random primers and the SSRs grouped the cocoyam into two clusters which corresponds with known morphological classification. The method would be used to screen large cocoyam germplasm (Quain et al., 2010b). Through this study genetic fingerprint of eight elite and one local check cocoyams has been documented. These fingerprinted cocoyam accessions are presently being evaluated on the field to establish their agronomic attributes and select some lines for release to farmers and the Ghanaian public for utilization. This will be the first time ever in Ghana that research output is releasing cocoyam varieties.

Genetic diversity of elite *Musa* cultivars and introduced hybrids in Ghana using SSR markers; in this study, molecular diversity was carried out on 10 *Musa* cultivars using SSR. *Musa* SSRs (49) marker was used for the diversity and NTSYS Data analysis used to establish conclusions on studies. Dendogram and similarity matrix generated, indicated that local false horn and intermediate French plantain are distantly related (16.78%). The closest related cultivars are two false horn (Apantu-Dichotomy and Osoboaso) at 70.32%. Similarity between introduced hybrids and local false horn plantains and local intermediate French plantains was in the range of 20.81–49.67% and 18.85–42.27% respectively. Apem (local intermediate French plantain) was distantly related to all the cultivars screened (16.78–36.84%). The information generated has documented diversity between the introduced hybrids and elite local cultivars and this will aid breeders mine for genes in the local cultivars that are responsible for earliness, peculiar taste and preferred cooking qualities (Quain et al., 2010c).

Genetic relationships between some released and elite Ghanaian cassava cultivars based on distance matrices has also been carried out (Acquah et al., 2010); Eleven (11) released and two local Ghanaian cassava cultivars were fingerprinted to estimate the genetic diversity among them using 35 SSR markers. Genomic DNA of thirteen cassava cultivars (*UCC,IFAD, Agelifiaa, Nyerikobga, Nkabom, Essam Bankye, Akosua Tumtum, Debor, Filindiakong, Afisiafi,Doku Duade, Bankye Hemaa* and *Bankye Botan*) were isolated and used as template for PCR amplification involving 35 SSR markers. The recorded gel bands (163 polymorphic bands) were subjected to NTSYS Version 2.1 software for cluster analysis and development of dendrogram to show the corresponding similarity coefficients. Genetic relationships between *Bankye Hemaa* and *Filindiakoh* and that between *Bankye Hemaa* and *Afisiafi* had similarity coefficients of 1.2%. The local cultivars, *Debor* and *Akosua Tumtum* were related at 52.31% similarity. *Filindiakoh* was found to be the relative to

Akosua Tumtum and *Debor* at 17.9 and 29.1% similarity, respectively. *Bankye Botan* and *Bankye Hemaa,* however, were distantly related to most of the cultivars, including the local varieties. *Bankye Botan* and *Bankye Hemaa* are distant relatives to most of the cultivars, including the local varieties which could however make these cultivars also very useful in breeding. This research work documented molecular information on released cassava varieties for the first time in Ghana. This information will contribute towards variety identification as they are released for utilization by farmers. The various research groups that have released cassava varieties over the years can also track the performance of their lines within the country and the subregion.

Groundnut is a member of the genus *Arachis* and the crop is divided into two subspecies and six botanical varieties based on morphological characteristics. A groundnut core collection of 831 accessions was developed from a total of 7432 US groundnut accessions based on morphological characteristics. Identification of DNA markers associated with the botanical varieties of groundnut would be useful in genotyping, germplasm management and evolutionary studies. A study was initiated to evaluate 22 groundnut genotypes representing six botanical varieties from a US groundnut core collection to determine their diversity using DNA microsatellites. Cluster analysis located the lines in their assigned specific botanical groups in agreement with available morphological classification for groundnut (Asibuo et al., 2010). Groundnut production and utilization in Ghana is presently very promising and selected groundnut varieties are produced for the confectionary industry. The output of this study will thus enable organizations that produce seed for farmers to cultivate scrutinize the stability and identity of their seeds.

2.3 Cryopreservation

Cryopreservation is a process of cooling cells, tissues or organs at ultralow temperature to preserve them indefinitely. The applied temperature is typically at −196 °C which is the boiling point of liquid nitrogen. Other temperatures applied in cryopreservation include -40°C, -70°C, -80°C, in programmable or ultra-low freezers, or in the vapor phase of liquid nitrogen at -150°C, or at -210°C in nitrogen slash. Technically, at these low temperatures, any biological, metabolic activity as well as biochemical reactions that would lead to the cell losing viability should cease and the preserved cell tissue or organ should be viable when retrieved from cooling. Due to the complex nature of cells, organs and tissues subjected to cryopreservation, other additives including cryoprotectants are used to prevent damage otherwise caused by cooling.

As the biotechnology research advanced, conservation of clonally propagated crops was identified as an important aspect. To facilitate the development of this technological tool, a PhD student worked on Complementary Conservation of Root and Tuber Genetic Resources - *Dioscorea* species and *Solenostemon rotundifolius.* The major focus of this study was the application of cryopreservation techniques to complement *in vitro* slow growth methods, since *in vitro* conservation under slow growth has been used for the conservation of clonally propagated crops. However, it demands periodic subculturing and regular attention and, with interruptions in electric power supply, conserved cultures are in danger of being lost. Presently the existing root and tuber germplasm conservation techniques serve a short to medium-term purpose only. The ultimate means of long-term conservation, which will complement all the existing modes being used and serve the purpose of base collections, is conservation in liquid nitrogen at −196°C (Engelmann, 2000). Storage of biological material at ultra-low temperatures, preferably that of liquid nitrogen, arrests all metabolic activities: consequently, no genetic changes occur, theoretically, permitting indefinite germplasm storage periods (Panis & Lambardi, 2005). However, in practice, although very long, indefinite (seed) storage may not be attainable (Walters et al., 2005) and this probably applies to other forms of germplasm as well (Benson & Bremner, 2004). The development of protocols tested three cryomodels: being vitrification–based, silica gel dehydration, encapsulation vitrification, as well as encapsulation desiccation. The study revealed that, the successful cryopreservation of *Dioscorea rotundata* which is an important crop producing mealy tubers is possible using a simple vitrification protocol. The procedure incorporates:

- pregrowth of the donor plant on 0.09 M sucrose-supplemented medium for five weeks
- preculture on 0.3 M sucrose supplemented medium for 5 d
- PVS2 solution for 40 min
- Rapid cooling in liquid nitrogen or slow cooling to -80°C

Through this study, *Dioscorea rotundata* accession "Pona" which is an elite variety in Ghana was successfully cryopreserved for the first time. The technique developed was simple, cost-effective and potentially reliable methodology that does not require sophisticated equipment. The procedures can be adapted for germplasm conservation of other species, using limited resources in laboratories in sub-Saharan Africa. It was also brought to the fore that to achieve an optimal recovery of cryopreserved explants, the donor plants should be adequately conditioned. *S. rotundifolius* (Frafra potato) was extremely sensitive to the vitrification based protocol. The nodal explants used in the experiments easily becomes hyperhydric, and are impossible to dehydrate sufficiently for cryopreservation, this provide a sound basis for further attempts to cryopreserve Frafra potato genetic resources. Presently, cryopreservation tools are being developed further for other vegetatively propagated crops in Ghana.

2.4 Marker assisted breeding

Marker assisted selection (MAS) is indirect selection process where a trait of interest is selected, not based on the trait itself, but on a marker linked to it (Ribaut and Hosington, 1998). During selection for tolerance to an abiotic or biotic stress by means of MAS, the plants not on basis of quantified but rather a marker allelomorph (allele) which is linked with the particular trait (disease) is used to determine the presence of the trait (disease). The assumption being that linked allele associates with the gene and/or quantitative trait locus (QTL) of interest. MAS can be useful for traits that are difficult to measure, exhibit low heritability, as well as those that are expressed late in development.

Over the years most breeding activities in the CSIR-CRI have been mainly through selection, based on agronomic traits. However, since 2005, due to training obtained from The International Centre for Tropical Agriculture (CIAT), Cali Columbia, the cassava breeding group has initiated development of crosses to select varieties. Some of the breeding efforts are towards pyramiding genes responsible for tolerance/resistance to cassava mosaic disease using wild relatives from the centre of origin in South America into local Ghanaian accessions through mechanical hybridization. Other similar crop development to select varieties that can withstand biotic and abiotic stresses include drought tolerance in maize, pod shattering in soybean and disease resistance in peanuts. All these processes are being hastened by the utilization of molecular markers to shorten the number of years used in breeding from 10 – 15 years to about four to six years.

2.5 Plant disease diagnostics

Effective crop management involves the early and accurate diagnosis of plant disease. Early introduction of effective control measures in plant development can facilitate plant diseases management. Reliance on symptoms is usually not satisfactory in this regard. This is because the disease may be well underway when symptoms first appear, and symptom expression can be highly variable. Biological techniques for disease diagnosis and pathogen detection are usually highly accurate but too slow and not amenable to large-scale application. Recent advances in molecular biology and biotechnology are being applied to the development of rapid, specific, and sensitive tools for the detection of plant pathogens. (Miller & Martin, 1988). Some Immunological and nucleic-acid hybridization-based methods available for pathogen detection in crop systems are listed below:

- Immunoassay Technology
  - Enzyme Linked Immunoabsorbent Assay (ELISA)
  - Colloidal Gold
  - Immunofluorescence Assay (IFA)
  - Radio Immunoassy (RIA)
- Nucleic – Acid hybridization – Based Pathogen detection
  - Nucleic Acid-Based Detection Technologies
    - Dot – Blot Assay
    - Nonradioactive Labels
    - Restriction Fragment Length Polymorphisms (RFLPS)
  - Nucleic Acid Probes
    - Uncloned Probes
    - Synthetic Probes
    - Cloned Probes and RFLPS
    - Viruses and Viroids
    - Mycoplasma – like organisms and bacteria
    - Fungi
    - Nematode

Molecular biology tools that have been explored so far in our Laboratory include the application of molecular markers to screen for occurrence of disease in crops. This is being applied for African Cassava Mosaic Virus (ACMV) in cassava, Tomato Yellow Leaf Curl Virus (TYLCV) and Root Knot diseases in tomatoes, yellow mottle virus in rice and Sweet Potato Virus Disease (SPVD) in sweetpotato is still at the developmental stages. In a recent study on cassava, disease observations in the field were confirmed with laboratory diagnostics using the polymerase chain reaction assay. African cassava mosaic virus (ACMV) and East African cassava mosaic virus were detected on all the cultivars either as single infections or as mixtures. The detection of EACMV on cassava at Fumesua and Ejura is the first to be reported in Ashanti region in Ghana. This study recommended that, with the advent and spread of the EACMV serotype of the mosaic virus in important cassava growing eco-zones and the emergence of some severe strains of the African cassava mosaic in the pathosystem highly resistant planting materials should be used for ratooning of mother plants as one of the methods to increase the production of clean planting materials for farmers. It was also indicated that, there is also the need to conduct an extensive survey in all the cassava growing areas in Ghana to determine the incidence and spread of emerging species of the cassava mosaic begomoviruses virus in order to develop better strategies to reduce the menace of the cassava mosaic disease in Ghana and sub-region as a whole (Lamptey et al., 2011).

2.6 Genetic engineering

Genetic transformation, also referred to as Modern Biotechnology, is the application of the techniques of molecular biology and/or recombinant DNA technology, or *in vitro* gene transfer, to develop products or impart specific capabilities to organisms. Although the various breeding programs in Ghana are still using conventional breeding tools, efforts are being made to improve plants through an additional technique known as genetic engineering or recombinant DNA (rDNA) science. This method does not rely on the pollination of flowers: it allows individual genes with desire traits to be moved directly from one organism into another living DNA of the same or different species. This technique is very vital for the improvement of most of our local staple crops which are vegetatively propagated. Genetic engineering was first accomplished in the laboratory in 1973 (Paarlberg, 2008). However, as of 2011, laboratories in Ghana have not started applying rDNA in crop research activities. In 2008, a biosafety legislative instrument was passed in the Ghanaian parliament to permit confined field trials and contained laboratory experiment by research scientists. In 2009, a project to Strengthen Capacity for the Safe Management of Biotechnology in Sub-Sahara Africa (SABIMA), was initiated by the Forum for Agricultural Research in Africa (FARA) with sponsored by the Syngenta Foundation for Sustainable Agriculture (SFSA). This project identified personalities who can champion the advancement of Biotechnology in Ghana (Champions) and also focal persons to run the project. Following advocacy and awareness creation workshops conducted by champions in the SABIMA project, with focus on policy makers since 2010, the policy makers were incited to pay critical attention to the biosafety bill that has been before parliament. The members of parliament subsequently, took the biosafety bill through consideration in parliament, in February 2011 and it was passed into law in June 2011. In Ghana, there already exist a National biosafety committee (NBC) which has started receiving applications for conducting confined field trials (CFT). The three proposals in the pipeline for consideration by the NBC are:

1. Introduction of BT-Cowpea in Ghana
2. Protein quality improvement in sweetpotato
3. Nitrogen use efficiency and salt tolerance in rice

Other research efforts have been initiated in collaboration with advanced laboratories using crops of local importance, and these include studies on the transgenic potential of *Dioscorea rotundata*, using agrobacterium-mediated genetic transformation (Quain et al., 2011). This study considered *D. rotundata* (yam) which is one of the important staples and sources of carbohydrate in the diet in Sub-Saharan African sub-region. The crop has several post harvest problems including poor storage of the tuber, availability of the edible planting materials and high cost of labour for cropping. The crop therefore needs to be improved using modern biotechnology methods. Studies were conducted to induce shoot regeneration in *D. rotundata* leaf petiole as explants. Explants (0.5 cm long) were cultured on MS medium supplemented with 0.2 mg L-1 2,4-D for 3 days, and transferred to MS medium containing TDZ alone or in combination with 2iP. Shoot regeneration was observed within 21 days on all media; however, the highest shoot regeneration, 7–9 shoots developing per explant were obtained on media supplemented with TDZ and 2iP. The shoots grew vigorously when transferred to MS medium supplemented with GA3. When petiole explants were subjected to *Agrobacterium*-mediated transformation using strain C58 and EHA101 harbouring a binary plasmid containing the β glucuronidase (*gusA* or *uidA*) intron gene under the transcriptional control of CaMV35S promoter, very high efficiency of transformation (25–65%) was obtained. This successful organogenesis and transformation protocol could be optimized and adapted in engineering of local yam quality and productivity for enhanced protein content, and longer shelf-life.

2.7 Advancements achieved in Ghana

To facilitate research in the application of biotechnology tools, Ghanaian researchers have partnered with regional, subregional and international programs. These include projects with financial and technical support from International Atomic Energy Agency (IAEA), Generation Challenge Program (GCP), United States Agency International Development (USAID), CORAF/WECARD, Syngenta foundation for Sustainable Agriculture, United States Department of Agriculture/Food and Agricultural Science (USDA/FAS), African Agricultural Technology Foundation (AATF), United Nations University – Institute for Natural Resources in Africa (UNU/INRA), World Bank, just to name a few. Collaboration with CGIAR centers as well as universities (Tuskegee University Alabama, USA, and Cornell University) has contributed immensely to recent research outputs, training human capacity, as well as technical and infrastructural capacities.

Presently, Ghanaian agricultural research organizations have graduated to the status of organizing national and regional training courses which hitherto were organized at Consultative Group (CG) centers in the subregion. This activity re-enforces the fact that both the human and infrastructural capacities have been developed to identify and solve regional problems.

Development of functional and sustainable Biotechnology systems needs to take into account `proper stewardship principle. Ghana is presently a participating country in the project to Strengthen Capacity for Safe Application of Biotechnology (SABIMA). This is an Excellence through stewardship based project ensuring the utilization of modern biotechnology tools is practiced in a responsible manner. Under this project, through stewardship training workshops, researchers are being trained to develop and properly document standard operating procedures as well as critical control points along the life cycle of product development. Biotechnology application policies are also being developed by the various institutes for implementation at the management level in the organizations.

Over the years, Ghanaian Universities had been training plant and animal breeders and researchers in related fields of applied and basic sciences, with limited or no knowledge in molecular biology. There is therefore a generation of Breeders and Research Scientists with little or no knowledge of molecular biology, tissue culture and modern biotechnology. However with the establishment of the West Africa Centre for Crop Improvement (WACCI), Alliance for Green Revolution in Africa (AGRA), Postgraduate Program at the University of Ghana at Legon, and Kwame Nkrumah University of Science and Technology (KNUST) respectively, a new generation of plant breeders with knowledge on the use of biotechnology in crop improvement are being turned out. This new generation of plant breeders is challenged to help solve African's problems on African's soil by developing crop varieties tolerant/resistant to biotic and abiotic stresses, and thus alleviate poverty in order to achieve the Millennium Challenge Goals. The Government of Ghana also assisted by providing funds for the establishment of Biotechnology laboratories in 2006, the CSIR – Crops Research institute and CSIR – Animal Research Institute. As a result, a new breed of lecturers, researchers and students are using conventional and biotechnological tools for crop and animal improvement. To us in the developing countries, some of the outmoded technological tools in the advanced countries are novel ideas. The use of laborious and conventional breeding methods with its attendant long duration have given way to evaluation and selection using marker assisted breeding, mutation breeding, use of tissue culture to select somaclonal variants and disease elimination, production of clean planting materials, and mass production of clonal planting material by means of tissue culture.

3. Conclusion

The application of modern biotechnology in developing countries especially in Africa, has great prospects. All the necessary efforts have to be employed in the form of financing, policies, technologies, collaboration etc. These will help us to realize the inherent potentials and immense contributions to the scientific advancement worldwide. All stakeholders are needed to play their various roles to ensure the responsible application of biotechnology in developing countries. The case of Ghana with respect to the CSIR-CRI alone stated above gives the clear indication that, consistency, great leadership, team work, human and infrastructural capacity building, good networking and collaboration are keys to establishing a sustainable system. Presently, under the West African Agriculture Productivity Program (WAAPP), with sponsorship from the World Bank, a multipurpose biotechnology facility is being constructed to facilitate root and tuber research activities in the sub-region. It is hopeful that the impetus will keep building up, and more innovative strategies will be put in place to harness utilization of biotechnology tools in the sub-region.

4. Acknowledgement

The authors wish to acknowledge the following persons for the crucial roles they played in training human resource and advocating for funds for the advancement of biotechnology in Ghana, name; Dr. Elizabeth Acheampong, Dr. M. Egnin, Dr. C. Bonsi, Prof. P. Berjak, Dr, Hans Adu-Dapaah, Prof. A Oteng-Yeboah, Dr. Y. Difie Osei, Prof. Walter Alhassan, Prof Boampensem, Dr. J. Asafo-Adjei, just to mention a few. The CGIARs are also duly acknowledged for their immense contribution in training human resource capacity, as well as improving infrastructural capacity.

5. References

Acquah, W. E.; Quain M. D.; Twumasi, P. (2011). Genetic relationships between some released and elite Ghanaian cassava cultivars based on distance matrices. *African Journal of Biotechnology* Vol. 10(6), pp. 913-921, Available online at
    http://www.academicjournals.org/AJB ISSN 1684-5315 © 2011 Academic Journals
Asibuo J. Y., He, G. ,Akromah, R.; Adu-Dapaah H.K. & Quain M.D. (2010). Genetic diversity of groundnut botanical varieties using simple sequence repeats. *.Aspects of Applied Biology* 96, Agriculture: Africa Engine for Growth- Plant science and biotechnology the key.
Ashun, M. D. (1991) – Effect of various Hormonal (Growth regulators) combinations on *in vitro* sprouting of various species of *Dioscorea* under light and dark conditions - B.Sc. dissertation submitted to University of Ghana - Legon.
Ashun, M. D. (1996)- *In vitro* studies on micropropagation of various yam species (*Dioscorea* species) M.Phil. Thesis submitted to University of Ghana - Legon.
Benson E.E., & Bremner, D. (2004). Oxidative stress in the frozen plant: a free radicle point of view. In: B.J. Fuller N. Lane E.E. Benson (eds). *Life in the frozen state*, CRC Press Boca Raton. pp 205 – 241.

Lamptey, J.N.L.; Quain, M.D.; Owusu Kyere, E.; Ribeiro,P.F.; Prempeh,R.; Afriyie-Debrah, C. & Abrokwa,L. (2011). Effect of Cassava Mosaic Disease on Successive Ratooning of Some Cassava Cultivars. The West Africa Root and Tuber Crops Conference, Abstract, pp 10.

Miller, S.A.; & Martin. R.R.; (1988) Molecular Diagnosis of Plant Disease. *Ann. Rev. Phytopathol.* 26 pp 409-32Molecular Biology Source
http://en.wikipedia.org/w/index.php?oldid=417169395

Murashige, T. & Skoog, F. (1962) A revised medium for rapid growth and bioassays with tobacco tissue cultures. *Physiol. Plant.* 15:473-497

Murch, S.J. & Saxena, P.K. (Eds.). (2005). Journey of a Single Cell to a Plant. Oxford & IBH Publishing Co. Pvt. Ltd., New Delhi, India

Otoo, J. A. & Quain, M. D. (2001) comparative studies of *in vitro* cleaned planting material and field selected apparently clean sweet potato planting material in proceedings of 8th International Society for Tropical Root crops - Africa Branch (ISTRC-AB) symposium at Cotonou 2001.

Paarlberg (2008). *Starved being for Science How Biotechnology is being kept out of Africa.* ISNB-13:978-0-674-02973-6

Panis B., & Lambardi, M. (2005), Status of Cryopreservation Technologies. *In Plants (Crops and Forest Trees). In The Role of Biotechnology* Villa Gualino, Turin, Italy–5-7 March, pp. 43–54.

Quain, M.D.; Egnin, M.; Bey, B.; Thompson, R. & Bonsi, C. (2011). Transgenic potential of *Dioscorea rotundata*, using agrobacterium-mediated genetic transformation. *Aspects of Applied Biology 110,* ISSN 0265-1491 pp71-79

Quain M. D., Adofo-Boateng, P.; Mensah Dzomeku, B.; & Agyeman, A. (2010a). Multiple Shoot Generation Media for Musa sapientum L. (False Horn, Intermediate French Plantain and Hybrid Tetraploid French Plantain) Cultivars in Ghana. *The African Journal of Plant Science and Biotechnology.* 4 (Special Issue 1), 102 – 106, © 2010 Global Science Books.

Quain, M. D.; Thompson, R.; Omenyo, E.L.; Asibuo, J.Y.; Appiah-Kubi, D.& Adofo-Boateng, P. (2010b). Assessing transferability of Sweetpotato EST SSR primers to cocoyam and micropropagation of nine elite cocoyam varieties in Ghana. *Aspects of Applied Biology* 96, Agriculture: Africa Engine for Growth - Plant science and biotechnology the key. pp. 269-276.

Quain, M. D.; Dzomeku, B.M.; Thompson,R.; Asibuo, J.Y.; Boateng, P.A. & Appiah-Kubi, D. (2010c). Genetic Diversity of Musa cultivars in Ghana and introduced hybrids. *Aspects of Applied Biology* 96, Agriculture: Africa Engine for Growth- Plant science and biotechnology the key. pp. 277-28

Quain, M. D. (2001.) Propagation of Root and Tuber Crops By Tissue Culture In Ghana. – *West Africa Seed and Planting Material Newsletter of the West Africa Seed Network (WASNET).* Issue No. 8

Quain, M. D. & Acheampong, E. (2001). Effect of Explant Age on *In vitro* Development of three *Dioscorea* species. Proceedings of 7th *International Society for Tropical Root crops - Africa Branch (ISTRC-AB)* symposium at Cotonou 2001.

Ribaut, J.M. & Hoisington, D. A. (1998). Marker assisted selection: new tools and strategies. *Trends Plant Sci.,* 3, 236–239

Secretariat of the Convention on Biological Diversity (2000). Cartagena Protocol on Biosafety to the Convention on Biological Diversity: text and annexes. Montreal:Secretariat of the Convention on Biological Diversity. ISBN: 92-807-1924-6

Walters, C.; Hiu, L.M. & Wheeler, L.M. (2005). Dying while dry: kinetics and mechanisms of deterioration in desiccated organisms. *Integrative Comparative Biology* 45, pp. 751-758.

Part 2

Novel Applications

7

Therapeutic Applications of Electroporation

Sadhana Talele
The University of Waikato
New Zealand

1. Introduction

Drug delivery is the method or process of administering a pharmaceutical compound to achieve a therapeutic effect in humans or animals. Most common routes of administration include the preferred non-invasive peroral (through the mouth), topical (skin), transmucosal (nasal, buccal/sublingual, vaginal, ocular and rectal) and inhalation routes. Many medications such as peptide, antibody, vaccine and gene based drugs, in general may not be delivered using these routes because they might be susceptible to enzymatic degradation or cannot be absorbed into the systemic circulation efficiently due to molecular size, to be therapeutically effective. For this reason such drugs have to be delivered by injection. For example, many immunizations are based on the delivery of protein drugs and are often done by injection. Current effort in the area of drug delivery include the development of targeted delivery in which the drug is only active in the target area of the body (for example, in cancerous tissues) and sustained release formulations in which the drug is released over a period of time in a controlled manner from a formulation Cemažar et al. (1998); Serša (2000); Serša et al. (1993). This is achieved by combining phenomenon of electroporation with the application of drugs.

This makes the area of drug delivery a study in which experts from most scientific discipline can make a significant contribution. To understand the barriers to drug delivery, it is useful to consider anatomical structures at a length scale suitable for variety of structures: a cell, a tissue, a muscle, organ. Better medical treatments may not always require stronger medicinal drugs. A better effect could be achieved by an effective method of administration. Often differences in the mode of drug administration produce substantial changes in drug concentration requirement, and thus can affect the unnecessary side effects of some of the drugs in favor of the patient. A good example of this is electrochemotherapy. The major disadvantage of clinically established chemotherapeutic drugs is their lack of selectivity of tumor cells. Therefore for a noticeable antitumor effect, high doses of the chemotherapeutic drugs are needed, which often cause systemic toxicity leading to severe side effects Serša et al. (1993). In cancer chemotherapy, some drugs do not exhibit anti-tumour effects because of impeded transport through the cell membrane Miklavĉiĉ and Kotnik (2004). To overcome this difficulty, a number of new approaches for drug delivery systems have been attempted. One of the approaches to better drug delivery is by making it topical and more effective at the tumor site with the use of electric field . This is electrochemotherapy. In this article a brief overview of electroporation and its use in electrotherapy is discussed.

2. Electrochemotherapy (ECT)

The combined use of chemotherapeutic drugs alongwith electroporation caused due to application of electric pulses is known as electrochemotherapy and is useful for local tumour control. Bleomycin and cisplatin which are commonly used drugs for chemotherapy have proven to be much more effective in electrochemotherapy than in standard chemotherapy when applied to tumour cell lines in vitro, as well as in vivo on tumours in mice Mir *et al.* (1991; 1995); Serŝa *et al.* (1995). Clinical trials have been carried out with encouraging results Glass *et al.* (1996); Gothelf *et al.* (2003); Kranjc *et al.* (2005); Serŝa *et al.* (2000); Snoj *et al.* (2005); Tozon *et al.* (2005). Especially, bleomycin has been reported to have shown a 700-fold increased cytotoxicity when used in ECT Cemaẑar *et al.* (1998); Serŝa (2000).This helps to achieve a substantial anti-tumour effect with a small amount of drug, that limits its side effects Serŝa *et al.* (1993).

2.1 Electroporation

It is possible to induce the formation of hole in a cell membrane by applying a sufficiently strong electric field pulse. This is known as electroporation. The effect is reversible when the cell membrane is temporarily permeated. Irreversible electroporation occurs when the cell membrane poration is of such a nature as to induce cell death. Polarization is one of the basic mechanisms of interactions of membranes with electric fields, leading to electroporation and related phenomenons of dielectrophoresis Neumann (1989); Pohl (1978) and electrofusion Neumann (1989); Zimmermann (1982).

2.1.1 Polarization of membranes

Polarization of membranes underlies their destabilization. Polarization is due to restricted motion of charges: electric fields exert forces on charges. These charges can either move if they are free (material is conductive) or accumulate if they are limited in their movement. This charge redistribution in a particular limited space leads to polarization. Figure 1 shows polarization of a single cell due to restriction by the cell membrane to the motion of ions.

2.2 Electric field interaction with polarized membranes and pore formation

The interaction of external electric field with the polarized membranes results in forces which can induce motions inside particles. This motion can result in structural rearrangement or fracture in the material. This can subsequently lead to electroporation and related phenomenon in case of cell membranes Dimitrov (1995); Pohl (1978). Membranes have low polarizability (relative dielectric constant about 5) and low conductivity (3×10^{-7} S/m) Kotnik *et al.* (1998). The cell membrane is generally bounded (externally and internally) by a medium of high dielectric constant (about 80) and a high conductivity (about 1.2 S/m). Application of external fields leads to accumulation of charge at the membrane surfaces; this creates an electric field inside the membrane that is much stronger than the surrounding field. The polarized membrane interacts with this field, resulting in structural rearrangements which can cause membrane poration.

It soon became apparent that a field-induced permeability increase is transient in nature although long-lived compared with the field duration. The term 'electropermeabilization' was used to explain the occurrence of permeability changes introduced by electrical impulses in vesicular membranes Neumann and Rosenheck (1972). It was later shown that the electric

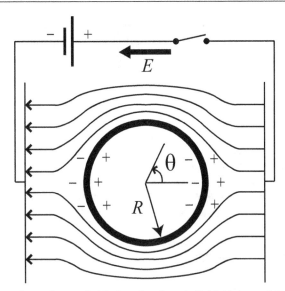

Fig. 1. Spherical particle in electric field, E is the electric field (Adapted from Chang *et al.* (1992); Dimitrov (1995)).

field induced change was transient Rosenheck *et al.* (1975) . The resistance changes in the membrane were attributed to dielectric breakdown Zimmermann *et al.* (1973).
Subsequent studies showed that the cell membranes of pulse treated cells were permeable to molecules of a size smaller than a certain limit, suggesting the creation of a porous membrane structure Kinosita and Tsong (1977b); Neumann and Rosenheck (1972); Zimmermann *et al.* (1973). It was also found that under appropriate conditions, the cells could recover, which implied that these electropores were resealable and could be induced without permanent damage to the cell Zimmermann *et al.* (1980), and the cytoplasmic macromolecular contents could be retained Kinosita and Tsong (1977a;b). Since then, a number of research groups have studied mechanisms of pore formation and detailed characteristics of the cell membranes modified by electric fields Abidor *et al.* (1979); Chernomordik *et al.* (1983); Glaser *et al.* (1988); Schwister and Deuticke (1985).
However, the pores themselves were not observed until the invention of rapid freezing electron microscopy in the 1990s. Chang et.al Chang and Reese (1990) were the first to observe them. Other aspects of electroporation, for example, vizualization of transmembrane potential and its evolution in space and time, resealing of pores and asymmetry in permeability of porated cells (sea urchin egg and liposomes) with the help of an optical microscope, were also reported Hibino *et al.* (1993); Kinosita *et al.* (1992). These microscopes have a time resolution of sub-microseconds suitable for studying electroporation.

2.2.1 Types of pores
The pores are assumed to be hydrophobic or hydrophilic. The hydrophobic pores, as shown in Figure 2a Abidor *et al.* (1979); Glaser *et al.* (1988); Neu and Krassowska (1999), are simply gaps in the lipid bilayer of the membrane, formed as a result of thermal fluctuations.

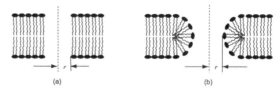

Fig. 2. Types of electropores: (a) Hydrophobic (nonconduting pore), (b) Hydrophilic pore (conducting pore).

The primary pores that participate in electrical behaviour and molecular transport are thought to be hydrophilic pores, with a minimum radius of about 1 nm, and a reasonable probability of various pore sizes much larger Weaver (1993). The 'hydrophilic' or 'inverted pores,' as shown in Figure 2b, have their walls lined with the water-attracting heads of lipid molecules. Hence, the hydrophilic pores allow the passage of water-soluble substances, such as ions, while the hydrophobic pores do not.

2.3 Facts about electrochemotherapy
Due to the availability of these electropores, electroporation can and has been used to deliver a variety of molecules for the purpose of DNA transfer, anesthesia, cosmetics, vaccination and chemotherapy. We discuss the electrochemotherapy details and results in brief as below.

1. Many studies reported that with belomycin doses far below toxicity, antitumour effectiveness of electrochemotherapy induced good responses of the tumours including tumour cures Serŝa (2000).

2. It has also been found that some tumours are more sensintive to one drug than to another used in electrochemotherapy Serŝa (2000).

3. Not all tumours have equal level of sensitivity to electrochemotherapy with bleomycin, but all tumor types (e,g., breast, colon, bladder, renal cell, malignant melanoma, basal cell carcinoma) have shown a response to electrochemotherapy Gehl (2008); Serŝa (2000).

4. Electrochemotherapy with bleomycin was performed on tumours in internal organs (brains and livers in rats) Serŝa (2000).

5. Complete eradication of treated nodules occurs in approximately 75% of the cases, and at least a partial remission occurs in 85-90% of the treated patients Gehl (2008).).

6. Mostly square pulses of duration 100 us, with electric field intensity of 1300V/cm or 1500V/cm, repetition frequency of 1 Hz is used. With higher amplitudes more cells in the tumour are permeabilized Gehl (2008).

7. Permeabilization of tumour cells is also dependent on the number of electric pulses, with eight electric pulses found optimum Jaroszeski *et al.* (2000).

8. Antitumour effectivenss is dependent on drug concentration in the tumour during application of electric pulses, a 3-minute interval between the treatments is optimal Jaroszeski *et al.* (2000).

9. The second most useful drug for electrotherapy has been found to be cisplatin. This is one of the drugs that induces resistance in cells, often early in the course of chemotherapy treatments. Electroporation has demonstrated itself to overcome this resistance of cells to cisplatin, at least to some degree Serŝa (2000).

10. Other attempts to determine whether other drugs would be effective in electrochemotherapy protocol in vivo do not prove to be good candidates because of their lipophilicity (being soluble in fat solvents) Serŝa (2000).

11. Achieving optimal electroporation during therapy without the need for repetitive treatment is an issue yet, however several ways of obtaining this are under research e.g. optimization of electric field by proper choice of value, number and duration of pulses and type of electric field, needle electrode design, rotating these needles between pulses Jaroszeski *et al.* (2000); Serŝa *et al.* (1993).

12. Impact of electrochemotherapy on the formation of metastates is yet to be established with preclinical and clinical studies, however studies have shown decreased number of metastases in rats Orlowski *et al.* (1998).

2.4 Noteworthy treatments of electroporation
There are many other special noteworthy applications of electrochemotherapy/ electroporation worth mentioning separately as below:

1. Treatment of human pancreatic tumours: cancers of pancreas is currently the fifth leading cause of cancer related deaths with a very low five year survival in the United States Jaroszeski *et al.* (2000). Since it is hard to detect in the early stages, it becomes difficult to treat. Conventional chemotherapeutic agents have not been vey effective for human pancreatic cancers Talele *et al.* (1991). A novel cancer treatment that uses a single intratumoral injection of bleomycin followed by application of the tumour site with square wave pulses has produced a large percentage of cures and a good number of partial regression in many different forms of cancers e.g. in human larynx Nanda *et al.* (1998b), human pancreas Dev *et al.* (1997); Nanda *et al.* (1998a).

2. Electrofusion (EF): Under appropriate physical conditions, delivery of electric pulses can lead to membrane fusion in close-contact adjacent cells. EF results in the encapsulation of both original cellsíntracellular material within a single enclosed membrane and can be used to produce genetic hybrids or hybridomas Zimmermann (1982). Hybridomas are hybrid cells produced by the fusion of an antibody secreting stimulated B-lymphocytes, with a tumour cell that grows well in culture. The hybridoma is then able to continue to grow in culture, and a large amount of specific desired antibodies can be recovered after processing. Electrofusion has proved to be a successful approach in the production of vaccines Orentas *et al.* (2001); Scott-Taylor *et al.* (2000), antibodies Schmidt *et al.* (2001), and reconstructed embryos in mammalian cloning Gaynor *et al.* (2005).

3. Transdermal drug delivery (TDD): Application of high-voltage pulses to the skin allows a large increase in induced ionic and molecular transport across the skin barrier Prausnitz *et al.* (1993). This has been applied for transdermal delivery of drugs, such as metoprolol Vanbever *et al.* (1994), and also works for larger molecules, for example, DNA oligonucleotides Vanbever *et al.* (1994).

4. Electroinsertion (EI): Another application of electroporation is insertion of molecules into the cell membrane. As the electric field induced membrane pores reseal, they entrap some of the transported molecules. Experiments on electroinsertion suggest the possibility of using the process to study certain physiological properties of these cells and understanding aspects of the lipid-protein interactions of the cell plasma membrane Mouneimne *et al.* (1992).

2.5 Known side effects of electrochemotherapy

1. During electrochemotherapy, the pulse delivery is usually painful for patients due to a a muscle contraction. Generally, a number of electric pulses are delivered, with a repetition frequncy of 1 Hz, which results in equal number of individual sensations and muscle contractions Zupanic *et al.* (2007). To reduce the number of individual muscle conractions, use of pulse frequncy larger than the frequncy of tetanic contraction has been suggested Miklavcic *et al.* (2005). It has been reported that increasing the pulse repetition frequncy to 5 kHz lowers the number of contractions, whereas clinical effectiveness remains same as that achieved by 1 Hz Marty *et al.* (2006); Snoj *et al.* (2007).

2. Just after the electric pulse delivery to the tumour nodule on the mice flanks, it is regularly observed a transient paralysis of the hind legs of the treated mouse, which lasts less than one minute and is always totally reversible Orlowski and Mir (2000).

3. Few days after electrochemotherapy, small scabs are often observed on the skin just at the level of electrode application. They always heal after a few days Orlowski and Mir (2000).

4. The delivery of electric pulses to tumours induces changes (reduced) in tumour blood flow. These changes have been observed to be sensitive to the frequency of applied electric field. The immediate reduction in tumour blood flow at 5 kHz was higher than the reduction at 1 Hz during the initial perid following pulse delivery however at longer times, the 5 kHz frequency had effects on tumour bood flow comparable to those abserved at 1 Hz Raeisi *et al.* (2010). Reduction in tumour blood flow may result in trapping of the drug in tumours, thus providing a longer time for the drug to act by decrease the drug washout from the tumour Mir (2006).

5. That antitumour effectiveness of electrochemotherapy is not only due to increased cytotoxicity of the drugs due to electroporation of tumour cells, but also due to reduced tumour blood flow and oxygenation Sersa *et al.* (2008).

3. Electrogenetherapy

Application of electroporation for transfer of DNA into cells to effect some form of gene therapy, often referred to as electrogenetransfection/electrogenetherapy, is currently being applied in some clinical trials. It is presently considered to have large potential as a non-viral method to deliver genetic material into cells, the process aimed at correcting genetic diseases Budak-Alpdogan *et al.* (2005). The genes being coiled up need a larger electropore for a longer time in order for it to enter the cells. Numerical modelling is useful to establish appropriate parameters to achieve this Krassowska and Filev (2007); Talele and Gaynor (2007); Talele *et al.* (2010). We have seen that due to electroporation the cells can be permeabilized such that the barrier function of the membrane is instantaneously compromised. During this time, genetic material may travel across the membrane. A successful gene transfer process is the one where the electrical and biological conditions of the cell are such that the barrier function of the cell membrane is rapidly restored for a cell survival. This process is termed a electrogenetransfer and when used for therapeutic purpose, electrogenetherapy. For gene therapy to be successful, the gene must be transferred efficiently to target cell without the cell damaging side effects. Most common method for gene transfer in the literature is the viral vector method to attach the gene of interest to enter in the target cell. This method may have detrimental effects of the virus Feuerbach and Crystal (1996) and thus alternative methods of gene transfer are necessary. The electrogenetherapy seems to be the most promising one since the side effects are close to none.

One very obvious fact is that the intake of genetic material by an electroporated cell is affected by the extent of cell membrane permeabilization. This can be dependent on several parameters like cell diameter, cell membrane thickness and capacitance, internal and external conductivities, the electric pulse parameters used for electroporating, the time duration they are used for and so on. Many of these parameters have an interdependent and a non linear effect on the end result and need complex mathematical use to be explained in full details. Without getting into these details, I would like to mention some prominent simple to understand comments about electrogenetherapy.

1. Once the cell is permeabilized by a pulse, more DNA enters the cells during the next pulse of lower field strength Sukharev *et al.* (1992). Multiple pulses and AC pulses seem to have better results Chang *et al.* (1991).

2. Adding the DNA immediately after the pulse usually results in a much lower transfection efficiency compared to adding DNA before the pulse Andreason and Evans (1989).

3. For a given pore forming pulse electric field, the transfection efficiency depends more on the total length of the pulses than of the time span when cells remain permeable, which suggests that uptake of DNA adsorbed on cell surfaces would also contribute to this efficiency Nickoloff and Reynolds (1992).

4. Other physical parameters such as geometry and concentration of DNA are also important e.g. bigger size DNA would need bigger pores on the cell wall and thus would be hard to enter Nickoloff and Reynolds (1992).

5. The transfection efficiency decreases with increasing gap between repeating pulses Chang *et al.* (1991). This also suggests that DNA is collected on the cell surface for subsequent push through the electropores.

6. If the electro pores are too large that cell membrane is unable to reseal then the cell dies.

7. Another reason the cells may not survive is the osmotic swelling. This is due to selective permeability of the membrane after electroporation. Once electroporated, the intracellular cell material molecules being large, these molecules cannot escape outside, however the small ions from outside can enter in, causing the cell to swell leading to bursting and death Baker and Knight (1983).

4. Conclusion

Electroporation may be widely used as a cancer treatment in near future with advantages of low toxicity and being topical and more effective at the tumor site. Newer drugs suitable for various types of cancers and an optimum methodology of application of the electric field is under extensive research. Electroporation may also be the popular phenomenon used for genetherapy without use of viral vectors.

5. References

Abidor, I. G., V. B. Arakelyan, L. V. Chernomordik, Y. A. Chizmadzhev, V. F. Pastushenko, and M. R. Tarasevich. Electric breakdown of bilayer membranes: I. The main experimental facts and their qualitative discussion. *Bioelectrochemistry and Bioenergetics*, 6, pp. 37–52 (1979).

Andreason, G. L. and G. A. Evans. Optimization of electroporation for transfection of mammalian cells. *Anal Biochem*, 180, pp. 269–275 (1989).

Baker, P. F. and D. E. Knight. High volatge techniques for gaining access to the interior of cells: application to the study of exocytosis and membrane turn-over. *Methods Enzymol*, 98, pp. 23–37 (1983).

Budak-Alpdogan, T., D. Banerjee, and J. R. Bertino. Hematopoietic stem cell gene therapy with drug resistance genes: An update. *Cancer Gene Therapy*, 12, pp. 849Ǔ–863 (2005).

Cemažar, M., D. Miklavĉiĉ, and G. Serŝa. Intrinsic sensitivity of tumor cells to bleomycin as an indicator of tumor response to electrochemotherapy. *Japanese Journal of Cancer Research*, 89, pp. 328–333 (1998).

Chang, D. C., B. M. Chassy, J. A. Saunders, and A. E. Sowers, editors. *Guide to Electroporation and Electrofusion*. Academic Press, San Diego, CA (1992).

Chang, D. C., P. Q. Gao, and B. L. Maxwell. High efficiency gene transfection by electropration using a radio-frequency electric field. *Biochim Biophys Acta*, 1992, pp. 153–160 (1991).

Chang, D. C. and T. S. Reese. Changes in membrane structure induced by electroporation as revealed by rapid-freezing electron microscopy. *Biophysical Journal*, 58, pp. 1–12 (1990).

Chernomordik, L. V., S. I. Sukharev, I. G. Abidor, and Y. A. Chizmadzhev. Breakdown of lipid bilayer membranesin an electric field. *Biochim Biophys Acta*, 736, pp. 203–213 (1983).

Dev, S. B., G. S. Nanda, Z. An, X. Wang, R. M. Hoffman, and G. A. Hofman. An effective electroporation therapy of human pancreatic tumours implanted in nude mice. *Drug Delivery*, 4, pp. 293–2996 (1997).

Dimitrov, D. S. Electroporation and electrofusion of membranes. In: *Handbook of Physics of Biological Systems*, edited by R. Lipowsky and E. Sackmann, volume 1, pp. 854–895. Elsevier (1995).

Feuerbach, F. J. and R. G. Crystal. Progress in human gene therapy. *Kidney Int.*, 49, pp. 1791–1794 (1996).

Gaynor, P., D. N. Wells, and B. Oback. Couplet alignment and improved electrofusion by dielectrophoresis for a zona-free high-throughput cloned embryo production system. *Medical and Biological Engineering and Computing*, 43, pp. 150–154 (2005).

Gehl, J. Electroporation for drug and gene delivery in the clinic:doctors go electric. In: *Electroporation protocols Preclinical and clinical gene medicine*, edited by S. Li, chapter 27, pp. 351–372. Humana Press, New Jersey (2008).

Glaser, R. W., S. L. Leikin, L. V. Chernomordik, V. F. Pastushenko, , and A. I. Sokirko. Reversible electrical breakdown of lipid bilayers: Formation and evolution of pores. *Biochim Biophys Acta*, 940, p. 275Ǔ287 (1988).

Glass, L. F., N. A. Fenske, M. Jaroszeski, R. Perrott, D. T. Harvey, D. S. Reintgen, and R. Heller. Bleomycin-mediated electrochemotherapy of basal cell carcinoma. *Journal of the American Academy of Dermatology*, 34, pp. 82–86 (1996).

Gothelf, A., L. M. Mir, and J. Gehl. Electrochemotherapy: Results of cancer treatment using enhanced delivery of bleomycin by electroporation. *Cancer Treatment Reviews*, pp. 1–17 (2003).

Hibino, M., H. Itoh, and K. J. Kinosita. Time courses of cell electroporation as revealed by submicrosecond imaging of transmembrane potential. *Biophysical Journal*, 64, pp. 1789–1800 (1993).

Jaroszeski, M. J., R. Gilbert, and R. Heller. *Methods in Molecular Medicine: Electrochemotherapy, Electrogenetherapy, and Transdermal Drug Delivery Electrically Mediated Delivery of Molecules to Cells*. Humana Press, Totowa, New Jersey (2000).

Kinosita, K. and T. Y. Tsong. Voltage-induced pore formation and hemolysis of human erythrocytes. *Biochim Biophys Acta*, 471, pp. 227–242 (1977a).

Kinosita, K. and T. Y. Tsong. Formation and resealing of pores of controlled sizes in human erythrocyte membrane. *Nature*, 268, pp. 438–441 (1977b).

Kinosita, K., Jr., M. Hibino, H. Itoh, M. Shigemori, K. Hirano, Y. Kirinoand, and T. Hayakawa. Events of membrane electroporation visualized on a time scale from microsecond to seconds. In: *Guide to Electroporation and Electrofusion*, edited by D. C. Chang, B. M. Chassy, J. A. . Saunders, and A. E. Sowers, chapter 3, pp. 29–46. Academic Press, San Diego, CA (1992).

Kotnik, T., D. Miklavĉiĉ, and T. Slivnik. Time course of transmembrane voltage induced by time-varying electric fields: a method for theoretical analysis and its application. *Bioelectrochemistry and Bioenergetics*, 45, pp. 3–16 (1998).

Kranjc, S., M. Cemaẑar, A. Grosel, M. Sentjurc, and G. Serŝa. Radiosensitising effect of electrochemotherapy with bleomycin in LPB sarcoma cells and tumors in mice. *BMC Cancer* (2005).

Krassowska, W. and P. D. Filev. Modeling electroporation in a single cell. *Biophysical Journal*, 92, pp. 404–417 (2007).

Marty, M., G. Sersa, J. R. Garbay, J. Gehl, C. G. Collins, M. Snoj, V. Billard, P. F. Geetsen, J. O. Larkin, D. Miklavcic, I. Pavlovic, S. M. Paulin-Kosir, M. Cemazar, N. Morsli, D. M. Soden, Z. Rudolf, C. Robert, G. OŝSullivan, and L. M. Mir. ElectrochemotherapyŮan easy, highly effective and safe treatment of cutaneous and subcutaneous metastases: results of esope (european standard operating procedures of electrochemotherapy) study. *European journal of cancer supplement*, 4, pp. 3–13 (2006).

Miklavĉiĉ, D. and T. Kotnik. Electroporation for electrochemotherapy and gene therapy. In: *Bioelectromagnetic Medicine*, edited by M. S. Markov, chapter 40, pp. 637–656. Marcel Dekker, New York (2004).

Miklavcic, D., G. Pucihar, M. P. S. Ribaric, M. Mali, A. M.-L. A, M. Petkovsek, J. Nastran, S. Kranjc, M. Cemazar, and G. Sersa. The effect of high frequency electric pulses on muscle contractions and antitumor efficiency in vivo for a potential use in clinical electrochemotherapy. *Bioelectrochemistry*, 65, pp. 121–128 (2005).

Mir, L. M. Bases and rationale of the electrochemotherapy. *European Journal of Cancer Suppl*, 4, pp. 38–44 (2006).

Mir, L. M., S. Orlowski, J. Belehradek, and C. Paoletti. Electrochemotherapy potentiation of antitumour effect of bleomycin by local electric pulses. *European Journal of cancer*, 27, pp. 68Ů–72 (1991).

Mir, L. M., S. Orlowski, J. J. Belehradek, J. Teissié, M. Rols, G. Serŝa, D. Miklavĉiĉ, R. Gilbert, and R. Heller. Biomedical applications of electric pulses with special emphasis on antitumor electrochemotherapy. *Bioelectrochemistry and Bioenergetics*, 38, pp. 203–207 (1995).

Mouneimne, Y., P. F. Tosi, R. Barhoumi, and C. Nicolau. Electroinsertion: An electrical method for protein implantation into cell membranes. In: *Guide to Electroporation and Electrofusion*, edited by D. C. Chang, B. M. Chassy, J. A. Saunders, and A. E. Sowers, chapter 20, pp. 327–346. Academic Press, San Diego, CA (1992).

Nanda, G. S., F. X. Sun, G. A. Hoffman, R. M. Hofman, and S. B. Dev. Electroporation enhances therapeutic efficiancy of anticancer drugs:treatment of human pancreatic tumour in animal model. *Anticancer Research*, 18, pp. 1361–1366 (1998a).

Nanda, G. S., F. X. Sun, G. A. Hoffman, R. M. Hofman, and S. B. dev. Electroporation therapy of human larynx tumours hep-2 implanted in nude mice. *Anticancer Research*, 18, pp. 999–1004 (1998b).

Neu, J. C. and W. Krassowska. Asymptotic model of electroporation. *Physical Review E*, 59, pp. 3471–3482 (1999).

Neumann, E. The relaxation hysteresis of membrane electroporation. In: *Electroporation and Electrofusion in Cell Biology*, edited by E. Neumann, A. E. Sowers, and C. A. Jordan, chapter 4, pp. 61–82. Plenum Press, New York (1989).

Neumann, E., S. Kakorin, and K. Toensing. Membrane electroporation and electro-mechanical deformation of vesicles and cells. *Faraday Discuss*, pp. 111–Ű125 (1998).

Neumann, E. and K. Rosenheck. Permeability changes induced by electric impulses in vesicular membranes. *Journal of Membrane Biology*, 10, pp. 279–290 (1972).

Nickoloff, J. A. and R. J. Reynolds. Electroporation mediated gene transfer efficiency is reduced by linear plasmid carrier dnas. *Anal. Biochem*, 205, pp. 237–243 (1992).

Okino, M. and H. Mohri. Effects of high voltage electrical impulse and an anticancer drug on in vivo growing tumours. *Japanese Journal of cancer research*, 78, pp. 1319–1321 (1987).

Orentas, R., D. Schauer, Q. Bin, and B. D. Johnson. Electrofusion of a weakly immunogenic neuroblastoma with dendritic cells produces a tumor vaccine. *Cellular Immunology*, 213, pp. 4–13 (2001).

Orlowski, S., D. An, J. J. Belehradek, and L. M. Mir. Antimetastatic effect of electrochemotherapy and histoincompatible interleukin-2-secreting cells in the murine lewis lung tumour. *Anticancer drugs*, 9, pp. 551–556 (1998).

Orlowski, S. and L. M. Mir. Treatment of multiple spontaneous breast tumors in mice using electrochemotherapy. In: *Electrochemotherapy, Electrogenetherapy, and Transdermal Drug Delivery : Electrically Mediated Delivery of Molecules to Cells (Methods in Molecular medicine)*, edited by M. J. Jaroszeski, R. Heller, and R. Gilbert, chapter 15, pp. 265–269. Humana Press, Totowa, New Jersey (2000).

Pohl, H. A. *Dielectrophoresis, the Behavior of Matter in Non-uniform Electric Fields*. Cambridge University Press, Cambridge (1978).

Prausnitz, M. R., V. G. Bose, R. Langer, and J. C. Weaver. Electroporation of mammalian skin: A mechanism to enhance transdermal drug delivery. *Proceedings of the National Academy of Sciences*, 90, pp. 10504–10508 (1993).

Raeisi, E., S. M. Firoozabadi, S. Hajizadeh, H. Rajabi, and Z. M. H. ZM. The effect of high-frequency electric pulses on tumor blood flow in vivo. *Journal of Membrane Biology*, 236, pp. 163–166 (2010).

Rosenheck, K., P. Lindner, and I. Pecht. Effect of electric fields on light-scattering and fluorescence of chromaffin granules. *Journal of Membrane Biology*, 20, pp. 1–12 (1975).

Schmidt, E., U. Leinfelder, P. Gessner, D. Zillikens, E. B. Brocker, and U. Zimmermann. CD19+ B lymphocytes are the major source of human antibody-secreting hybridomas generated by electrofusion. *Journal of Immunological Methods*, 255, pp. 93–102 (2001).

Schwister, K. and B. Deuticke. Formation and properties of aqueous leaks induced in human erythrocytes by electrical breakdown. *Biochim Biophys Acta*, 816, pp. 332–348 (1985).

Scott-Taylor, T. H., R. Pettengell, I. Clarke, G. Stuhler, M. C. L. Barthe, P. Walden, and A. G. Dalgleish. Human tumour and dendritic cell hybrids generated by electrofusion: Potential for cancer vaccines. *Biochim Biophys Acta*, 1500, pp. 265Ű–267 (2000).

Serŝa, G. Electrochemotherapy. In: *Electrochemotherapy, Electrogenetherapy, and Transdermal Drug Delivery : Electrically Mediated Delivery of Molecules to Cells (Methods in Molecular medicine)*, edited by M. J. Jaroszeski, R. Heller, and R. Gilbert, chapter 6, pp. 119–133. Humana Press, Totowa, New Jersey (2000).

Serŝa, G., M. Cemažar, and D. Miklavĉiĉ. Antitumor effectiveness of electrochemotherapy with cisdiamminedichloroplatinum(II) in mice. *Cancer Research*, 55, pp. 3450Ű–3455 (1995).

Sersa, G., D. Miklavcic, M. Cemazar, Z. Rudolf, G. Pucihar, and M. Snoj. Electrochemotherapy in treatment of tumors. *Eur J Surg Oncol*, 34, pp. 232–240 (2008).

Serŝa, G., S. Novaković, and D. Miklavĉiĉ. Potentiation of bleomycin antitumor effectiveness by electrotherapy. *Cancer Letters*, 69, pp. 81–84 (1993).

Serŝa, G., B. Stabuĉ, M. Cemažar, D. Miklavĉiĉ, and Z. Rudolf. Electrochemotherapy with cisplatin: the systemic antitumour effectiveness of cisplatin can be potentiated locally by the application of electric pulses in the treatment of malignant melanoma skin metastases. *Melanoma Research*, 10, pp. 381–385 (2000).

Snoj, M., M. Cemazar, B. S. Kolar, and G. Sersa. Effective treatment of multiple unresectable skin melanoma metastases by electrochemotherapy. *Croat Med Journal*, 48, pp. 391–395 (2007).

Snoj, M., Z. Rudolf, M. Cemažar, B. Jancar, and G. Serŝa. Successful sphincter-saving treatment of anorectal malignant melanoma with electrochemotherapy, local excision and adjuvant brachytherapy. *Anti-Cancer Drugs*, 16, pp. 345–348 (2005).

Sukharev, S. I., S. I. Klenchin, V. A. Serov, L. V. Chernomordik, and Y. A. Chizmadzhev. Electroporation and electrophoretic dna tranfer into cells: The effect of dna interaction with electropores. *Biophysics Journal*, 63, pp. 1320–1327 (1992).

Talele, S. and P. Gaynor. Nonlinear time domain model of electropermeabilization: Response of a single cell to an arbitrary applied electric field. *Journal of Electrostatics*, 65, pp. 775–784 (2007).

Talele, S. and P. Gaynor. Non-linear time domain model of electropermeabilization: Effect of extracellular conductivity and applied electric field parameters. *Journal of Electrostatics*, 66, pp. 328–334 (2008).

Talele, S., P. Gaynor, J. van Ekeran, and M. J. Cree. chemotherapy in pancreatic cancer;a rational persuit. *Anticancer Drugs*, 2, pp. 3–10 (1991).

Talele, S., P. Gaynor, J. van Ekeran, and M. J. Cree. Modelling single cell electroporation with bipolar pulse parameters and dynamic pore radii. *Journal of Electrostatics*, 68, pp. 261–274 (2010).

Tozon, N., V. Kodre, G. Serŝa, and M. Cemažar. Effective treatment of perianal tumors in dogs with electrochemotherapy. *Anticancer Research*, 25, pp. 839–946 (2005).

Vanbever, R., N. Lecouturier, and V. Preat. Transdermal delivery of metoprolol by electroporation. *Pharmacological Research*, 11, pp. 1657Û–1662 (1994).

Weaver, J. C. Electroporation: a general phenomenon for manipulating cells and tissues. *Journal of Cellular Biochemistry*, 51, pp. 426Û–435 (1993).

Zimmermann, U. Electric field-mediated fusion and related electrical phenomena. *Biochim Biophys Acta*, 694, pp. 227–277 (1982).

Zimmermann, U., J. Schulz, and G. Pilwat. Transcellular Ion Flow in Escherichia coli B and Electrical Sizing of Bacterias. *Biophysical Journal*, 13(10), pp. 1005–1013 (1973).

Zimmermann, U., J. Vienken, and G. Pilwat. Development of drug carrier systems: electric field induced effects in cell membranes. *Journal of Electroanalytical Chemistry*, pp. 553–574 (1980).

Zupanic, A., S. Ribaric, and D. Miklavcic. Increasing the repetition frequency of electric pulse delivery reduces unpleasant sensations that occur in electrochemotherapy. *Neoplasma*, 54, pp. 246–250 (2007).

8

Surface Aspects of Titanium Dental Implants

Kinga Turzo
University of Szeged, Faculty of Dentistry,
Hungary

1. Introduction

This book chapter presents a brief description of a new emerging field of science, the biological surface science and stresses its importance in the field of alloplastic materials and dental implants. It is not intended to present a comprehensive review of the field, but rather to indentify some important trends and directions in the surface modifications of titanium (Ti) dental implants targeting the improvement of their bio/osseointegration (second subchapter). The third subchapter outlines the impact of fluoride on surfaces of titanium implants or other dental devices. The fourth one will give an overview of the effects of some chemical cleaning agents on titanium implant surfaces. The interaction between Ti and fluoride containing prophylactic agents or chemical cleaning agents can result in a beneficial or/and destructive alteration of the surface of Ti dental appliances. The objective of our studies was to characterize these specific modifications and alterations of Ti surfaces. The fifth subchapter focuses on the relation between biological surface science and dental implants related research.

2. Improving osseointegration of titanium implants by surface modifications - latest trends

These days much effort goes into the design, synthesis, and fabrication of Ti dental implants to obtain a long term (lifelong) secure anchoring in the bone. First of all, implants must carry and sustain the dynamic and static loads they are subjected to. The bulk structure of the material governs this ability. Evidently, it is important to achieve a proper function with the shortest possible healing time, with a very low failure rate, and with minimal discomfort for the patient. These factors are important for cost reasons, too.

The success and the long-term prognosis of dental implants depend mainly on three factors: 1) on the anchorage of the artificial root in the host bone, i.e. on the osseointegration; 2) on the peri-implant mucosal seal; 3) finally on the adequate loading of the implant, transmitted by the abutment, i.e. the biomechanical factor (Figure 1.) (Adell et al., 1981; Brånemark, 1983a; Davies, 1998).

During osseointegration, which is the formation of a direct connection between the living bone and the surface of the load-carrying implant, strong links must be formed between the biomaterial and the surrounding bone tissue (Binon et al., 1992; Cochran, 1999; Morra & Cassinelli, 1997; Olefjord & Hansson, 1993).

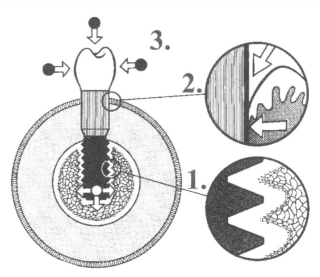

Fig. 1. Scheme of biointegration of a dental implant, representing osseointegration (1), mucosal seal (2) and biomechanical forces (3).

The family of Ti and its alloys represent a major class of materials successfully applied in prosthetic dentistry, dental implantology, and orthopaedics just because they meet the most important requirements of alloplastic materials (Brånemark et al., 1983b; Meffert et al., 1992).

Ti has been used in dentistry for over 30 years; its use in surgery was reported even earlier: in 1947 J. Cotton introduced Ti and its alloys as implants with medical applications. It is the sevenths most frequent metal in the earth's crust and it is a quite light material. Its density is 4.5 g/cm³, considerably less than that of other metals used in dentistry, like gold (19.3 g/cm³) or CoCrMo alloy (8.5 g/cm³). In its unalloyed condition, Ti is as strong as steel, but it is 45% lighter (density of stainless steel is 7.9 g/cm³). Its melting point is 1672-1727°C, and its other thermal properties (like thermal conductivity) are similar to those of the dental tissues. The Vickers Hardness Number (VHN) of Ti is 210, similar to gold alloys type III, IV (hard): 135-250. Ti6Al4V alloy has a VHN of 320 which is close to the value of CoCr alloys: 350-390 (Wang & Fenton, 1996; O'Brien, 2002).

Commercially pure (CP) Ti is available in four grades, which vary according to the oxygen (0.18 to 0.40 weight percent) and iron (0.20 to 0.50 weight percent) contents. These slight concentration differences have a substantial effect on the physical properties of the metal. Oxygen, in particular, has a great influence in the ductility and strength of Ti (Park & Kim, 2000; O'Brien, 2002).

To alter its properties, Ti can be alloyed with a wide variety of elements (Al, V, Nb, Zr), that can improve its strength, high temperature performance, creep resistance, weldability, response to ageing heat treatments, and formability.

Ti is a dimorphic allotrope: while at room temperature CP Ti has α-phase (HCP-hexagonally closed packed), above 883°C a body centred cubic (BCC), the β–phase will form (allotropic phase transformation). The β–form is stronger but more brittle than α–phase (Lautenschlager & Monaghan, 1993).

The two most useful properties of this metal are exceptional corrosion resistance and the highest strength-to-weight ratio of any metal (Lautenschlager & Monaghan, 1993; O'Brien, 2002; Park & Kim, 2000; Wang & Fenton, 1996).

Ti and its alloys are resistant to corrosion because of the formation of an insoluble and continuous titanium oxide layer on the surface (Figure 2) having one of the highest heats of reaction: $\Delta H = -912$ kJ/mol. In air, the oxide (usually TiO_2), begins to form within nanoseconds (10^{-9} s) and reaches a thickness of 20–100 Å already in 1 s. It is very adherent to the parent Ti, protects the metal from other impurities and it is impenetrable to oxygen. (Lautenschlager & Monaghan, 1993). TiO_2 may be catalytically active for a number of organic and inorganic chemical interactions influencing biological processes at the implant interface: the TiO_2 oxide film permits a compatible layer of biomolecules to attach. Explanations for that are still largely unidentified, but the high dielectric constant ($\varepsilon = 50\text{-}170$) of TiO_2 versus 4-10 for alumina and dental porcelain can outcome in considerably stronger van der Waals bonds between molecules and TiO_2 than other oxides. It is the nature of this surface layer that is thought to give titanium its excellent biocompatibility (Lautenschlager & Monaghan, 1993).

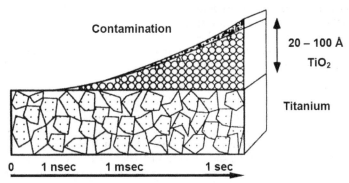

Fig. 2. Formation of an insoluble 20–100 Å thick TiO_2 layer on the surface of metal titanium.

Both CP Ti and Ti6Al4V own exceptional corrosion resistance for a wide range of oxide states and pH levels (Wang & Fenton, 1996). However, even in its passive condition, Ti is not totally inert. Its ions are released due to the chemical dissolution of titanium dioxide. Elevation of implant elements in blood can be observed for Ti6Al4V (measured in the fibrous membrane encapsulating implants), but they are non toxic: 21 ppm Ti, 10.5 ppm Al, 1 ppm V (Puleo & Nanci, 1999).

The strength of the material varies between a much lower value than that of 316 stainless steel or the CoCr alloys and a value about equal to that of annealed stainless steel of the cast CoCrMo alloy. But comparing its specific strength (yield strength per density) Ti alloys exceed any other implant materials. Ti, however, has poor shear strength making it less advantageous for bone screws, plates and similar applications (Park & Kim, 2000).

Ti also has the advantage that its mechanical properties (like elastic modulus) are closer to those of bone than are those of other metals, like stainless steel or CoCr alloys. Although its shear strength is too low for use in major load-bearing applications, it remains the material of choice for dental implants.

Although the bulk properties (mechanical and thermal characteristics) of biomaterials are important with respect to their biointegration, the biological responses of the surrounding tissues to dental implants are controlled mostly by their surface characteristics (chemistry and structure) because biorecognition takes place at the interface of the implant and host tissue. Biological tissues interact mainly with the outermost atomic layers of an implant which is about 0.1-1 nm thick. The molecular and cellular events at the bone-implant interface are well described in Puleo & Nanci, 1999 and Kasemo, 2002 but many crucial aspects are still far from being understood (Kasemo, 2002; Puleo & Nanci, 1999). Although our knowledge regarding the molecular structure of the bone-implant interface has evolved much in the last decade there are still many uncertainties (Klinger et al., 1998).

Justification of the surface modification of implants is therefore straightforward: to retain the key physical properties of an implant while modifying only the outermost surface to control the bio-interaction. As a result, a lot of research work is devoted to elaborate methods of modifying surfaces of existing implants (biomaterials) to achieve the desired biological responses.

These responses can be several: in case of a healthy patient a regular osseointegration process, but in case of elder or even ill patients a smaller bone quantity or not ideal bone quality means a handicap in biointegration. These cases are often avoided by patient selection. As the average human lifespan is growing, ever more people need tooth replacement using Ti dental implants. The demand is increasing to speed up the otherwise long osseointegration period (3-6 months) to rehabilitate the damaged chewing apparatus of the patients as soon as possible even for people in a worse than average health status.

The question of optimally functionalized Ti implant surface is very complex, not only for the above mentioned problems, but also because a dental implant has several different functional parts (root and neck), which are in contact with different biological tissues: alveolar bone, connective and epithelial tissue (Figure 3.). Usually a smooth surface is developed for epithelial attachment and to prevent plaque formation, a machined or rough oblique part for proper connective tissue attachment, and a rough surface for anchorage in the bone (Figure 3.).

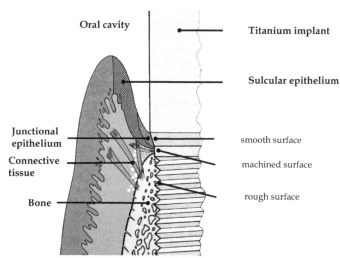

Fig. 3. The different types of biological tissues (epithelial, connective tissue, alveolar bone) in contact with the Ti implant will determine the ideal features of the surface of the implant at its given position.

The optimal implant surface is different for any given purposes, thus, when the goal is to develop an implant surface, then the targeted functional part and the purpose of the modification has to be specified.

For dental implants, likewise to biomaterials, bio- and osseointegration processes can be controlled at molecular and cellular level by the modification of the implant surface. There are several possible surface modifications, usually classified as **physicochemical and biochemical methods** (Puleo & Nanci, 1999).

Many of the surface modifications are in experimental stage and the *in vitro, in vivo* or clinical studies are still ahead. It is our belief that these surfaces will represent a huge positive contribution to clinical implant science, especially if we target the elder or ill patients.

2.1 Physicochemical methods

The most common physicochemical treatments are chemical surface reactions (e.g.: oxidation, acid-etching), sand blasting, ion implantation, laser ablation, coating the surface with inorganic calcium phosphate, etc. These methods are altering the energy, charge, and composition of the existing surface but they will provide surfaces with modified roughness and morphology as well (Ratner & Hoffman, 1996).

Surface energy/charge/composition/morphology is amongst the physicochemical characteristics, which can be altered in numerous ways. The surface energy (or surface wetting capability) plays an important role not only in protein adsorption, but also with respect to cell attachment and spreading (Baier & Meyer, 1988). This physical property can be determined by measuring the contact angles formed with the surface by different liquids. The surface charge influences both the molecular or cellular orientation and the cellular metabolic activity (Meyle, 1999).

One of the best examples demonstrating the importance of hydrophilic and hydrophobic properties of surfaces is the SLActive surface of Straumann implant. Neutralization of the implant in nitrogen atmosphere and storage in NaCl solution assures a hydrophilic surface with good clinical results. These implants present higher implant stability in the critical treatment period of 2 to 4 weeks (Eriksson et al. 2004; Buser et al. 2004).

Ion implantation methods are generally used to improve the mechanical quality of an implant. For example, iridium was ion implanted in Ti6Al4V alloy to improve its corrosion resistance and implanting nitrogen into Ti reduces wear significantly (Buchanan et al., 1990; Sioshansi, 1987).

The roughness (R_a) of the implant surface plays a significant role in anchoring cells and connecting the surrounding tissues, thereby leading to a shorter healing period. These surfaces display advantages over smooth ones as the area of contact with the bone is enlarged by micro-structuring the implant surface. Creation of mechanical interlocking accelerates bone ingrowths (Cochran et al., 1998; Joob-Fancsaly et al., 2004; Santis et al., 1996). The influence of R_a on growth of the different cells has been studied by many authors and it is known that epithelial cells do not attach so strongly to acid-etched or sand-blasted surfaces as to smooth (polished, $R_a < 0.5$ μm) surfaces, while fibroblasts adhere as well to rough (as machined) and even smooth surfaces (Klinge & Meyle, 2006). Surfaces with a smooth topography promote epithelial cell growth, spreading, and the production of focal contacts on Ti surfaces (Baharloo et al., 2005).

Larsson et al. carried out implantation in rabbit bone, concluding that the surface roughness and the oxide thickness affect the rate of bone adhesion in the early stages of implantation (1-7 weeks) (Larsson et al., 1994; Larsson et al., 1996).

Other authors suggest that the metabolic activity (the production of osteocalcin, prostaglandin E2 (PGE2) and transforming growth factor-β1 (TGF-β1) or alkaline phosphatase activity) of osteoblast-like cells is significantly increased on rough (sand-blasted, etched or plasma-sprayed) surfaces. It has been concluded that the surface roughness may modulate the activity of cells interacting with an implant and thereby affect the bone-healing process (Boyan et al., 1998; Meyle, 1999)

Acid-etching, sand blasting and Ti plasma-spraying are typical methods for developing rough surfaces. These are well documented using *in vitro* and *in vivo* methods and are already applied in the production of dental implants (Buser et al. 1991; Wennerberg et al., 1997; Wong et al., 1995).

To increase the roughness of solid surfaces, a number of laser-based techniques have been applied in the last decade (Bauerle, 2000; Joob-Fancsaly et al., 2000). The advantages of using lasers for ablation of surfaces are the precise control of the frequency of the light, the wide range of frequencies available, the high energy density, the ability to focus and raster the light and the ability to pulse the source and control reaction time. Lasers commonly used for surface modification include ruby, Nd:YAG, argon, CO_2 and excimer. Recent studies on the laser machining of dental implants revealed that an appropriate structure with minimum contamination could be achieved by means of laser treatment (Gaggl et al., 2000; Pető et al., 2001). After multipulse irradiation with a focused Nd:YAG laser beam, a crown-like structure formation was observed on the Ti surface (György et al., 2002). An efficient oxidation of Ti by Nd:YAG laser irradiation was reported (Nánai et al. 1997; Perez del Pino et al. 2002. In addition to the prompt, intense heating of the surface, excimer laser illumination may also enhance the sterilizing effect as a consequence of the high dose of UV light (Bereznai et al., 2003).

Our group has developed two kinds of physicochemical surface modifications of Ti implants: laser surface modifications (ablation and polishing) and octacalcium phosphate layer deposition on the surface of implants (Bereznai et al., 2003; Szekeres et al., 2005). The properties of these modified surfaces were investigated by modern surface science techniques (atomic force microscopy (AFM), scanning electron microscopy (SEM), and X-ray photoelectron spectroscopy (XPS)). These experiments demonstrated that excimer laser ablation effectively increased the surface area available for the attachment of bone-cells and also raised the thickness of the TiO_2 layer (Figure 4.). In addition, XPS measurements showed that the UV light of the excimer laser effectively sterilized the surface. Amongst the most promising surface modifications that could enhance the osseointegration of dental implants are the laser modifications, as proven by our preliminary *in vitro* (cell culturing) results.

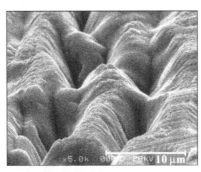

Fig. 4. SEM image of ablated holes formed on a titanium surface with 0.5 ps pulses of a KrF excimer laser. The laser fluence was 2.4 J/cm^2 and 1000 shots were applied (Bereznai et al., 2003).

The importance of these studies is further enhanced by the fact, that they involve laser technologies, which already have numerous industrial applications. However, even these techniques must still be refined, since medical applications require high accuracy in determination of both mechanical and chemical characteristics of the surface.

Despite of some positive indications from *in vitro* studies, the effectiveness of these two latter mentioned methods (ion implantation and laser-based methods) have not yet been investigated sufficiently under *in vivo* conditions.

Inorganic materials, such as bio-reactive calcium phosphate (or hydroxyl-apatite – HAP) coatings, have been applied extensively because of their chemical similarity to bone minerals. Several studies showed that these coatings achieve a very intimate contact between the implant and bone (Hench, 1996; Rohanizadeh et al., 2005; Sun et al., 2001; Szabo et al., 1995). Clinical investigations reported a high degree of success with HAP-coated implants, reducing the healing period (Block et al, 1996). However, in other studies HAP-coated implants showed signs of peeling off the covering material from the implant surface which may induce foreign body reactions (Buser et al. 1991; Matsui et al. 1994).

Furthermore, the long-term clinical study of Wheeler SL. (1996) on HAP-coated oral implants reported a significantly lower survival rate (77.8% after 8 years) for HAP-coated implants as compared to the Ti-plasma-sprayed (TPS)-coated implants (92.7%) (Wheeler, 1996). Biodegradation of these coatings may be the reason why HAP-coated implants are no longer the surface modifications of choice.

2.2 Biochemical methods

In addition to the physicochemical methods, biochemical techniques have currently generated a great deal of interest. Today we struggle to produce biomaterials that interact with specific targets within the body or mimic tissue architecture. It is well known that biological systems have a highly developed ability to recognize special features of the surface on the molecular scale. We look to nature when we design "biomimetic" materials to understand how cells interact with other cells, extracellular proteins, and tissues. This knowledge is then utilized to develop bio-mimetic strategies for functional, interactive biomaterials. These strategies include biodegradable and "smart" materials used in targeted drug delivery systems and tissue engineering, as well as biochemical modifications of biomaterial surfaces for implant or wound-healing applications (Dillow & Lowman, 2002; de Jonge et al., 2008).

In case of implants the goal of biochemical methods is to immobilize peptides, proteins, enzymes on the surface in order to induce specific cell and tissue responses (adhesion, signaling, stimulation) and to control the tissue-implant interface with molecules delivered directly there (Hoffman, 1996; Ito et al., 1991; Puleo & Nanci, 1999).

A lot of different, biologically functional molecules can be immobilized onto Ti surfaces to enhance bone regeneration at the interface of implant devices. An essential aspect is to maintain the bioactivity (or the recognizable binding site) of these molecules while incorporating them into a biomimetic coating (Dillow & Lowman, 2002).

Immobilization of bio-molecules can be achieved by physical absorption (van der Waals or electrostatic interactions), physical entrapment (use of barrier systems) and covalent attachment. The selection of the immobilization method depends on the working mechanism of the specific bio-molecules which dictates, for instance, a short-term, transient immobilization for growth factors and a long-term immobilization for adhesion molecules and enzymes.

Anchoring of proteins to Ti surfaces was previously achieved by direct adsorption onto the surface, but such physical adsorption frequently induces denaturation and loss of the functional activity of the protein (Hoffman, 1996). To eliminate these problems, our group developed a self-assembled polyelectrolyte (PE) multilayer film (Pelsőczi et al, 2005). The film is made by alternating adsorption of poly-cations (poly-L-lysine (PLL)) and poly-anions (poly-L-glutamic acid (PGA)) from aqueous solution onto a charged, solid surface (in our case, Ti surface, Figure 5.). PE film coatings modify the solid/liquid interface in such a way as to ensure a proper environment for the adsorption of proteins. The alternating adsorption technique has been successfully applied in different fields of science, as a consequence of its numerous practical applications. It can be automated, it involves the use of aqueous solutions, it is environment-friendly, and various substrates can be covered with films of readily variable thickness.

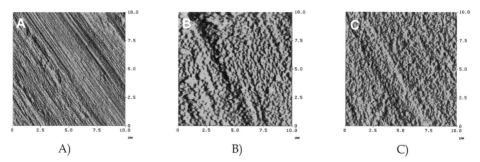

A)   B)   C)

Fig. 5. Typical AFM deflection images of Ti substrate and PE layers on Ti (*in situ* measurements). A) Bare Ti, $z = 500$ nm; B) (PLL/PGA)$_6$ multilayer, $z = 800$ nm; and C) (PLL/PGA)$_8$ film, $z = 1.5$ μm (Pelsőczi et al, 2005).

The inorganic calcium phosphate coatings as well as the purely organic components of bone can serve as carrier systems for osteogenic drugs, rendering them osteoinductive as well as osteoconductive. The most promising candidates for osteogenic agents are the members of the transforming growth factor β (TGF-β) superfamily, such as bone morphogenic proteins (BMPs). BMP-2 has been successfully coprecipitated with the inorganic components and, thus incorporated, retains its biological activity *in vitro* (Liu et al, 2004).

At present four major strategies exist for organic coating approaches: immobilization of extra cellular matrix (ECM) proteins (collagen, etc.) or peptide sequences as modulators for bone cell adhesion, deposition of cell signaling agents (bone growth factors) to trigger new bone formation, immobilization of DNA for structural reinforcement and enzyme-modified Ti surfaces for enhanced bone mineralization (de Jonge et al., 2008).

The cell membrane receptor family of integrins is involved in cell adhesion to ECM proteins. These integrins bind to specific amino acid sequences within ECM molecules, in particular to RGD (arginine-glycine-aspartic) sequence. For this reason the most commonly used peptide sequence for surface modification is the above mentioned cell adhesion motive (Ferris et al, 1999; Schliephake et al, 2005).

The structural properties of DNA show high potential for application as biomaterial coating, regardless of its genetic information. Additionally, DNA can be used as a drug delivery system since its functional groups allow incorporation of growth factors. The studies of Beucken et al., 2007 proved that DNA-based coatings improved the deposition of CaP (van der Beucken et al., 2007).

A relatively new approach for surface modification is the enzyme-modified titanium surface to enhance bone mineralization along the implant surface. Especially, the enzyme alkaline phosphatase (AP) is known to increase the local concentration of inorganic phosphate, and to decrease the concentration of extracellular pyrophosphate, a potent inhibitor of mineralization (Golub & Boesze-Battaglia, 2007).

In the last decade another viable bio-mimetic strategy appeared, the organic-inorganic composite coatings. These mimic the bone structure, which is composed of an organic matrix (90% of which are collagenous proteins) and an inorganic CaP phase. Collagen-CaP (Morra et al., 2003), growth-factor-CaP (Liu et al., 2007) and polyelectrolyte multilayers-CaP (Sikirić et al., 2008) composite coatings were developed until now with promising *in vitro* and *in vivo* experimental results.

As in the case of physicochemical surface modifications, many of the biochemical methods described above are in experimental stage yet and their *in vivo* or clinical studies are still ahead.

3. Impact of fluoride on surfaces of titanium implants or other dental devices

Endosseous dental implants and surgical implants for fixating or replacing hard tissue are made from "commercially pure" Ti (CP Ti) and the most common Ti alloy, Ti6Al4V (Mändl et al., 2005; Park & Kim, 2000). Ti is also used in prosthetic dentistry to manufacture crowns and multiple-unit fixed restorations (Huget, 2002; Wang & Fenton, 1996), and in orthodontic dentistry to produce Ti brackets (Harzer et al., 2001).

Dental arch wires and orthopedic braces are usually made from the special TiNi shape memory alloy (Mändl et al., 2005; Park & Kim, 2000).

Ti and its alloys are resistant to corrosion because of the formation of an insoluble TiO_2 layer on the surface, as described earlier. Oxidative agents are well known to exert a corrosive effect on the alloys used in dentistry, with the exceptions of Ti and other bioinert materials. Indeed, oxidative processes can thicken and condense the TiO_2 layer on the surface, improving the corrosion stability of the underlying Ti. On the other hand, reductive agents, such as fluoride (F^-), may have the opposite effect and attack this layer. Strietzel et al., demonstrated that Ti ion release was enhanced in the presence of F^-, and this effect was even further accelerated at low pH (Strietzel et al., 1998). High F^- concentrations and an acidic pH are known to impair the corrosion resistance of Ti (Toumelin-Chemla et al., 1996), and as a result crevice and pitting corrosion occur (Reclaru & Meyer, 1998; Schiff, et al., 2002).

Patients regularly use different oral care products containing F^-, such as toothpastes, rinsing solutions, or prophylactic gels. The Ti alloys applied in the form of orthodontic wire (Huang, 2007; Walker et al., 2005) or as the framework of a prosthesis, therefore come into contact with a wide range of preventive agents and these F^--containing materials can attack the surface of Ti (Boere, 1995; Siirilä & Könönen, 1991).

SEM investigations have revealed that topical F^- solutions can cause stress corrosion cracking on CP Ti (Könönen et al., 1995) Galvanic corrosion has been reported to occur between orthodontic wires and brackets (NiTi and CuNiTi) immersed in fluoride mouthwashes (Schiff et al., 2006).

Such corrosion has two undesirable consequences: the mechanical performance of the wire-bracket system deteriorates, and the risk of local Ni^{2+} release is increased.

Moreover, such F^--containing agents may come into contact with the neck part of dental Ti implants, which may extend into the oral cavity (Fig. 3). The long-term success of dental implants depends to a large extent on the gingival attachment to the neck of an implant. This mucosal seal ensures protection against bacterial attack and other injurious effects exerted by the oral environment. The epithelial attachment (junctional epithelium) may be anchored onto a rough or a smooth surface by hemidesmosomes through a preformed glycoprotein layer. A rough surface is more favorable for the plaque accumulation in the peri-implant crevices of the gingiva, which is an undesired effect in this very sensitive region of the implant. Accordingly, to avoid pathogenic plaque accumulation, the neck of an implant must be polished (Bollen et al., 1996; Vogel, 1999). From this respect, it is easy to realize the great importance of the maintenance of the continuity of these surfaces.

In 1999, Nakagawa et al. found a relation between the F^- concentration and the pH at which the corrosion of CP Ti occurred (Nakagawa et al., 1999). The results of their anodic polarization and immersion tests indicated that the corrosion of Ti in a F^--containing solution depends on the concentration of hydrofluoric acid (HF). The passivation film on Ti was destroyed when the HF concentration in the solution was > 30 ppm. In 1995, Boere had demonstrated that the corrosion of Ti is enhanced in an acidic environment, because F^- in solution combines with H^+ to form HF, even if the NaF concentration is low (Boere, 1995). Nakagawa et al. investigated the corrosion behaviour of Ti alloys: Ti6Al4V, Ti6Al7Nb, and the new alloy Ti-0.2Pd (Nakagawa et al., 2001). Their experimental results demonstrated that even a low F^- concentration causes corrosion in an acidic environment. If Ti alloy contains at least 0.2% Pd, this process does not take place. The high corrosion resistance of this alloy is because of the surface enrichment of Pd promoting the repassivation of Ti.

The studies by Huang (Huang, 2002) indicated that, when the NaF concentration was >0.1%, the protectiveness of TiO_2 on Ti was destroyed by F^-, leading to the severe corrosion of Ti. In 2003, Huang investigated the effects of F^- and albumin concentrations on the corrosion resistance of Ti6Al4V in acidic (pH 5) artificial saliva (Huang, 2003). The XPS results showed that when the NaF concentration was >0.1%, a hexafluorotitanate complex (Na_2TiF_6) was formed on the Ti surface, which destroyed the stable TiO_2 layer.

As the pH of the rinses and gels used for caries prevention in dentistry ranges from 3.5 up to neutral, and the F^- concentration in these materials is between 1000 and 10,000 ppm (Nakagawa et al., 1999), it is essential for the dental practitioner to know whether a F^--containing material can attack the Ti surface or can modify the corrosion resistance of the Ti surface of a dental implant, a prosthesis, or the wires of orthodontic braces. Besides 0.1–0.15% (1000–1500 ppm) F^-, toothpastes contain other constituents, such as rubbing, cleaning, foaming materials, and calcium complexes, which reduce the effectiveness of toothpastes by 25–50% (Neubert & Eggert, 2001).

Although all the above-mentioned studies point to the deleterious effect of F^--containing prophylactic gels, there are a huge number of data documenting that F^- exerts a bone-promoting activity. Ellingsen et al. proved that, when F^- is incorporated in the TiO_2 layer, the retention of implants is significantly increased, even as compared with rough surface implants (Ellingsen, 1995; Ellingsen et al., 2004). The success of a TiO_2-blasted surface with a F^--modified TiO_2 layer (OsseoSpeed implants, Astratech) is because of the ability of the F^- coating to stimulate the bone response, leading to binding between Ti and the phosphate from tissue fluids. The free F^- catalyzes this reaction and induces the formation of fluoridated hydroxyapatite and fluorapatite in the surrounding bone (Ellingsen, 1995). The studies by Cooper et al., demonstrated that the F^- modification of TiO_2 grit-blasted CP Ti surfaces enhanced osteoblastic differentiation and interfacial bone formation (Cooper et al., 2006).

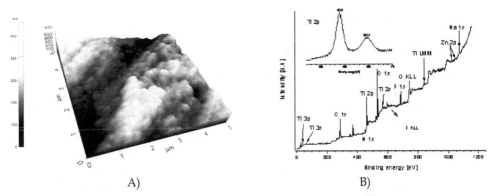

A)   B)

Fig. 6. 3D AFM picture (A) of a Ti disc treated with gel (12,500 ppm F^-, pH 4.8), reveals deep corrosive regions and granular forms. Image size: 5 x 5 µm. Typical XPS spectra of gel-treated Ti discs (B). Three new peaks can be observed on the spectra originating from Na_2TiF_6, which modifies the TiO_2 layer of the surface (Stájer et al., 2008).

In our study (Stájer et al., 2008), the effects of different F^--containing caries-preventive prophylactic rinses and gels on the surface structure and roughness of CP Ti were investigated, through the use of XPS and AFM. A further aim was to survey the attachment and proliferation of human epithelial cells after treatment of the Ti surface with an acidic NaF solution, a widely used F-containing mouthwash or a gel. The epithelial cell attachment and proliferation were examined by means of dimethylthiazol-diphenyl tetrazolium bromide (MTT) and protein content assays. For the visualization of cells, SEM was applied. The results of this study proved that aqueous 1% NaF solution (3800 ppm F-, pH 4.5) or high (12,500 ppm) F- content gel (pH 4.8) strongly corroded the surface and modified its composition (Figure 6.). XPS revealed formation of a strongly bound F--containing complex (Na_2TiF_6). AFM indicated an increase in roughness (R_a) of the surfaces: 7-fold for the NaF solution and smaller for the gel or a mouthwash (250 ppm F-, pH 4.4). MTT revealed that cell attachment was significantly increased by the gel, but was not disturbed by either the mouthwash or the NaF. Cell proliferation determined by MTT decreased significantly only for the NaF treated samples; protein content assay experiments showed no such effect. This study indicates that epithelial cell culturing results can depend on the method used, and the adverse effects of a high F-concentration and low pH should be considered when prophylactic gels are applied by patients with Ti implants or other dental devices.

4. Effects of chemical cleaning agents on titanium implant surfaces

The failure of dental implants is caused mainly by the inflammatory processes affecting the soft and hard tissues of the oral cavity, as the lifetime of such implants rely on the responses of the various surrounding tissues (the alveolar bone, or the conjunctive and epithelial parts of the mucosa, Figure 3.) (Norowski & Bumgardner, 2009).

Peri-implant infections involve peri-implant mucositis, defined as a reversible inflammatory change of the peri-implant soft tissues without bone loss, and peri-implantitis, an inflammatory process resulting in loss of supporting bone and associated with bleeding and suppuration (Norowski & Bumgardner, 2009; Renvert et al., 2008b; Zitzmann & Berglundh, 2008).

Several studies have evaluated peri-implant infections, but only a few were cross-sectional and provide information on the prevalence of peri-implant diseases among patients with implants functioning for approx. 10 years. The incidence of peri-implant mucositis has been reported to be in the range of 60% of implant recipients and in 48% of implants (Renvert et al., 2007; Roos-Jansaker et al., 2006a). The prevalence of peri-implantitis was found to be around 15, 16 and 28% with respect to the recipients (Fransson et al., 2005; Renvert et al., 2007; Roos-Jansaker et al., 2006a), and 7 and 12% regarding implant sites (Fransson et al., 2005; Roos-Jansaker et al., 2006a). The differences in the prevalence of peri-implantitis may be explained by differing criteria used for the diagnosis of peri-implantitis, as well as variations in maintenance procedures (Fransson et al., 2008; Zitzmann & Berglundh, 2008).

The etiology of marginal peri-implantitis is based mainly on an infectious factor and a biomechanical factor (Uribe et al., 2004). Although the causes may differ in both cases, microbial colonization occurs on the surface of the implant (Renvert et al., 2008b; Kotsovilis et al., 2008). If the conditions become pathogenic, bacteria start to proliferate, leading to inflammation around the implant. Peri-implant diseases have been primarily linked to Gram-negative anaerobic microflora (Leonhardt et al. 1999). The process is aggravated by microorganism colonization and their toxins, and extensive bone destruction will occur. The inflammation spreads apically thus, in very severe cases, therefore, the patient may lose the implant. Methods which remove the bacteria and the toxins from the surface of challenged implants would prevent or terminate the development of peri-implant bony defects.

The therapy of peri-implantitis in the surgical phase is a complex process, starting with surgical debridement of devitalized peri-implant tissue and continuing with decontamination of the exposed implant surface. The implant surface can be cleaned by mechanical (an air-powder abrasive) or chemical (citric acid, H_2O_2, chlorhexidine digluconate (CHX) or EDTA) procedures or with laser irradiation (CO_2, diode, Er:YAG or Nd:YAG) (Roos-Jansaker et al., 2003; Schwarz et al., 2006). To support antimicrobial treatment, topical and/or systemic antibiotics may be administered (Roos-Jansaker et al., 2003). After removal of damaged tissues from the peri-implant pocket, surgical treatment (guided tissue regeneration with or without the use of bone grafts and barrier membranes) promotes regeneration of any bone defect (Khoury & Buchmann, 2001; Roos-Jansaker et al., 2003).

For the chemical detoxification of implants, various cleaning solutions are used: CHX, H_2O_2, citric acid, phosphoric acid gel, delmopinol, ListerineR, iodine, saline irrigation, beta-isodona, chloramine-T, etc. Besides these chemical agents, a number of systemic antibiotics can be applied to support the therapy: e.g. tetracycline, amoxicillin, augmentin, metronidazol, penicillin, etc. (Roos-Jansaker et al., 2003; Zitzmann & Berglundh, 2008).

CHX is a commonly administered antimicrobial agent with a wide range of medical applications. It is used in dentistry as a mouthwash and topical antimicrobial. In the treatment of peri-implantitis it can serve as a rinsing solution (Abu-Ta'a et al., 2008; Roos-Jansaker et al., 2006b) or more often as an implant irrigation solution, in combination with systemic antibiotics (Khoury & Buchmann, 2001; Roos-Jansaker et al., 2003). Renvert et al. investigated the difference in effectiveness of minocycline microspheres and CHX gel, and concluded that the adjunctive use of these microspheres led to improved probing depths and bleeding scores, CHX alone resulting in only a limited reduction of the bleeding scores (Renvert et al., 2006; Renvert et al. 2008a). CHX is also effective in the surgical treatment of late peri-implant defects using guided tissue regeneration (Hämmerle et al., 1995; Schou et al., 2003a).

Recognizing the increasing interest in the functionalization of dental implant surfaces with antimicrobial agents prior to implantation, Barbour et al. investigated the adsorption of CHX to TiO_2 crystals of anatase and rutile (Barbour et al., 2007). Their results proved that CHX in 4-morpholinoethanesulphonic acid (MES) and phosphate-buffered saline (PBS) buffers adsorbed rapidly to anatase and rutile TiO_2, equilibrium being attained in less than 60 sec, with gradual desorption over a period of several days. More CHX adsorbed to anatase than to rutile, and the CHX desorbed more rapidly from anatase than from rutile, depending on the buffer used.

The study by Burchard et al. revealed that fibroblasts adhere more readily to surfaces exposed to CHX or saline than to those exposed to stannous fluoride (Burchard et al., 1991).

Saturated citric acid can also be applied for the decontamination of Ti surfaces in the surgical treatment of peri-implantitis with bone grafts and membranes (Deporter & Todescan, 2001; Schou et al., 2003b). In a comparison of the effects of citric acid and 10% H_2O_2, Alhag et al. demonstrated that rough surfaces (with an enhanced TiO_2 layer and textured surface; Nobel Biocare AB®, Gothenburg, Sweden) which were plaque-contaminated and cleaned with either solution, can re-osseointegrate (Alhag et al., 2008). H_2O_2 can be used successfully at a concentration of 3% in the surgical treatment of peri-implantitis, employing bone substitutes with, or without, resorbable membranes (Roos-Jansaker et al. 2007a; Roos-Jansaker et al., 2007b).

Some authors, including Khoury & Buchmann have even used a combination of these three different cleaning solutions in the surgical therapy of peri-implantitis (Khoury & Buchmann, 2001). After removal of the granulomatous tissue, the surgical site was repeatedly rinsed with CHX, after which citric acid (pH = 1) was applied for 1 min to decontaminate the implant surface, this then being rinsed with H_2O_2 and 0.9% saline.

Dennison et al. found that machined implants (without a surface coating) are decontaminated by a variety of methods (air-powder abrasive, citric acid solution, or CHX) more readily than hydroxyapatite-coated surfaces (Dennison et al., 1994).

The above-mentioned chemical agents are commonly applied in the therapy of peri-implantitis, but only investigations relating to the adsorption of CHX on different TiO_2 crystals (anatase and rutile) appear to have been conducted. When used for implant surface decontamination, these materials may alter the morphology and chemical structure of the surface. The aim of our investigation (Ungvári et al., 2010), therefore, was to study the effects of three cleaning solutions in clinical use for peri-implantitis therapy (H_2O_2, citric acid and CHX gel) on Ti.

In vitro studies are essential in the development of such treatments, as these are the basic steps with which to reveal the action of cleaning solutions on the implant surface. Additionally, fewer animal experiments would be required.

Commercially pure (grade 4) machined Ti discs (CAMLOG Biotechnologies AG, Switzerland) were treated with 3% H_2O_2 (5 min), saturated citric acid (pH = 1) (1 min) or CHX gel (5 min), and their surface properties were examined by AFM and XPS. A further aim was to survey the response of the biological environment to these changes, by examining the attachment and proliferation of human epithelial cells after treatment of the Ti surfaces with these solutions. The epithelial cell attachment and proliferation was examined by means of MTT and protein-content assays (the latter with bicinchoninic acid).

Our results revealed no significant difference in the roughness (AFM measurements) of the three treated surfaces. XPS confirmed the constant presence of typical surface elements and an intact TiO_2 layer on each surface. The XPS peaks after CHX gel treatment demonstrated C-O and/or C=O bond formation, due to CHX infiltrating the surface. MTT and BCA assays indicated similar epithelial cell attachments in the three groups; epithelial cell proliferation being significantly higher after H_2O_2 than after CHX gel treatment (not shown by BCA assays). In conclusion these agents do not harm the Ti surface. Cleaning with H_2O_2 slightly enhances human epithelial cell growth, in contrast to CHX gel.

5. Biological surface science in dental implants research

Biological surface science was defined in 2002 by Bengt Kasemo from Göteborg University, Sweden, as a broad interdisciplinary area in which the properties and processes at interfaces between alloplastic materials (biomaterials) and biological environments are studied and also biofunctional surfaces are fabricated (Kasemo, 2002).

Alloplastic or biomaterials are synthetic materials used in devices replacing parts of living systems or to function in intimate contact with the living tissues for any period of time. Beside this, alloplastic materials must not have any adverse or damaging effect on the body as a whole. The successful biointegration of biomaterials may depend on several factors related to the material, like the bulk and surface characteristics of the material, the design (construction) and the biocompatibility of the material. Naturally, the applied surgical technique and the general health condition and life-quality of the patient are also important features. Biocompatibility is defined as the acceptance of an artificial implant by the surrounding tissues and by the body as a whole (Park, 2000).

Achievement of biointegration of alloplastic materials is one of the most important targets of research in the medical, dental, and biological sciences. For developing a viable biomaterial, the knowledge of several disciplines have to be integrated, like science and engineering (e.g. materials science: structure-property relationship of synthetic and biological materials), biology and physiology (e.g. cell and molecular biology, anatomy, animal and human physiology, histopathology, experimental surgery, immunology) and clinical sciences (e.g. dental and maxillofacial implantology, orthopedics, neurosurgery, obstetrics, plastic, reconstructive, cardiovascular surgeries, etc.) (Park, 2000). Medical implants in the human body, biomimetic materials, tissue engineering, biosensors and biochips for diagnostics, bioelectronics and artificial photosynthesis are research areas constituting a strong driving force for the current rapid development of biological surface science (Kasemo, 2002).

The development of different surface modifications with the ultimate goal of improving the biointegration process of alloplastic materials is stimulating the progress in surface specific biotechnologies, like fabrication of high-tech, sophisticated surfaces (e.g. self-organizing monolayers) and in nano- and microfabrication (Kasemo, 2002).

The importance of bio- and alloplastic materials' knowledge in dentistry is evident as one of the goal of dentistry is to maintain and improve the health of the human teeth (oral cavity) in order to improve the quality of life of the dental patient. All of these activities require the replacement or alteration of the existing tooth structure and also the development of auxiliary dental appliances using alloplastic materials. As healthcare improves and people tend to live longer, materials with specific biomedical applications become more and more important. The most frequently used medical implants are dental implants that serve to substitute human teeth. In dentistry the main challenges for centuries have been the development and selection of biocompatible prosthetic materials that can withstand the adverse conditions of the oral environment. Oral cavity represents a multivariate external environment with a wide range of circumstances, like foods, abrasion, acidic pH, temperatures from 5 to 55°C, high masticator forces, bacteria, etc. (Lemons, 1996).

There is an increasing need to develop materials that can be implanted into the maxillofacial area in order to rehabilitate the damaged chewing apparatus due to loss of natural teeth. Multiplicity of dental applications requires more than one type or class of material because no material has yet been developed that can fulfil the varying requirements.

The main research topics are investigations of the biointegration of alloplastic materials and studies of how the chemical and surface microstructural modifications of the titanium dental implants influence their biointegration. These studies relate to replacements of body structures in case the biological function requires a significant load-bearing capability. Examples for this are dental implants and artificial hip-joint replacements. These devices have several common aspects: in general they are made of Ti, and their biological integration depends strongly on the surface structure of the metal. These studies have general interest in basic research but are highly applicable for biomedical and industrial uses as well. As experienced by many groups, including ours, the practical/clinical applications of the findings of studies in cooperation with representatives of other basic sciences, are welcomed on both sides. The multi -or interdisciplinary aspects of these topics are obvious, and without the results of basic science the field of alloplastic materials and biological surface science could not have developed so extensively. Nevertheless, without the experience and observations of clinical scientists, these studies would be purposeless.

6. Conclusion

Since the interactions between medical implants and their biological surroundings are controlled mostly by their surface characteristics (chemistry and structure), characterization of these properties is of main importance. Biorecognition takes place at the interface of the implant and host tissue and the most relevant molecular and cellular events are also localized at this bone-implant interface. The osseointegration of dental implants is relatively long (3-6 months); therefore the surface modifications that can shorten this process will achieve a decreased healing time, a lower failure rate and minimal discomfort for the patients. These improved/modified surfaces will open the possibility of implantation for people in relatively poor health status. A lower bone quantity and/or quality of elder or even ill patients is frequently a handicap in biointegration, therefore these cases are often rejected at patient selection.

The surface modifications outlined above retain the key physical properties of the implants and modify only their outermost surfaces with the ultimate goal of achieving the desired biological responses. The advantages and disadvantages of different physicochemical and biochemical surface modifications were presented in this chapter. Most of these modifications are still in experimental stage, their *in vivo* or clinical studies are still ahead. However, the tools offered by the biological surface science will certainly provide a huge contribution to the development of this field. These methods will help us in understanding the biological events occurring at the bone-implant interfaces and in the development of optimally functionalized implant surfaces as well.

Studies of the impact of fluoride on Ti implant surfaces are becoming more and more important because the chance for such interactions potentially degrading artificial surfaces is quickly growing with the increasing use of biomaterials in the oral cavity. The oral hygiene is enhanced considerably by the use of prophylactic rinses and gels. For both the dental practitioner and the patient it is essential to know whether or not a F^--containing material has the potential to destruct the Ti surface of a dental implant, prosthesis, or the wires of orthodontic braces or at least to decrease its corrosion resistance.

Failure of dental implants is mostly caused by the inflammatory processes affecting the soft and hard tissues. Peri-implant infections involve peri-implant mucositis, defined as a reversible inflammatory change of the periimplant soft tissues without bone loss, and peri-implantitis, an inflammatory process resulting in a loss of supporting bones and associated with bleeding and suppuration. The treatment of peri-implantitis, that causes tissue deterioration surrounding the osseointegrated implants, involves surface decontamination and cleaning. However, chemical cleaning agents may alter the structure of implant surfaces. This sub-chapter presented an overview of the current literature and pointed out the importance of further *in vitro* and *in vivo* studies for the safe application of these decontaminating agents on titanium implants.

7. Acknowledgment

This book chapter was supported by the János Bolyai Research Scholarship of the Hungarian Academy of Sciences and the ETT-248/2009 grant of the Hungarian Ministry of Health. The author thanks Prof. András Fazekas for the valuable discussions, Prof. Zoltán Rakonczay (University of Szeged, Faculty of Dentistry) and Assoc. Prof. Gábor Laczkó (University of Szeged, Faculty of Science and Informatics) for the thorough revision. The author is grateful for the support of Prof. Katalin Nagy (dean of the Faculty of Dentistry) and for the experimental work of our PhD students: Dr. István Pelsőczi K., Dr. Danica Matusovits, Dr. Anette Stájer and Dr. Krisztina Ungvári.

8. References

Abu-Ta'a, M., Quirynen, M., Teughels, W. & van Steenberghe, D. (2008). Asepsis during periodontal surgery involving oral implants and the usefulness of peri-operative antibiotics: a prospective, randomized, controlled clinical trial. *J Clin Periodontol*, Vol.35, pp.58-63

Adell, R., Leckholm. U., Rockler. B. & Brånemark, P.I. (1981). A 15-year study of osseointegrated implants in the treatment of the edentulous jaw. *Int J Oral Surg*, Vol.10, pp. 387-416

Alhag, M., Renvert, S., Polyzois, I. & Claffey, N. (2008). Re-osseointegration on rough implant surfaces previously coated with bacterial biofilm: an experimental study in the dog. *Clin Oral Impl*, Vol.19, pp.182-187

Anusavice, K.J. (1996). Dental Implants, In: *Phillips' Science of Dental Materials*, K.J. Anusavice, (Ed.), 655-662, W.B. Saunders Company, Tenth Edition, ISBN 0-7216-5741-9, Philadelphia, Pennsylvania, USA

Baharloo, B., Textor, M. & Brunette, D.M. (2005). Substratum roughness alters the growth, area, and focal adhesions of epithelial cells, and their proximity to titanium surfaces. *J Biomed Mater Res A*, Vol.74, pp.12–22

Baier, R.E. & Meyer, A.E. (1988). Implant surface preparation. *Int J Oral Maxillofac Implants*, Vol.3, pp.9-20

Barbour, M.E., O'Sullivan, D.J. & Jagger, D.C. (2007). Chlorhexidine adsorption to anatase and rutile titanium dioxide. *Colloids and Surfaces A: Physicochem Eng Aspects*, Vol.307, pp.116-120

Bauerle, D. (2000). *Laser processing and chemistry*. Springer, Berlin, Heidelberg, New York, Tokyo ISBN 978-3-540-66891-6

Bereznai, M., Pelsőczi, I., Tóth, Z., Turzó, K., Radnai, M., Bor, Z. & Fazekas, A. (2003). Surface modifications induced by ns and sub-ps excimer laser pulses on titanium implant material. *Biomaterials*, Vol.24, pp.4197-4203

van den Beucken, J.J., Walboomers, X.F., Leeuwenburgh, S.C., Vos, M.R., Sommerdijk, N.A., Nolte, R.J. & Jansen, J.A. (2007). Multilayered DNA coatings: *in vitro* bioactivity studies and effects on osteoblast-like cell behavior. *Acta Biomater*, Vol.3, pp.587-596

Binon, P.P., Weir, D.J. & Marshall, S.J. (1992). Surface analysis of an original Brånemark implant and three related clones. *Int J Oral Max Imp*, Vol.7, pp.168-175

Block, M., Gardiner, D., Kent, J., Misiek, D., Finger, I. & Guerra, L. (1996). Hydroxyapatite-coated cylinder implants in the posterior mandible: 10 years observations. *Int J Oral Maxillofac Implants*, Vol.11, pp.626-633

Boere, G. (1995). Influence of fluoride on titanium in an acidic environment measured by polarization resistance technique. *J Appl Biomater* , Vol.6, pp.283–288

Boyan, B.D., Batzer, R., Kieswetter, K., Liu, Y., Cochran, D.L., Szmuckler-Moncler, S.S., Dean, D.D. & Schwartz, Z. (1998). Titanium surface roughness alters responsiveness of MG63 osteoblast-like cells to 1 alpha,25-(OH)(2)D-3. *J Biomed Mater Res*, Vol.39, pp.77–85

Brånemark, P.I. (1983a). Osseointegration and its experimental background. *J Prosthet Dent*, Vol.50, pp. 399-410

Brånemark, P.I., Adell, R., Albrektsson, T., Lekholm, U., Lundkvist, S. & Rockler, B. (1983b). Osseointegrated titanium fixtures in the treatment of edentulousness. *Biomaterials*, Vol.4. pp. 25–28

Buchanan, R.A., Lee, I.S. & Williams, J.M. (1990). Surface modification of biomaterials through noble metal ion implantation. *J Biomed Mater Res*, Vol.24, pp.309–318

Burchard, W.B., Cobb, C.M., Drisko, C.L. & Killoy, W.J. (1991). Effects of chlorhexidine and stannous fluoride on fibroblast attachment to different implant surfaces. *Int J Oral Maxillofac Implants*, Vol.6, pp.418-426

Buser, D., Schenk, R.K., Steinmann, S., Fiorellini, J.P., Fox, C.H. & Stich, H. (1991). Influence of surface characteristics on bone integration of titanium implants – a histomorphometric study in miniature pigs. *J Biomed Mater Res*, Vol.25, pp.889–902

Buser, D., Broggini, N., Wieland M., Schenk, R., Denzer, A., Cochran, D., Hoffman, B., Lussi, A. & Steinemann, S.G. (2004). Enhanced bone apposition to a chemically modified SLA titanium surface. *J Dent Res*, Vol.83, No.7, pp.529-533.

Bollen, C.M.L., Papaioannou, W., van Eldere, J., Schepers, E., Quirinen, M. & van Steenberghe, D. (1996). The influence of abutment surface roughness on plaque accumulation and peri-implant mucositis. *Clin Oral Implants Res*, Vol.7, pp.201–211

Cochran, D.L. (1999). A comparison of endosseous dental implant surfaces. *J Periodontology*, Vol.70, pp.1523-1539

Cochran, D.L., Schenk, R.K., Lussi, A., Higginbottom, F.L. & Buser, D. (1998). Bone response to unloaded and loaded titanium implants with a sandblasted and acid-etched surface: a histometric study in the canine mandible. *J Biomed Mater Res*, Vol.40, pp.1-11

Cooper, L.F., Zhou, Y., Takebe, J., Guo, J., Abron, A., Holmén, A. & Ellingsen, J.E. (2006). Fluoride modification effects on osteoblast behaviour and bone formation at TiO_2 grit-blasted c.p. titanium endosseous implants. *Biomaterials*, Vol.27, pp.926–936

Davies, J.E. (1998). Mechanisms of endosseous integration. *Int J Prosthodont*, Vol.11, pp. 391-401

Dennison, D.K., Huerzeler, M.B., Quinones, C.R.G. (1994). Contaminated implant surfaces: an *in vitro* comparison of implant surface coating and treatment modalities for decontamination. *J Periodontol*, Vol.65, pp.942-948

Deporter, A.D. & Todescan, R. Jr. (2001). A possible "rescue" procedure for dental implants with a textured surface geometry: a case report. *J Periodontol*, Vol.72, pp.1420-1423

Dillow, A.K. & Lowman, A.M. (2002). *Biomimetic Materials and Design*, Dillow, A.K. & Lowman, A.M. (Eds.), Marcel Dekker, ISBN 0-8247-0791-5, New York, USA

Ellingsen, J.E. (1995). Pre-treatment of titanium implants with fluoride improves their retention in bone. *J Mater Sci: Mater Med*, Vol.6, pp.749–753

Ellingsen, J.E., Johansson, C.B., Wennerberg, A. & Holmen, A. (2004). Improved retention and bone-to-implant contact with fluoride modified titanium implants. *Int J Oral Maxillofac Implants*, Vol.19, pp.659–666

Eriksson, C., Nygren, H. & Ohlson K. (2004). Implantation of hydrophilic and hydrophobic titanium discs in rat tibia: cellular reactions on the surfaces during the first 3 weeks in bone. *Biomaterials*, Vol.25, No.19, pp.4759-4766.

Ferris, D.M., Moodie, G.D., Dimond, P.M., Gioranni, C.W., Ehrlich, G.M. & Valentini, R.F. (1999). RGD-coated titanium implants stimulate increased bone formation *in vivo*. *Biomaterials*, Vol.20, pp.2323-2331

Fransson, C., Lekholm, U., Jemt, T. & Berglundh, T. (2005). Prevalence of subjects with progressive bone loss at implants. *Clin Oral Impl Res*, Vol.16, pp.440-446

Fransson, C., Wennström, J. & Berglundh, T. (2008). Clinical characteristics at implants with a history of progressive bone loss. *Clin Oral Impl Res* , Vol.19, pp.142-147

Gaggl, A., Schultes, G., Müller, W.D. & Kärcher, H. (2000). Scanning electron microscopical analysis of laser-treated titanium implant surfaces - a comparative study. *Biomaterials*, Vol.21, pp.1067–1073

Golub, E.E. & Boesze-Battaglia, K. (2007). The role of alkaline phosphatase in mineralization. *Curr. Opin. Orthop*, Vol.18, pp.444-448

György, E., Mihailescu, I.N., Serra, P., Pérez del Pino, A. & Morenza, J.L. (2002). Crown-like structure development on titanium exposed to multipulse Nd:YAG laser irradiation. *Appl Phys A*, Vol.74, pp.755–759

Harzer, W., Schroter, A., Gedrange, T., Muschter, F. (2001). Sensitivity of titanium brackets to the corrosive influence of fluoride-containing toothpaste and tea. *Angle Orthod*, Vol.71, pp.314–323

Hämmerle, C., Fourmousis, I., Winkler, J.R., Weigel, C., Brägger, U. & Lang, N.P. (1995). Successful bone fill in late peri-implant defects using guided tissue regeneration. A short communication. *J Periodontol*, Vol.66, pp.303-308

Hench, L.L. (1996). Ceramics, glasses and glass-cements. In: *Biomaterials Science: An Introduction to Materials in Medicine*, B.D. Ratner, A.S. Hoffman, F.J. Schoen, J.E. Lemons (Eds.). 309-312, Academic Press, ISBN 0-12-582461-0, San Diego, California, USA

Hoffman, A.S. (1996). Biologically functional materials. In: *Biomaterials Science: An Introduction to Materials in Medicine*, B.D. Ratner, A.S. Hoffman, F.J. Schoen, J.E. Lemons (Eds.). 309-312, Academic Press, ISBN 0-12-582461-0, San Diego, California, USA

Huang, H. (2002). Effects of fluoride concentration and elastic tensile strain on the corrosion resistance of commercially pure titanium. *Biomaterials*, Vol.23, pp.59–63

Huang H. (2003). Effect of fluoride and albumin concentration on the corrosion behavior of Ti-6Al-4V alloy. *Biomaterials*, Vol.24, pp.275–282

Huang, H.H. (2007). Variation in surface topography of different NiTi orthodontic archwires in various commercial fluoride-containing environments. *Dent Mater*, Vol.23, pp.24–33

Huget, E.F. (2002). Base metal casting alloys. In: *Dental Materials and Their Selection* O'Brien, W.J. (Ed.), 3. ed. Quintessence, ISBN 0-86715-406-3

Ito, Y., Kajihara, M. & Imanishi, Y. (1991). Materials for enhancing cell adhesion by immobilization of cell-adhesive peptide. *J Biomed Mater Res*, Vol.25, pp.1325-1337 de Jonge, L.T., Leeuwenburgh, S.C.G., Wolke, J.G.C. & Jansen, J.A. (2008). Organic-inorganic surface modifications for titanium implant surfaces. *Pharmaceutical Research* Vol.25, No.10, pp.2357-2369

Joob-Fancsaly, A., Divinyi, T., Fazekas, A., Peto, G. & Karacs, A. (2000). Surface treatment of dental implants with high-energy laser beam. *Fogorv Sz*, Vol.93, pp.169-180

Joob-Fancsaly, A., Huszar, T., Divinyi, T., Rosivall, L. & Szabo, G. (2004). The effect of the surface morphology of Ti-implants on the proliferation activity of fibroblasts and osteoblasts. *Fogorv Sz*, Vol.97, pp.251-255

Kasemo, B. (2002). Biological surface science. *Surface Science*, Vol.500, pp. 656-677

Khoury F. & Buchmann, R. (2001). Surgical therapy of peri-implant disease: a 3-year follow-up study of cases treated with 3 different techniques of bone regeneration. *J Periodontol*, Vol.72, pp.1498-1508

Klinge, B. & Meyle, J. (2006). Soft-tissue integration of implants. Consensus report of Working Group 2. *Clin Oral Impl Res*, Vol.17, pp.93–96

Klinger, M.M., Rahemtulla, F., Prince, C.W., Lucas, L.C. & Lemons, J.E. (1998). Proteoglycans at the bone-implant interface. *Crit Rev Oral Biol Med*, Vol.9, No.4, pp.449-463

Kotsovilis, S., Karoussis, I.K., Trianti, M. & Fourmousis, I. (2008). Therapy of peri-implantitis: a systematic review. *J Clin Periodontol*, Vol.35, No.7, pp.621-629

Könönen, M.H.O., Lavonius, E.T. & Kivilahti, J.K. (1995). SEM observations on stress corrosion cracking of commercially pure titanium in a topical fluoride solution. *Dent Mater*, Vol.11, pp.269–272

Larsson, C., Thomsen, P., Lausmaa, J., Rodahl, M., Kasemo, B. & Ericson, L.E. (1994). Bone response to surface modified titanium implants: studies on electropolished implants with different oxide thickness and morphology. *Biomaterials*, Vol.15, pp.1062-1074

Larsson, C., Thomsen, P., Aronsson, B.O., Rodahl, M., Lausmaa, J., Kasemo, B. & Ericson L.E. (1996). Bone response to surface modified titanium implants: studies on the early tissue response to machined and electropolished implants with different oxide thicknesses. *Biomaterials*, Vol.17, pp.605-616

Lautenschlager, E.P. & Monaghan, P. (1993). Titanium and titanium alloys as dental materials. *Int Dent J*, Vol.43, pp. 245–253

Lemons, J.E. (1996). Dental Implants, In: *Biomaterials Science: An Introduction to Materials in Medicine*, B.D. Ratner, A.S. Hoffman, F.J. Schoen, J.E. Lemons (Eds.). 309-312, Academic Press, ISBN 0-12-582461-0, San Diego, California, USA

Leonhardt, A., Renvert, S. & Dahlén, G. (1999). Microbial findings at failing implants. *Clin Oral Impl Res*, Vol.10, pp.339-345

Liu, Y., Hunziker, E.B. & de Groot, K. (2004). BMP-2 incorporated into biomimetic coatings retain its biological activity. *Tissue Eng*, Vol.10, pp.101-108

Liu, Y., Huse, R.O., de Groot, K., Buser, D. & Hunziker, E.B. (2007). Delivery mode and efficacy of BMP-2 in association with implants. *J Dent Res*, Vol.86, pp.84-89

Matsui, Y., Ohno, K., Michi, K. & Yamagata, K. (1994). Experimental study of high velocity flame-sprayed hydroxyapatite coated and noncoated titanium implants. *Int J Oral Maxillofac Implants*, Vol.9, pp.397-404

Mändl, S., Gerlach, J.W. & Rauscenbach, B. (2005). Surface modification of NiTi for orthopaedic braces by plasma immersion ion implantation. *Surf Coat Technol*, Vol.196, pp.293–297

Meffert, R.M., Langer, B. & Fritz, M.E. (1992). Dental implants: a review. *J Periodontol*, Vol.63, pp. 859–870

Meyle, J. (1999). Cell adhesion and spreading on different implant surfaces. *Proceedings of the 3rd European Workshop on Periodontology*, Lang, N., Karrig. T. & Lindhe, J. (Eds.), ISBN 3-87652-306-0 Quintessenz Verlags-GmbH, Berlin, Germany

Morra, M. & Cassinelli, C. (1997). Organic surface chemistry on titanium surfaces via thin film deposition. *J Biomed Mater Res*, Vol.37, pp.198-206

Morra, M., Cassinelli, C., Cascardo, G., Cahalan, P., Cahalan, L., Fini, M. & Giardino R. (2003). Surface engineering of titanium by collagen immobilization. Surface characterization and *in vitro* and *in vivo* studies. *Biomaterials*, Vol.24, pp.4639-4654

Nakagawa, M., Matsuya, S., Shiraishi, T. & Ohta, M. (1999). Effect of fluoride concentration and pH on corrosion behavior of titanium for dental use. *J Dent Res*, Vol.78, pp.1568–1572

Nakagawa, M., Matsuya, S. & Udoh, K. (2001). Corrosion behavior of pure titanium and titanium alloys in fluoride-containing solutions. *Dent Mater J* , Vol.20, pp.305–314

Nánai, L., Vajtai, R. & George, T.F. (1997). Laser-induced oxidation of metals: state of art. *Thin Solid Films*, Vol.298, pp.160-164

Neubert, R. & Eggert, F. (2001). Fluoridhaltige Zahnpasten. *Dtsch Apoth Ztg*, Vol.141, pp.42–45

Norowski, P.A., Jr. & Bumgardner, J.D. (2009). Review. Biomaterial and Antibiotic Strategies for peri-implantitis. *J Biomed Mater Res Part B: Appl Biomater*; Vol.88B, pp.530-543

Olefjord, I. & Hansson, S. (1993). Surface analysis of four dental implant systems. *Int J Oral Max Imp*, Vol.8, pp.32-40

O'Brien, W.J. (2002). *Dental Materials and Their Selection*, 3. ed. Quintessence, ISBN 0-86715-406-3

Park, J.B. (2000). Biomaterials, In: *The Biomedical Engineering Handbook*, 2nd ed., Vol. I, Bronzino, J.D., (Ed.), IV-1-IV-5, CRC Press and IEEE Press, ISBN 0-8493-0461-X, Boca Raton, Florida, USA

Park, J.B. & Kim, Y.K. (2000). Metallic biomaterials. In: *The Biomedical Engineering Handbook* 2nd ed., Vol. I, Bronzino, J.D., (Ed.), 37-1-37-20, CRC Press and IEEE Press, ISBN 0-8493-0461-X, Boca Raton, Florida, USA

Pelsőczi, I., Turzó, K., Gergely, Cs., Fazekas, A., Dékány, I. & Cuisinier, F.: Structural characterization of self-assembled polypeptide films on titanium and glass surfaces by atomic force microscopy. *Biomacromolecules*, Vol.6, No.6, pp.3345-3350

Perez del Pino, A., Serra, P. & Morenza, J.L. (2002). Oxidation of titanium through Nd:YAG laser irradiation. *Appl Surf Sci*, Vol.197-198, pp.887-890

Pető, G., Karacs, A., Pászti, Z., Guczi, L., Divinyi, T. & Joób, A. (2001). Surface treatment of screw shaped titanium dental implants by high intensity laser pulses. *Appl Surf Sci*, Vol.7524, pp.1-7.

Puleo, D.A. & Nanci, A. (1999). Understanding and controlling the bone-implant interface. *Biomaterials*. Vol.20, No.23-24, (December), pp.2311-2321

Ratner, B.D. & Hoffman, A.S. (1996). Thin films, Grafts, and Coatings, In: *Biomaterials Science: An Introduction to Materials in Medicine*, B.D. Ratner, A.S. Hoffman, F.J.

Schoen, J.E. Lemons (Eds.). 309-312, Academic Press, ISBN 0-12-582461-0, San Diego, California, USA

Reclaru, L. & Meyer, J.M (1998). Effects of fluorides on titanium and other dental alloys in dentistry. *Biomaterials*, Vol.19, pp.85–92

Renvert, S., Lessem, J., Dahlén, G., Lindahl, C. & Svensson, M. (2006). Topical minocycline microspheres versus topical chlorhexidine gel as an adjunct to mechanical debridement of incipient peri-implant infections: a randomized clinical trial. *J Clin Periodontol*, Vol.33, pp.362-369

Renvert, S., Roos-Jansaker, A.M., Lindahl, C., Renvert, H. & Persson, G.R. (2007). Infection at titanium implants with or without a clinical diagnosis of inflammation. *Clin Oral Impl Res*, Vol.18, pp.509-516

Renvert, S., Lessem, J., Dahlén, G., Renvert, H. & Lindahl, C. (2008a). Mechanical and repeated antimicrobal therapy using a local drug delivery system in the treatment of peri-implantitis: A Randomized Clinical Trial. *J Periodontol*, Vol.79, pp.836-844

Renvert, S., Roos-Jansaker, A-M. & Claffey N. (2008b). Non-surgical treatment of peri-implant mucositis and peri-implantitis: a literature review. *J Clin Periodontol*, Vol.35, pp.305-315

Rohanizadeh, R., LeGeros, R.Z., Harsono, M. & Bendavid, A. (2005). Adherent apatite coating on titanium substrate using chemical deposition. *J Biomed Mater Res A*, Vol.72, pp.428-438

Roos-Jansaker, A-M., Renvert, S. & Egelberg, J. (2003). Treatment of peri-implant infections: a literature review. *J Clin Periodontol*, Vol.30, pp.467-485

Roos-Jansaker, A.M., Lindahl, C., Renvert, H. & Renvert, S. (2006a). Nine- to fourteen year follow-up of implant treatment. Part II: presence of peri-implant lesions. *J Clin Periodontol*, Vol.33, pp.290-295

Roos-Jansaker, A-M., Renvert, H., Lindahl, C. & Renvert S. (2006b). Nine- to fourteen year follow-up of implant treatment. Part III: factors associated with periimplant lesions. *J Clin Periodontol*, Vol.33, pp.296-301

Roos-Jansaker, A-M., Renvert, H., Lindahl, C. & Renvert, S. (2007a). Surgical treatment of periimplantitis using a bone substitute with or without a resorbable membrane: a prospective cohort study. *J Clin Periodontol*, Vol.34, pp.625-632

Roos-Jansaker, A-M., Renvert, H., Lindahl, C. & Renvert S. (2007b). Submerged healing following surgical treatment of peri-implantitis: a case series. *J Clin Periodontol*, Vol.34, pp.723-727

Santis, D., Guerriero, C., Nocini, P.F., Ungersbock, A., Richards, G., Gotte, P. & Armato, U. (1996). Adult human bone cells from jaw bones cultured on plasma-sprayed or polished surfaces of titanium or hydroxyapatite discs. *J Mater Sci Mater Med*, Vol.7, pp.21-28

Schiff, N., Grosgogeat, B., Lissac, M. & Dalard, F. (2002). Influence of fluoride content and pH on the corrosion resistance of titanium and its alloys. *Biomaterials*, Vol.23, pp.1995–2002

Schiff, N., Boinet, M., Morgon, L., Lissac, M., Dalard, F. & Grosgogeat, B. (2006). Galvanic corrosion between orthodontic wires and brackets in fluoride mouthwashes. *Eur J Orthod*, Vol.28, pp.298–304

Schliephake, H., Scharnweber, D., Dard, M., Sewing, A., Aref, A. & Roessler, S. (2005). Functionalization of dental implant surfaces using adhesion molecules. *J Biomed Mater Res B Appl. Biomater*, Vol.73, pp.88-96

Schou, S., Holmstrup, P., Jorgensen, T., Stoltze, K., Hjorting-Hansen, E. & Wenzel, A. (2003a). Autogenous bone graft and ePTFE membrane in the treatment of peri-implantitis. I. Clinical and radiographic observation in cynomolgus monkeys. *Clin Oral Impl Res*, Vol.14, pp.391-403

Schou, S., Holmstrup, P., Jorgensen, T., Skovgaard, L.T., Stoltze, K., Hjorting-Hansen, E. & Wenzel, A. (2003b). Implant surface preparation in the surgical treatment of experimental peri-implantitis with autogenous bone graft and ePTFE membrane in cynomolgus monkeys. *Clin Oral Impl Res*, Vol.14, pp.412-422

Schwarz, F., Bieling, K. & Bonsmann, M. (2006). Nonsurgical treatment of moderate and advanced periimplantitis lesions: a controlled clinical study. *Clin Oral Invest* Vol.10, pp.279-288

Sikirić, M.D., Gergely, C., Elkaim, R., Wachtel, E., Cuisinier, F.J. & Füredi-Milhofer. H. (2008). Biomimetic organic-inorganic nanocomposite coatings for titanium implants. *J Biomed Mater Res A* Vol.89, No.3, pp.759-771

Sioshansi, P. (1987). Surface modification of industrial components by ion implantation. *Mater Sci Eng*, Vol.90, pp.373-383

Siirilä, H.S. & Könönen, M. (1991). The effect of oral topical fluorides on the surface of commercially pure titanium. *Int J Oral Maxillofac Implants*, Vol.6, pp.50–54

Stájer, A., Ungvári, K., Pelsőczi, K.I., Polyánka, H., Oszkó, A., Mihalik, E., Rakonczay, Z., Radnai, M., Kemény, L., Fazekas, A., Turzó, K. (2008). Corrosive effects of fluoride on titanium: investigation by X-ray photoelectron spectroscopy, atomic force microscopy and human epithelial cell culturing. *J Biomed Mater Res A*. Vol.87, No.2, pp.450-458

Strietzel, R., Hösch, A., Kalbfleish, H. & Buch, D. (1998). *In vitro* corrosion of titanium. *Biomaterials*, Vol.19, pp.1495–1499

Sun, L., Berndt, C.C., Gross, K.A. & Kucuk, A. (2001). Material fundamentals and clinical performance of plasma-sprayed hydroxyapatite coatings: a review. *J Biomed Mater Res*, Vol.58, pp.570-592.

Szabo, G., Kovacs, L. & Vargha, K. (1995). Possibilities for improvement of the surface properties of dental implants (2). The use of ceramic oxides in surface coating for titanium and tantalum implants. *Fogorv Sz*, Vol.88, pp.73-77

Toumelin-Chemla, F., Rouelle, F. & Burdairon, G. (1996). Corrosive properties of fluoride-containing odontologic gels against titanium. *J Dent*, Vol.24, pp.109–115

Ungvári, K., Pelsőczi, K.I., Kormos, B., Oszkó, A., Rakonczay, Z., Kemény, L., Radnai, M., Nagy, K., Fazekas, A., Turzó, K. (2010). Effects on titanium implant surfaces of chemical agents used for the treatment of peri-implantitis. *J Biomed Mater Res B Appl Biomater*, Vol.94, No.1, pp.222-229

Uribe, R., Penarrocha, M., Sanchis, J.M. & Garcia, O. (2004). Marginal peri-implantitis due to occlusal overload. A case report. *Med Oral*, Vol.9, pp.159-162

Vogel, G. (1999). Biological aspects of a soft tissue seal. In: *Proceedings of the, 3rd European Workshop on Periodontology*, Lang, N.P., Karring, T., Lindhe, J. (Eds.). ISBN: 3-87652-306-0 Quintessenz Verlags- GmbH, Berlin, Germany

Wang R.R. & Fenton A. (1996). Titanium for prosthodontic applications: A review of the literature. *Quintessence Int*, Vol.27, pp. 401–408

Walker, M.P., White, R.J. & Kula, K.S. (2005). Effect of fluoride prophylactic agents on the mechanical properties of nickel-titanium-based orthodontic wires. *Am J Orthod Dentofacial Orthop*, Vol.127, pp.662–669

Wennerberg, A., Ektessabi, A., Albrektsson, T., Johansson, L. & Andersson, B. (1997). A 1-year follow-up of implants of differing surface roughness placed in rabbit bone. *Int J Oral Max Implants*, Vol.12, pp.486–494

Wheeler, S.L. (1996). Eight-year clinical retrospective study of titanium plasma-sprayed and hydroxyapatite-coated cylinder implants. *Int J Oral Maxillofac Implants*, Vol.1, pp.340-350

Wong, M., Eulenberger, J., Schenk, R. & Hunziker E. (1995). Effect of surface topology on the osseointegration of implant materials in trabecular bone. *J Biomed Mater Res*, Vol.29, pp.1567–1575

Zitzmann, N.U. & Berglundh, T. (2008). Definition and prevalence of peri-implant diseases. *J Clin Periodontol*, Vol.35, Suppl.8, pp.286-291

9

Antihypertensive Peptides Specific to *Lactobacillus helveticus* Fermented Milk

Taketo Wakai and Naoyuki Yamamoto
Microbiology & Fermentation Laboratory,
Calpis., Ltd., Fuchinobe, Chuo-ku, Sagamihara-shi, Kanagawa,
Japan

1. Introduction

Peptides are well known as nitrogen sources to supply various amino acids for many different organisms, and also have many hormonal functions in our body. Previous studies have reported a secondary role for peptides with specific amino acid sequences that possess biological function *in vivo* (1-8). To prepare biologically active peptides (bioactive peptides), food ingredients containing protein are generally hydrolyzed by some proteolytic enzymes. In particular, milk proteins such as bovine casein and whey proteins have been used for preparations of bioactive peptides because inexpensive and safe sources are readily available. Various physiologically functional peptides, such as immunostimulating peptides (1), antimicrobial peptides (2), opioid peptides (3), mineral soluble peptides (4) and antihypertensive peptides (5-8) and have been isolated from enzymatic hydrolyzates of raw food materials and fermented food products.

Among these bioactive peptides, antihypertensive peptides have been extensively studied and reviewed (5-8). Hypertension is a major risk factor in cardiovascular disease, such as heart disease and stroke. In order to reduce the incidence of disease, pharmacological substances can be used to decrease high blood pressure to within the normal range. Angiotensin I-converting enzyme (kininase II; EC 3.4.15.1) (**ACE**) is predominantly expressed as a membrane-bound ectoenzyme in vascular endothelial cells and several other cell types, including absorptive epithelia, neuroepithelia and male germinal cells (9, 10). A dipeptidyl carboxypeptidase, ACE catalyzes the production of a vasoconstrictor, angiotensin II, and inactivates a vasodilator, bradykinin (11, 12). The first competitive inhibitors to ACE were reported as naturally occurring peptides isolated from snake venom (13, 14). Then, inhibitory activities on ACE, which plays an important role in blood pressure regulation are generally assessed for preparation of antihypertensive peptides.

Among lactic acid bacteria, *Lactobacillus helveticus* had the highest extracellular proteinase activity and the highest ability to release the specific antihypertensive peptides in the fermented milk (15). So, this paper mainly reviews processing of the antihypertensive peptides, Val-Pro-Pro (VPP) and Ile-Pro-Pro (IPP), by proteolytic enzymes of *L. helveticus*. The antihypertensive effects of these *in vitro* and *in vivo* studies, clinical study, and the mode of action are also reviewed. The antihypertensive effect of the *L. helveticus* fermented milk compared to that produced by various lactic acid bacteria and proteolytic systems of *L.*

helveticus is reviewed for the discussion in more detail. Finally, processing of VPP and IPP specific to the *L. helveticus* is discussed in the context of comparative genome analyses of corresponding proteolytic enzymes in various lactic acid bacteria.

1.1 Antihypertensive peptide in fermented milk

1.1.1 ACE inhibitory peptides from milk proteins

Many kinds of ACE inhibitory (**ACEI**) peptides have been reported from enzymatic hydrolyzates of milk protein, as well as synthetic peptides and fermented products (16-24) (Table 1). Spontaneously hypertensive rat (**SHR**) is a useful animal model to evaluate the antihypertensive activity of ACEI peptides because the systolic blood pressure of SHR reaches over 230 mmHg and is powerful tool for detection of the *in vivo* effect. Some of the orally administered ACEI peptides have demonstrated strong antihypertensive effects in SHR (Table 1).

Peptide	Source	Preparation	*IC50 (µM)	Dose (mg/kg)	***SBP (mm Hg)
<Enzymatic hydrolysate>					
FFVAPFPEVFGK	αs1-casein	Trypsin	77	100	-13.0
AVPYPQR	β-casein	Trypsin	15	100	-10.0
TTMPLW	αs1-casein	Trypsin	16	100	-13.6
LKPNM	Aldolase	Thermolysin	2.4	60	-23
LKP	Aldolase	Chicken muscle	0.32	60	-18
IPA	β-lactogloblin	Proteinase K	141	8	-31
VYPFPG	β-casein	Proteinase K	221	8	-22
GKP	β-microglobulin	Proteinase K	352	8	-26
FP	β-casein, albumin	Proteinase K	315	8	-27
YKVPQL	αs1-casein	Proteinase	22	1	-12.5
<Fermented products>					
RF	Sake lees	Brewing	ND	100	-17
VW	Sake lees	Brewing	1.4	100	-10
YW	Sake lees	Brewing	10.5	100	-28
VY	Sake	Brewing	7.1	100	-31
IYPRY	Sake	Brewing	4.1	100	-19
VPP	β-casein	Fermentation	9	1.6	-20
IPP	β- and κ -casein	Fermentation	5	1	-15.1
YP	αs1, β- and κ-casein	Fermentation	720	1	-27.4

ND: Not described

*IC50: Peptide concentration that shows 50% inhibition of ACE activity

***SBP: systolic blood pressure of spontaneously hypertensive rat

Table 1. Antihypertensive peptides derived from caseins by proteolytic action.

Lactic acid bacteria have proteolytic systems that can hydrolyze milk protein and have been reported to utilize the peptides released from the milk protein casein (25-27). Among lactic acid bacteria, *Lactobacillus helveticus* had the highest extracellular proteinase activity and the ability to release the largest amount of peptides in the fermented milk (Table 2). As a result, among various kinds of fermented milk, the antihypertensive effect was specific to the L. helveticus fermented milk (15). In our study, an antihypertensive effect related to ACEI peptides was found in sour milk produced by L. helveticus (19, 28). Two ACE inhibitory peptides were purified from sour milk and identified as VPP and IPP. The ACEI activity of the two peptides was very high, 9 μM and 5 μM, compared to other reported peptides (Table 1). The amino acid sequences of VPP and IPP were found in the primary structure of bovine β-casein (84-86) (74-76) and κ-casein (108-110), respectively. These peptides were produced during fermentation (16), but were not found in the hydrolyzate of casein after digestion with an extracellular proteinase of L. helveticus (29). Oral administration of L. helveticus fermented milk containing VPP and IPP to SHR, with a single dose of 5 ml/kg body weight, significantly decreased systolic blood pressure between 4 and 8 h after administration (28). The antihypertensive effect of these two chemically synthesized peptides was also observed between 2 and 8 h after administration and the effects were dose-dependent. Furthermore, a dose-dependent antihypertensive effect of these two chemically synthesized peptides was also observed from 0.1 to 10 mg/kg of body weight (28).

Strain	Peptide conc. (%)	Proteinase act. (U/ml)	*ACEI act. (U/ml)	***Change in SBP (mmHg)
Non-fermented milk	0.00	-	0	- 5.0 ± 7.3
(Lactobacilli)				
L. helveticus CP790	0.19	230	58	- 27.4 ± 13.3 * *
L. helveticus CP611	0.25	367	70	- 20.0 ± 9.6 * *
L. helveticus CP615	0.18	420	51	- 23.0 ± 13.4 * *
L. helveticus JCM1006	0.15	182	26	- 15.2 ± 9.3 *
L. helveticus JCM1120	0.10	112	34	- 6.5± 10.8
L. helveticus JCM1004	0.21	186	48	- 29.3 ± 13.6 * *
L. delbrueckii subsp. bulgaricus CP973	0.19	105	22	- 0.8 ± 8.2
L. delbrueckii subsp. bulgaricus JCM1002	0.11	124	28	- 4.5 ± 4.0
L. casei CP680	0.01	35	3	- 0.2 ± 6.6
L. casei JCM1134	0.00	28	9	- 7.0 ± 11.2
L. casei JCM1136	0.09	25	18	- 9.6 ± 7.2
L. acidophilus JCM1132	0.00	28	8	- 8.7 ± 7.8
L. delbrueckii subsp. lactisJCM1105	0.08	18	16	- 3.3 ± 3.5
(Streptococci)				
S. thermophilus CP1007	0.02	25	3	- 2.4 ± 8.1
(Lactococci)				
L. lactis subsp. lactis CP684	0.00	35	4	- 7.3 ± 10.5
L. lactis subsp. cremoris CP312	0.02	18	4	- 5.8 ± 13.9

Significant differences from the control, **p< 0.01, *p< 0.05.

*ACEI activity: Peptides that show 50% inhibition of ACE activity was defined as one unit.

***SBP: systolic blood pressure of spontaneously hypertensive rat

Table 2. Antihypertensive effects in spontaneously hypertensive rats and ACE inhibitory activities of various fermented milk.

1.2 Clinical effects of the fermented milk

Hypertension is a major risk factor in cardiovascular diseases, such as heart disease and stroke. In order to reduce the incidence of disease, pharmacological substances can be used to decrease high blood pressure to within the normal range. In the first Japanese study with the fermented milk, hypertensive subjects were randomly assigned to two groups: the one group ingested 95 ml of the milk, containing 3.4 mg of VPP and IPP, daily for 8 wk; the other group ingested the same amount of artificially acidified milk as a placebo, for 8 wk (30). In the fermented milk group, systolic blood pressure decreased significantly between 4 and 8 wk after the beginning of ingestion, but not in the placebo group (30). Moreover, clinical tests were performed for Japanese subjects with different blood pressure levels, which confirmed the mild and prolonged effects for the hypertensive subjects following oral administration of bioactive milk (30-33) (Fig. 1). In a pilot study conducted in Finland, the antihypertensive effect was also observed in the group ingesting *L. helveticus* fermented milk containing the two tripeptides (34, 35). Moreover, a recent study indicated significant beneficial effects of hypertensive patients ingesting the *L. helveticus* fermented milk over a long period of time for 21 wk (35). There was a significant decrease in systolic blood pressure (6.7 ± 3.0 mmHg) by comparative study with placebo group.

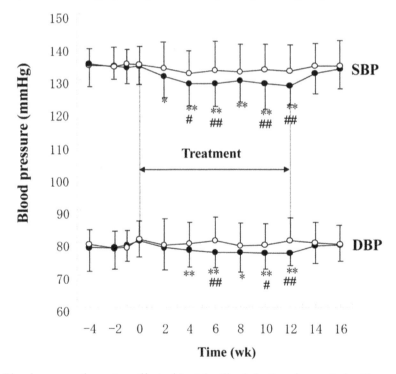

Fig. 1. Blood pressure lowering effect of *Lactobacillus helveticus* fermented milk product for subjects with high-normal hypertension (Nakamura *et al.*, J. Nutritional Food (in Japanese) 2004, 7, 123-137). Significant difference from initial value (*t*-test): $^{*}p$ <0.05, $^{**}p$ <0.01. Significant difference from placebo group (Bonferroni test): # p<0.05, ## p<0.01.

1.3 Processing of antihypertensive peptides in *L. helveticus*

1.3.1 Proteolytic system in lactic acid bacteria

Many kinds of proteolytic enzymes have been reported from lactic acid bacteria, and were reviewed extensively (26, 27, 36, 37). The components of the proteolytic systems of lactic acid bacteria are divided into three groups, including the extracellular proteinase that catalyzes casein breakdown to peptides, peptidases that hydrolyze peptides to amino acids and a peptide transport system. The number of proteinases was reported from lactococci, which are mainly used in cheese making. The extracellular proteinase activity is linked to cell growth in milk and seemed to be essential for utilization of milk protein for growth. The gene encoding the proteinases, named *prtP, prtH* and *prtH2*, has been sequenced and characterized from several of *L. lactis* strains (38-40). These proteinases are processed to active enzymes by removal of N-terminal polypeptides from the pre-proteinase with the help of the maturation protein, PrtM.

1.4 Processing by a cell wall-associated proteinase

The first step in casein decomposition is typically caused by an extracellular proteinase, and further digestion to amino acids is catalyzed by many kinds of intracellular peptidases (25, 36). Among the lactic acid bacteria, *L. helveticus* has the highest extracellular proteinase activity and the ability to release the highest amount of peptides in fermented milk (Table 2). *L. helveticus* strains were classified into 2 types based on differences in the extracellular proteinase (41). One type has an enzyme with a molecular weight of 170 kDa with homology to the lactococcus enzyme, and other has an enzyme with a molecular weight of 45 kDa (29, 42). Polycloncal antibodies raised against the mature 170 kDa proteinase reacted not only with the 170 kDa enzyme but also the 53 kDa protein (Table 3). The 53 kDa protein was thought to be a degradation product from the 170 kDa active enzyme. A gene encoding a proteinase with a homology to the *Lactococcus* proteinase was cloned and sequenced from *L. helveticus* CNRZ32 (43). A gene encoding *PrtM* gene was also cloned from *L. helveticus* CNRZ32. On the other hand, the gene encoding a small type of proteinase with a molecular weight of 45 kDa was cloned from *L. helveticus* CP790 strain (44). A 46 kDa pre-proteinase was activated to the 45 kDa active proteinase by release of 7 amino acids from the N-terminus with the help of a maturation protein (44, 45).

Moreover, a slight difference in the specificity of the two types of proteinases toward casein was suggested for the two types of *L. helveticus* strains (46). However, there were no clear relationships between proteinase specificity and the antihypertensive effects and the ACEI activities of the fermented milk, which depended on the strain of *L. helveticus* used. The extracellular proteinase activity of each *L. helveticus* strain was almost correlated with ACEI activity in the fermented milk (Table 1). These results strongly suggest that the proteolysis of casein by the extracellular proteinase is the most important parameter in the processing of active components. This possibility is also supported by the fact that the *L. helveticus* CM4, which was selected for its strong proteinase activity compared to other *L. helveticus* strains, has the ability to release ACE inhibitory peptides in the fermented milk (47). The proposed importance of the proteinase was also supported by the fact that a proteinase negative mutant was not able to generate antihypertensive peptides in the fermented milk, whereas the wild-type strain had the ability to release strong antihypertensive peptides in the fermented milk (16). Strain CM4, which had the highest antihypertensive peptide

No	Strain	Subspecies	Reactivity (CP790)1	Reactivity (CP53)2	Type
1	*L. helveticus* CP39	J3	45 kDa	45 kDa	A
2	*L. helveticus* CP53	H4	ND5	53, 170 kDa	B
3	*L. helveticus* CP209	J	45 kDa	45 kDa	A
4	*L. helveticus* CP210	J	45 kDa	45 kDa	A
5	*L. helveticus* CP293	J	45 kDa	45 kDa	A
6	*L. helveticus* CP510	J	45 kDa	45 kDa	A
7	*L. helveticus* CP611	J	45 kDa	45 kDa	A
8	*L. helveticus* CP615	J	45 kDa	45 kDa	A
9	*L. helveticus* CP617	J	45 kDa	45 kDa	A
10	*L. helveticus* CP789	J	45 kDa	45 kDa	A
11	*L. helveticus* CP790	J	45 kDa	45 kDa	A
12	*L. helveticus* JCM1004	H	ND	53, 170 kDa	B
13	*L. helveticus* JCM1006	J	45 kDa	45 kDa	A
14	*L. helveticus* JCM1007	J	ND	53, 170 kDa	B
15	*L. helveticus* JCM1062	J	ND	53, 170 kDa	B
16	*L. helveticus* JCM1103	H	ND	53, 170 kDa	B
17	*L. helveticus* JCM1120	H	ND	53, 170 kDa	B

(Yamamoto et al., 1998, Biosci. Biotech. Biochem. 58, 776-778)

[1]Monoclonal antibody to the proteinase from L. helveticus CP790

[2]Polyclonal antibody to the proteinase from L. helveticus CP53

[3]Classified as L. helveticus biovar jugurti

[4]Classified as L. helveticus biovar helveticus

[5]Not detected

Table 3. Immunological Difference of Proteinases of *L. helveticus* Strains with Two Types of Antibodies.

production in fermented milk, seemed to have the potential for use as a functional food product. Currently, we completed the whole genome sequence of *L. helveticus* CM4, and the results revealed the presence of 2,171 open reading frames in 2,028,493 bp of whole DNA sequence (unpublished data). Then, whole genome sequences of CM4 and DPC4571 (48) were compared as shown in Table 4 for full understanding of the intracellular processing to VPP and IPP (49). As shown in Table 4, three genes for cell-wall associated proteinase genes, *prtY*, *prtH2* and *prtM2*, and 23 kinds of intracellular peptidases were detected in the CM4 sequence. The genes of *prtH2* and *prtM2* were detected both in CM4 and DPC4571, but the *prtM2* was considered to be pseudo-genes for extracellular proteinase in the previous study (50). However, no *prtH1* and *prtM1* genes reported in CNRZ32 (51) were detected in sequences in both CM4 and DPC4571 strains. The genes of *prtH2* and *prtM2* were detected both in CM4 and DPC4571, but the *prtM2* was considered to be pseudo-genes for extracellular proteinase in the previous study. However, no *prtH1* and *prtM1* genes reported in CNRZ32 (51) were detected in sequences in both CM4 and DPC4571 strains. On the other hand, one of the cell wall-associated proteinase gene (*prtY*) corresponding to a 45 kDa proteinase previously detected in *L. helveticus* CP790 strain (41, 42) was detected in CM4 strain but not in DPC4571 strain (Table 1). These results reveal the cell wall-associated extracellular proteinase which plays key role in decomposition of casein might be different among detected proteolytic enzymes between CM4 and DPC4571. On the other hand, the large size of the proteinase gene (*prtH*) corresponding to a 200 kDa proteinase and its maturation gene (*prtM*) reported by Pederson *et al.* (43) were not detected in the CM4 and DPC4571 genes. This observation supports our previous results that there are two types of extracellular proteinases, and *L. helveticus* can be classified by the proteinase type as shown in Table 4. This result also suggested that the extracellular proteinase might play a key role to release high amount of VPP and IPP from the comparative analysis of both proteolytic genes (Table 4).

| Proteolytic enzyme | Gene | Molecular weight (kDa) | | Protein ID | Identity |
		CM4	DPC4571	DPC4571	(%)
Proteinase	*prtY*	47.0	ND	-	-
	prtH2	181.6	180.871*	-	99.2
	prtM	33.7	32.7	ABX27563	98.0
Aminopeptidase	*pepC1*	51.4	51.4	ABX26582	98.9
	pepC2	52.9	50.2	ABX27065	98.6
	pepN	95.8	95.9	ABX27731	99.4
	pepN2	57.2	57.2	ABX27544	100.0
	pepA	41.3	40.1	ABX26758	99.4
XPDAP	*pepX*	90.5	90.6	ABX27419	99.6
Endopeptidase	*pepE*	50.0	50.0	ABX26466	99.8
	pepE2	51.4	50.3	ABX26457	99.8
	pepF	68.1	68.1	ABX27686	99.3
	pepO	73.6	73.5	ABX27358	99.4
	pepO2	73.8	73.5	ABX27211	98.6
	pepO3	73.1	72.6	ABX26433	99.7
Tripeptidase	*pepT*	47.1	46.7	ABX27305	99.3
	pepT2	48.8	48.4	ABX27165	99.5
Dipeptidase	*pepD1*	54.0	54.1	ABX27625	99.4
	pepD2	54.9	54.9	ABX27375	99.8
	pepD3	53.5	53.5	ABX27723	100.0
	pepV	51.5	51.5	ABX27224	98.9
	pepDA	53.5	53.5	ABX26492	99.6
Prolidase	*pepQ*	41.2	41.2	ABX26664	99.5
	pepQ2	41.4	41.1	ABX27405	99.7
Prolinase	*pepPN*	35.0	35.0	ABX27633	99.7
Proline iminopeptidase	*pepI*	33.9	33.8	ABX26375	99.3

ND: Not detected, *size of the reported pseudo-gene

Table 4. Genes encoding proteolytic enzymes reported in *Lactobacillus helveticus* strains

By the proteolytic action of the extracellular proteinase in CP790 (and CM4), a long β-casein peptide with a 28 amino acid residue including VPP and IPP sequences was generated (29) (Fig. 2). The proteinase activity is easily repressed by accumulated peptides by the proteinase in the fermented milk. Moreover, the enzyme activity is inactivated by pH drop during the fermentation. So, the first degradation of casein by the extracellular proteinase would be occurred mostly at the beginning of the fermentation.

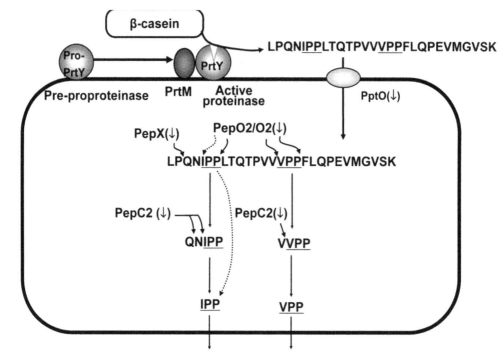

Fig. 2. Postulated proteolytic system for Val-Pro-Pro and Ile-Pro-Pro processing in *Lactobacillus helveticus*. PrtY: cell-wall proteinase, PrtM: maturation protein, PptO: oligopeptide transporter, PepO: endopeptidase O, PepO2: endopeptidase O2, PepC2: aminopeptidase C2, PepX: X-prolyl dipeptidyl aminopeptidase. Up- and down-regulation and amount of release are indicated by arrows as (↑) and (↓), respectively.

1.4.1 Intracellular processing by some peptidases

Next, the long peptide was thoroughly hydrolyzed to shorter peptides by intracellular peptidases. Intracellular peptidases of *L. helveticus* and these genes were well reviewed, previously (37), however, there was no clear explanation for the processing of VPP and IPP. Long peptide containing two kinds of tri-peptide sequences released by the extracellular proteinase from β-casein will be incorporated into the cell by the oligopeptide transporter (**PptO**), and processed intracellularly to VPP and IPP by some peptidases (Fig. 2). Carboxyl peptidase most likely needs for the release of C-terminal amino acid from Pro-Pro-X sequence. However, recently, a key enzyme that can catalyzed carboxyl terminal processing to produce VPP and IPP was detected and purified from the CM4 strain (Fig. 2) (52). The enzyme had a homology to an endopeptidase (**PepO**) from *L. helveticus* CNRZ32 by amino terminal sequence analysis of the purified enzyme (52) and a homology with the deduced amino acid sequence of the gene (53). The enzyme can catalyze C-terminal processing of VPPFL and IPPLT to VPP and IPP.

Based on the previous reported characteristics of many peptidases from *L. helveticus*, processing of the N-terminal sequence of VPP and IPP are presumed and summarized in

Fig. 2. Generally, amino peptidase shows broad specificity toward amino acid at N-terminal end, however, amino terminal processing seems to terminate if proline residue is present at the N-terminal end in the peptide. However, X-prolyl dipeptidyl aminopeptidase (**XPDAP**) is able to release the di-peptide with a sequence of X-Pro, from the N-terminus. On the other hand, aminopeptidase may stop the hydrolysis at a Xaa-Pro-Pro- sequence if it is present. Therefore, the N-terminal processing to release VPP and IPP may be catalyzed by specific aminopeptidases, such as pepC2 and XPDAP as shown in Fig. 2. However, for more detailed understanding of these peptide processing in *L. helveticus*, productivities of these peptides in transformant strains expressing the each postulated peptidase gene or disrupting of these corresponding peptidase genes from wild type strain should be studied.

1.5 Comparison of the *L. helveticus* proteolytic system to those in other lactic acid bacteria

The unprocessed proteinase of *L. lactis* consists of about 1950 amino acid residues, and the mature proteinase of *L. lactis* is a serine type enzyme with a molecular mass between 180-190 kDa (38-40). The gene encoding the proteinase, named *prtP*, has been cloned and sequenced from several of *L. lactis* strains (38-40). These proteinases are processed to active enzymes by removal of N-terminal polypeptides from the pre-proteinase with the help of the maturation protein, PrtM. Recent decay, whole genome sequences for more than 36 kinds of lactobacilli have been reported. The comparative analysis of those genes to other species revealed that the CM4 peptidases were more homologous to those in *Lactobacillus acidophilus* NCFM strain (54)(Table 5). Moreover, homologies of CM4 peptidases were detected with peptidases in *Lactobacillus gasseri* ATCC33323 (55) and *Lactobacillus johnsonii* (NC533) (56). These observations suggest that the *L. acidophilus* group has similar peptidases, endopeptidase, XPDAP, and aminopeptidase to *L. helveticus* CM4 and might have the ability to process VPP and IPP if the initial decomposition of casein by an extracellular proteinase is accelerated in the fermentation process.

1.6 Repression of proteolytic systemg

For growth of lactic acid bacteria in milk, the proteolytic system is activated in the milk medium because of a limited amount of amino acids. However, during fermentation in the milk medium, the proteolytic system of lactic acid bacteria is repressed by accumulated peptides in the fermented milk. The amount of VPP and IPP in the *L. helveticus* fermented milk was also repressed if amino acids were added to the fermented milk (57). Microarray analysis of the whole *L. helveticus* CM4 genome suggested extracellular proteinase, endopeptidases, XPDAP, some aminopeptidases and some kinds of peptide transporters might be suppressed by addition of amino acids and be involved in the processing of the VPP and IPP (Fig. 2). Regulatory systems (57) that repress the proteolytic enzymes and transporters in the presence of amino acids was reported as codY system in *Lactococcus lactis* (58, 59). However, there is no codY-like protein in the *Lactobacillus* genome that has been reported. Thus, a novel type regulatory system for the proteolytic system must exist in lactobacilli and strongly affect on the release of VPP and IPP in *L. helveticus*.

	Proteinase		Aminopeptidase			XPDAP	Endopeptidase					
	prtV	prtH	pepN	pepC1	pepC2	pepX	pepE	pepE2	pepF	pepO	pepO2	pepO3
Lb. acidophilus NCFM	-	-	90	91	87	91	89	90	88	85	61	-
Lb. gasseri ATCC 33323	-	-	67	83	76	72	70	83	77	63	-	77
Lb. johnsonii NCC533	-	-	66	82	75	72	69	73	76	65	-	78
Lb. delbrueckii subsp. *bulgaricus* ATCC BAA-365	-	-	71	-	53	70	71	-	-	58	-	68
Lb. casei subsp. *casei* ATCC334	-	-	62	59	-	-	-	-	54	-	-	-
Lc. lactis IL1403	-	-	-	-	-	-	-	-	-	-	-	-

Lb: Lactobacillus; Lc: Lactococcus

All values are shown as % homology. Values below 50% were omitted (-).

Table 5. Proteolytic enzymes in various lactic acid bacteria with homology to enzymes in *Lactobacillus helveticus* CM4, which are expected to have roles in processing Val-Pro-Pro and Ile-Pro-Pro

2. Conclusion

In this paper, we showed the potential of bioactive peptides to maintain blood pressure in the normal range. About 30% of Japanese people are estimated to be at risk for hypertension. Generally, hypertension has been improved by medication and partly by controlling the diet. Recently, some food products containing antihypertensive peptides and proven antihypertensive effects in clinical studies were recognized as functional foods, Foods for Specified Health Use (**FOSHU**) in Japan. Biologically functional peptides exerting a mild influence on hypertensive subjects without adverse effect have enormous potential in reducing the risk of cardiovascular disease.

Among many kinds of commercially available lactic acid bacteria, L. helveticus has ability to release functional peptides such as antihypertensive peptides in the fermented milk. These potential is strongly depends on the activities of proteolytic enzymes of L. helveticus. So, the isolation and mutation breeding of a new L. helveticus strain having strong ability to release peptides in the fermented milk is very important to develop industrially useful fermented milk. For the understanding of the antihypertensive peptides, based on comparative analysis with other Lactobacillus proteolytic enzymes, the extracellular proteinase and endopeptidase enzymes seemed to be unique to L. helveticus and seemed at least partially explain the L. helveticus specific release of VPP and IPP from fermented milk. For a full understanding of protein processing, the genomic information and the analysis of the L. helveticus will be a very useful tool and might be needed in future studies.

3. References

[1] Migliore-Samour, D., Floćh, F. and Jollès, P.: Biologically active casein peptides implicated in immunomodulation. J. Dairy Res., 56, 357-362 (1989).
[2] Bellamy, W., Takase, M., Yamauchi, K., Shimamura, S.and Tomita, M.: Identification of the bactericidal domain of lactoferrin. Biochimica et Biophysica Acta, 1121, 130-136 (1992).
[3] Teschemacher, H., Koch, G. and Brantl, V.: Milk protein-derived opioid receptor ligands. Biopolymers, 43, 99-117 (1997).
[4] FitzGerald, R. J.: Potential uses of caseinophosphopeptides. Int. Dairy J., 8, 451-457 (1998).
[5] Yamamoto, N.: Antihypertensive peptides derived from food proteins. Biopolymers, 43, 129-134 (1997).
[6] Meisel, H. and Bockelmann, W.: Bioactive peptides encrypted in milk proteins: proteolytic activation and thropho-functional properties. Antonie Van Leeuwenhoek, 76, 207-215 (1999).
[7] Yamamoto, N. and Takano, T.: Antihypertensive peptides derived from milk proteins. Nahrung, 43, 159-164 (1999).
[8] Takano, T.: Anti-hypertensive activity of fermented dairy products containing biogenic peptides., Antonie Van Leeuwenhoek, 82, 333-340 (2002).
[9] Caldwell, P. R., Seegal, B. C., Hsu, K. C., Das, M. and Soffer, R. L.: Angiotensin-converting enzyme: vascular endothelial localization. Science, 191, 1050-1051 (1976).
[10] El-Dorry, H. A., Bull, H. G., Iwata, K., Thornberry, N. A., Cordes, E. H. and Sofffer, R. L.: Molecular and catalytic properties of rabbit testicular dipeptidyl carboxypeptidase., J. Biol. Chem., 257, 14128-14133 (1982).

[11] Skeggs, L. T. J., Kahn, J. R. and Shumway, N. P.: The preparation and function of the hypertensin-converting enzyme. J. Exp. Med., 103, 295-299 (1956).

[12] Yang, H. Y. T., Erdos, E. G. and Levin, Y.: A dipeptidyl carboxypeptidase that converts angiotensin I and inactivates bradykinin. Biochimica et Biophysica Acta, 214, 374-376 (1970).

[13] Ferreira, S. H., Bartelt, D. C. and Greene, L. J.: Isolation of bradykinin-potentiating peptides from Bothrops jararaca venom. Biochemistry, 9, 2583-2259 (1970).

[14] Ondetti, M. A., Williams, N. J., Sabo, E. F., Plušcec, J., Weaver, E. R. and Kocy, O.: Angiotensin-converting enzyme inhibitors from the venom of Bothrops jararaca. Isolation, elucidation of structure, and synthesis. Biochemistry, 10, 4033-4039 (1971).

[15] Yamamoto, N., Akino, A. and Takano, T.: Antihypertensive effect of different kinds of fermented milk in spontaneously hypertensive rats. Biosci. Biotech. Biochem., 58, 776-778 (1994).

[16] Yamamoto, N., Akino, A. and Takano, T.: Antihypertensive effect of the peptides derived from casein by an extracellular proteinase from Lactobacillus helveticus CP790. J. Dairy Sci., 77, 917-922 (1994).

[17] Maruyama, S., Mitachi, H., Tanaka, H., Tomizuka, N. and Suzuki, H.: Study on the active site and antihypertensive activity of angiotensin I-converting enzyme inhibitors derived from casein. Agric. Biol. Chem., 51, 1581-1586 (1987).

[18] Kohmura, M., Nio, N., Kubo, K., Minoshima, Y., Nunekata, E. and Ariyoshi, Y. Inhibition of angiotensin-converting enzymes by synthetic peptides of human beta-casein. Agric. Biol. Chem., 53, 2107-2114 (1989).

[19] Nakamura, Y., Yamamoto, N., Sakai, K., Okubo, A., Yamazaki, S. and Takano, T.: Purification and characterization of angiotensin I-converting enzyme inhibitors from sour milk. J. Dairy Sci., 78, 777-783 (1995).

[20] Yamamoto, N., Maeno, M. and Takano, T.: Purification and characterization of an antihypertensive peptide from a yogurt-like product fermented by Lactobacillus helveticus CPN4. J. Dairy Sci., 82, 1388-1393 (1999).

[21] Maeno, M., Yamamoto, N. and Takano, T.: Identification of an antihypertensive peptide from casein hydrolysate produced by a proteinase from Lactobacillus helveticus CP790. J. Dairy Sci., 79, 1316-1321 (1996).

[22] Abubakar, A., Saito, T., Kitazawa, H., Kawai, Y. and Itoh, T.: Structural analysis of new antihypertensive peptides derived from cheese whey protein by proteinase K digestion. J. Dairy Sci., 81, 3131-3138 (1998).

[23] Pihlanto-Leppala, A., Koskinen, P., Piilola, K., Tupasela, T. and Korhonen, H.: Angiotensin I-converting enzyme inhibitory properties of whey protein digests: concentration and characterization of active peptides. J. Dairy Res., 67, 53-64 (2000).

[24] Murakami, M., Tonouchi, H., Takahashi, R., Kitazawa, H., Kawai, Y. and Negishi, H.: Saito, T. Structural analysis of a new anti-hypertensive peptide (beta-lactosin B) isolated from a commercial whey product. J. Dairy Sci., 87, 1967-1974 (2004).

[25] Smid, E. J., Poolman, B. and Konings, W. N.: Casein utilization by lactococci. Appl. Environ. Microbiol., 57, 2447-2452 (1991).

[26] Pritchard, G. G. and Coolbear, T.: The physiology and biochemistry of the proteolytic system in lactic acid bacteria. FEMS Microbiol. Rev., 12, 179-206 (1993).

[27] Tan, P. S., Poolman, B. and Konings, W. N.: Proteolytic enzymes of Lactococcus lactis. J. Dairy Res., 60, 269-286 (1993).

[28] Nakamura, Y., Yamamoto, N., Sakai, K. and Takano, T.: Antihypertensive effect of sour milk and peptides isolated from it that are inhibitors to angiotensin I-converting enzyme. J. Dairy Sci., 78, 1253-1257 (1995).

[29] Yamamoto, N., Akino, A. and Takano, T.: Purification and specificity of a cell-wall-associated proteinase from *Lactobacillus helveticus* CP790. J. Biochem., 114, 740-745 (1993).

[30] Hata, Y., Yamamoto, M., Ohni, M., Nakajima, K., Nakamura, Y. and Takano, T.: A placebo-controlled study of the effect of sour milk on blood pressure in hypertensive subjects. Am. J. Clin. Nutr., 64, 767-771 (1996).

[31] Kajimoto, O., Aihara, K., Hirata, H., Takahashi, R. and Nakamura, Y.: Hypotensive Effects of Tablets Containing "lactotripeptides (VPP, IPP)". J. Nutr. Food (in Japanese), 4, 51-61 (2001).

[32] Kajimoto, O., Aihara, K., Hirata, H., Takahashi, R. and Nakamura, Y.: Safety evaluation of excessive intake of the tablet containing "lactotripeptides (VPP, IPP)" in healthy volunteers. J. Nutr. Food (in Japanese), 4, 37-46 (2001).

[33] Nakamura, Y., Kajimoto, O., Aihara, K., Mizutani, J., Ikeda, N., Nishimura, A. and Kajimoto, Y.: Effects of the liquid yogurts containing "lactotripeptide (VPP,IPP)" on high-normal blood pressure. J. Nutr. Food (in Japanese), 7, 123-137 (2004).

[34] Seppo, L., Jauhiainen, T., Poussa, T. and Korpela, R.: A fermented milk high in bioactive peptides has a blood pressure-lowering effect in hypertensive subjects. Am. J. Clin. Nutr., 77, 326-330 (2003).

[35] Seppo, L., Kerojoki, O., Suomalainen, T. and Korpela, R.: The effect of a *Lactobacillus helveticus* LBK-16 H fermented milk on hypertension - a pilot study on humans. Milchwissenschaft, 57, 124-127 (2002).

[36] Kunji, E. R., Mierau, I., Hagting, A., Poolman, B. and Konings, W. N.: The proteolytic systems of lactic acid bacteria. Antonie Van Leeuwenhoek, 70, 187-221 (1996).

[37] Christensen, J. E., Dudley, E. G., Pederson, J. A. and Steele, J. L.: Peptidases and amino acid catabolism in lactic acid bacteria. Antonie Van Leeuwenhoek, 76, 217-246 (1999).

[38] Kok, J. and Venema, G.: Genetics of proteinases of lactic acid bacteria. Biochimie, 70, 475-488 (1988).

[39] Kiwaki, M., Ikemura, H., Shimizu-Kadota, M. and Hirashima, A.: Molecular characterization of a cell wall-associated proteinase gene from *Streptococcus lactis* NCDO763. Mol. Microbiol., 3, 359-369 (1989).

[40] de Vos, W. M., Vos, P., de Haard, H. and Boerrigter, I.: Cloning and expression of the *Lactococcus lactis* subsp. *cremoris* SK11 gene encoding an extracellular serine proteinase. Gene, 85, 169-176 (1989).

[41] Yamamoto, N., Ono, H., Maeno, M. and Takano, T. Classification of *Lactobacillus helveticus* strains by Immunological differences in extracellular proteinases. Biosci. Biotech. Biochem., 62, 1228-1230 (1998).

[42] Yamamoto, N., Akino, A., Takano, T. and Shishido, K.: Presence of active and inactive molecules of a cell wall-associated proteinase in *Lactobacillus helveticus* CP790. Appl. Environ. Microbiol., 61, 698-701 (1995).

[43] Pederson, J. A., Mileski, G. J., Weimer, B. C. and Steele, J. L.: Genetic characterization of a cell envelope-associated proteinase from *Lactobacillus helveticus* CNRZ32. J. Bacteriol., 181, 4592-4597 (1999).

[44] Yamamoto, N., Shinoda, T. and Takano, T.: Molecular cloning and sequence analysis of a gene encoding an extracellular proteinase from *Lactobacillus helveticus* CP790. Biosci. Biotech. Biochem., 64, 1217-1222 (1999).

[45] Yamamoto, N. and Takano, T.: Maturation protein need for activation of an extracellular proteinase in *Lactobacillus helveticus* CP790. J. Dairy Sci., 80, 1949-1954 (1997).

[46] Ono, H., Yamamoto, N., Maeno, M. and Takano, T.: Purification and characterization of a cell-wall associated proteinase from *Lactobacillus helveticus* CP53. Milchwissenschaft, 52, 373-377 (1997).

[47] Yamamoto, N., Ishida, Y., Kawakami, N. and Yada, H.: *Lactobacillus helveticus* bacterium having high capability of producing tripeptide, fermented milk product, and process for preparing the same. EU Patent, 1016709A1 (1991).

[48] Callanan, M., P. Kaleta, J. O'Callaghan, O. O'Sullivan, K. Jordan, O. McAuliffe, A. Sangrador-Vegas, L. Slattery, G. F. Fitzgerald, T. Beresford, and R. P. Ross. Genome sequence of *Lactobacillus helveticus*, an organism distinguished by selective gene loss and insertion sequence element expansion. 190:727-735 (2008).

[49] Shinoda, T., Wakai, T., Uchida, Hattori, M., Nakamura, Y. and Yamamoto, N.: Comparative Analysis of Proteolytic Enzymes Need for Processing of Antihypertensive Peptides between *Lactobacillus helveticus* CM4 and DPC4571. In preparatoin for the submission to publicatoin (2011).

[50] Genay, M., L. Sadat, V. Gagnaire, and S. Lortal.: *prtH2*, not *prtH*, is the ubiquitous cell wall proteinase gene in *Lactobacillus helveticus*. 75:3238-3249 (2009).

[51] Pederson, J. A., Mileski, G. J., Weimer, B. C., Steele, J.L.: Genetic characterization of a cell envelope-associated proteinase from *Lactobacillus helveticus* CNRZ32. J. Bacteriol. 181: 4592-4597 (1999).

[52] Ueno, K., Mizuno, S., Yamamoto, N.: Purification and characterization of an endopeptidase has an important role in the carboxyl terminal processing of antihypertensive peptides in *Lactobacillus helveticus* CM4. Lett. Appl. microbiol., 39, 313-318 (2004).

[53] Yamamoto, N., Shinoda, T. and Mizuno, S.: Cloning and expression of an endopeptidase gene from *Lactobacillus helveticus* CM4 involved in processing antihypertensive peptides. Milchwissenschaft, 59, 593-597 (2004).

[54] Altermann, E., Russell, W. M., Azcarate-Peril, M. A., Barrangou, R., Buck, B. L., McAuliffe, O., Souther, N., Dobson, A., Duong, T., Callanan, M., Lick, S., Hamrick, A., Cano, R. and Klaenhammer, T. R.: Complete genome sequence of the probiotic lactic acid bacterium *Lactobacillus acidophilus* NCFM. Proc. Natl. Acad. Sci. U. S. A., 102, 3906-3912 (2005).

[55] DOE Joint Genome Institute:, *Lactobacillus gasseri* whole genome shotgun sequencing project. (2002).

[56] Pridmore, R. D., Berger, B., Desiere, F., Vilanova, D., Barretto, C., Pittet, A. C., Zwahlen, M. C., Rouvet, M., Altermann, E., Barrangou, R., Mollet, B., Mercenier, A., Klaenhammer, T., Arigoni, F. and Schell, M. A.: The genome sequenc e of the probiotic intestinal bacterium *Lactobacillus johnsonii* NCC 533. Proc. Natl. Acad. Sci. U. S. A., 101, 2512-2517 (2004).

[57] Wakai, T., and Yamamoto. N.: Repressive release of antihypertensive peptides, Val-Pro-Pro and Ile-Pro-Pro by peptides in *Lactobacillus helveticus* fermented milk. In preparation for the submission to publication.

[58] den Hengst C. D., Curley, P., Larsen, R., Buist, G., Nauta, A., van Sinderen, D., Kuipers, O. P., Kok, J.: Probing direct interactions between CodY and the oppD promoter of *Lactococcus lactis*. J Bacteriol. 187: 512-521 (2005).

[59] den Hengst, C. D., van Hijum, S. A., Geurts, J. M., Nauta, A., Kok, J., Kuipers, O.P.: The *Lactococcus lactis* CodY regulon: identification of a conserved cis-regulatory element. J Biol Chem. 280: 34332-34342 (2005).

10

Synthetic PEG Hydrogels as Extracellular Matrix Mimics for Tissue Engineering Applications

Georgia Papavasiliou*, Sonja Sokic and Michael Turturro
Illinois Institute of Technology, Department of Biomedical Engineering,
USA

1. Introduction

In recent years the field of tissue engineering, or regenerative medicine, has developed from the need to replace damaged and/or diseased tissues and organs by combining biomaterial scaffolds, biological signaling molecules, and cells. The regeneration of tissues may be achieved using either one of two principle approaches: 1) the *in vitro* construction or 2) the *in vivo* induction of tissue. In the first approach, biomaterial scaffolds are combined with biofunctional signaling molecules and cells and a fully functional tissue is grown *in vitro* which can then be implanted into the host. In the second approach, scaffolds are tailored with the desired biochemical composition as well as physical and mechanical properties of the target tissue, implanted into the host, and the body is used as a bioreactor to regenerate the tissue of interest. Therefore, biomaterial scaffolds play a central role in regenerative medicine as physical and biochemical milieus that dictate cell behavior, function, and tissue regeneration (Lutolf & Hubbell, 2005). While both natural and synthetic biomaterials have been extensively explored as scaffolds for tissue regeneration, polymeric materials from synthetic sources are advantageous due to their tunable mechanical properties and ability to systematically and selectively incorporate biological signals of the natural extracellular matrix (ECM) enabling controlled study of cell-substrate interactions. Over the last several decades synthetic crosslinked hydrogels of poly(ethylene) glycol have been extensively investigated for numerous biomedical applications including drug delivery, immunoisolation, and as matrices for engineering tissues. PEG hydrogels are biocompatible, hydrophilic polymers composed of 3D interstitial crosslinks that swell extensively in aqueous environments with water content similar to soft tissues. These biomaterials are inherently resistant to non-specific cell adhesion and protein adsorption, thus providing a blank slate upon which ECM-derived signals can be systematically introduced as well as spatially and temporally manipulated to control cell behavior and tissue regeneration. The continued enhancement of PEG-based biomaterial strategies towards the rational design of scaffolds is highly dependent on their ability to independently control the incorporation of multiple biofunctional signaling molecules from alterations in hydrogel degradation kinetics and mechanical properties, to temporally and spatially tune the presentation of mechanical and biofunctional signals, and to promote rapid and guided neovascularization (new blood vessel formation) prior to complete material degradation. The combination of the above-

---

*Corresponding Author

mentioned strategies may ultimately result in PEG-based scaffolds that contain the necessary cues that recapitulate the dynamic environment of the ECM ultimately leading to the regeneration of tissues. Here we address common approaches and polymerization techniques used to fabricate PEG scaffolds with 3D spatial and/or temporal variations in physical, mechanical, and biofunctional cues of the native ECM and highlight the important signals embedded in these scaffolds that support cell behavior and tissue regeneration.

2. The molecular constituents and role of the extracellular matrix

Since cells receive a myriad of signals from and interact highly with their immediate extracellular microenvironment, the extracellular matrix (ECM), the effective design and improvement of biomaterial scaffold strategies is highly dependent on incorporated knowledge of ECM structure and function. Once believed to function as a passive scaffold for the maintenance of tissue and organ structure, it is now well recognized that the ECM is a dynamic construct that upon its interaction with cells undergoes constant remodeling (i.e. assembly and degradation of its constituents) particularly during the normal physiological process of development, differentiation, and wound healing (Daley et al., 2008). The ECM is composed of macromolecular constituents that are primarily produced locally by the cells that surround it. In most connective tissues, ECM macromolecules are secreted primarily by fibroblast cells, while in connective tissues of bone and cartilage the macromolecular constituents are secreted by osteoblasts and chondrocytes, respectively (Alberts et al, 2008). The macromolecular constituents of the ECM are composed of a three-dimensional array of protein fibers and filaments embedded in a hydrated gelatinous network of polysaccharide chains of glycosaminoglycans (GAGs) and proteoglycans (Fig. 1).

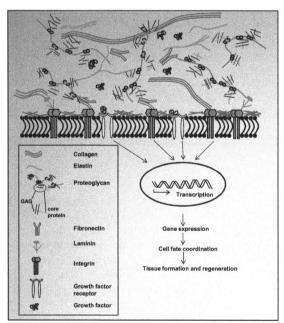

Fig. 1. Schematic illustration of key ECM components and their interactions with cells.

The ECM also contains growth factors and other bioactive molecules, as well as binding sites for cell-surface molecules exposed upon proteolysis (Somerville et al., 2003). Macromolecular GAGs are hydrophilic, unbranched, negatively charged polysaccharide chains that form stiff and highly extended conformations. Their negative charges attract counterions inducing an osmotic effect that enables the matrix to occupy large volumes of water. The four main groups of GAGs are hyaluronic acid, chondroitin sulfate, keratin sulfate, and heparin sulfate. Most GAGs are sulfated, covalently linked to protein via linker proteins and are synthesized intracellularly and released via exocytosis. The exception among the GAGs is hyaluronic acid which is released directly from the cell surface by an enzyme complex embedded in the plasma membrane, is unsulfated, not covalently linked to proteins, and free of sugar groups (Alberts et al., 2008; Saltzman, 2004). Hyaluronic acid is found in abundant quantities during embryogenesis and wound healing where its presence modulates cell migration in the extracellular space by controlling the level of hydration in tissues. Proteoglycans have extremely high sugar content ($\sim$ 95%) and are composed of a core protein to which GAGs are attached via linker proteins (Fig. 1). The GAG chains of proteoglycans enable them to form gels of varying porosity and charge density allowing them to serve as filters that regulate the molecular diffusion of molecules and cells. This structural and chemical organization allows proteoglycans to function as mediators of cell adhesion, regulators of growth factor secretion, and activators of secretion of proteolytic enzymes and protease inhibitors that are involved in both the assembly and degradation of ECM components.

The proteins of the ECM can be classified into structural (i.e., collagen and elastin) and adhesive (i.e., fibronectin and laminin) types. The most abundant structural protein in the ECM is collagen, which is secreted in large quantities primarily by cells of connective tissues. Various chemically distinct forms of collagen exist, each with a basic macromolecular unit comprised of an α triple helical structure composed of three polypeptide chains of ~1000 amino acids each that are specific to the type of collagen (Patino et al., 2002). Collagen assembles into different supramolecular structures allowing it to have extraordinary functional diversity. The most common forms of collagen found in the ECM are types I, II, III, and IV. Collagens I, II, and III are of the fibrillar type where upon secretion into the ECM organize into higher order polymer structures composed of long (several microns in length) and thin collagen fibrils (10-300 nm in diameter). Collagen assembly into fibrils and fibers is stabilized by covalent crosslinks formed by lysine residues between constituent collagen molecules (Saltzman, 2004). These crosslinks provide collagen fibrils with the tensile strength required to resist tensile forces. Fibrillar collagens interact with cells through integrin receptors on the cell membrane (Fig. 1) to induce cell differentiation and migration during embryonic development. Collagen type IV is a network forming collagen resulting in a mesh-like lattice forming a major portion of the basal lamina which separates epithelial sheets from other tissues and binds to cells via indirect binding to laminin adhesive proteins (Fig. 1). In healthy tissues, collagen is continuously degraded and replaced with turnover times ranging from months to years depending on the tissue type (Alberts et al., 2008).

Elasticity of the ECM is provided from elastic fibers that are primarily composed of elastin molecules that form crosslinked networks of fibers and sheets that enable them to stretch and relax upon deformation. Similar to collagen IV, these crosslinked structures are formed via covalent bond interactions of lysine residues between individual elastin molecules. This anisotropic structure of protein fibers and filaments mechanically and biochemically influences cell behavior. Cells within this microenvironment sense the mechanical properties of the ECM and convert mechanical signals into biochemical signals due to the direct interaction of the ECM with the cytoskeleton via cell-surface receptors (Lutolf & Hubbell, 2005).

Proteins of the adhesive type possess multiple domains that contain specific binding sites for ECM macromolecules as well as for integrins on the surface of cells. Integrins are a class of cell-surface receptors consisting of two non-covalently associated transmembrane α and β heterodimers that contain an extracellular ligand binding domain, a transmembrane domain, and a cytoplasmic domain (Steffensen et al., 2011). Integrin-ECM ligand binding interactions provide communication between the intracellular and extracellular environments which in turn affect cellular fate processes such as proliferation, migration, and differentiation (Fig. 1). Cells adhere by integrin-mediated interactions with ECM adhesion proteins such as fibronectin and laminin (Patterson et al., 2010). During this cell-matrix binding event, the extracellular domain of the integrin receptor binds to adhesive protein oligopeptide regions and the cytoplasmic portion binds to focal adhesions linking the ECM to the cytoskeleton resulting in signal transduction. The adhesive protein fibronectin is a large dimeric glycoprotein composed of two subunits each containing approximately 2500 amino acids linked together by disulfide bonds. Integrin receptors on cell surfaces bind to a fibronectin domain containing the well-known tripeptide adhesion sequence arginine-glycine-aspartic acid (RGD) and the neighboring "synergy site" while other distinct protein domains bind to collagen and heparin (Daley et al., 2008). While most cell types bind to the RGD sequence, an additional region (IIICS) within fibronectin has been identified to contain the peptide sequences REDV and LDV that permit the adhesion of specific cell types such as neural cells and lymphocytes. Laminin, a large cross-shaped adhesion protein composed of α (400 kDa), β (215 kDa), and γ (205 kDa) polypeptide subunits linked by disulfide bonds, is primarily associated with basement membranes. Different types of isoforms of laminin exist which are created by the different types of α, β, and γ polypeptide chain interactions. Multiple functional regions have been identified in laminin such as those containing the RGD and YIGSR sequences resulting in cell adhesion, the IKVAV peptide motif involved in neurite growth, and a heparin and collagen IV binding region (Saltzman, 2004).

Within a single tissue the ECM is continuously being remodeled by cells via the degradation of ECM components resulting from the secretion of proteolytic enzymes followed by the reassembly of newly synthesized protein components secreted by cells. Cells adhered to the ECM have to change from an adhesive phenotype to a migratory phenotype before they can migrate within its three-dimensional structure (Vu & Werb, 2000). During the process of cell migration, cells secrete proteolytic enzymes (proteases) that cleave a variety of ECM substrates to break down physical barriers that inhibit cell locomotion. While there are several families of proteases secreted by cells that are involved in ECM degradation, proteases of the matrix metalloproteinase (MMP) family have been shown to be central to tissue homeostasis and ECM remodeling (Steffensen et al., 2011). MMPs are a family of structurally related endopeptidases comprising 23 enzymes in humans all acting to degrade parts of the ECM. Most MMPs are secreted in latent form as pro-enzymes. The latency of the zymogen is dependent on a "cysteine switch" formed by the interaction of a conserved cysteine on the pro-domain with a catalytic zinc atom within the active site of the enzyme blocking access to the catalytic site. Enzyme activation results from disruption of the cysteine-zinc pairing due to displacement of the pro-domain by conformational change or proteolysis induced by the protease plasmin or by other MMPs (Singer & Clark, 1999). There are four main classes of MMPs: collagenases (MMPs -1, 8, and 13) that degrade fibrillar type collagens I, II and III in their triple helical domains to form gelatin, gelatinases (MMPs- 2 and 9) that degrade gelatin as well as basement membrane proteins and elastin, stromelysins (MMP-3, 10, and 11) which have broad substrate specificity but possess the ability to activate other MMPs, and membrane-type MMPs (MT-MMPs) that are bound to the cell membrane and similar to stromelysins, can activate other MMPs in addition to their ability to degrade ECM components (Somerville et al., 2003).

The ECM also modulates tissue dynamics through its ability to bind, store, and release growth factors (Lutolf et al., 2003; Somerville et al., 2003). Growth factors may be released by cells for immediate signaling or embedded within ECM components and growth factor binding proteins and released upon ECM proteolytic degradation (Chen & Mooney, 2003). Controlled growth factor release within the ECM is balanced by extracellular degradation. During processes such as wound healing, angiogenesis, and tissue repair, ECM-bound growth factors are released upon MMP proteolytic activity. MMPs can also cleave growth factor receptors on cell surfaces to release growth factor receptor ectodomains contributing to mechanisms by which cells can target growth factor activity (Steffensen et al., 2011). Studies using this approach have shown that cleavage of the ectodomain of acidic fibroblast growth factor receptor 1 (FGFR-1) by MMP-2 releases the entire extracellular domain of the receptor which then enables the binding and sequestration of fibroblast growth factor (FGF) in the tissue, suggesting that soluble forms of FGF-1 produced upon MMP-2 proteolysis contribute to angiogenic and mitogenic activities of FGF (Levi et al., 1996).

In summary, the complex structure and function of the ECM as well as the interaction of its molecular constituents with cells regulate numerous cellular fate processes such as adhesion, proliferation, migration, and differentiation. This dynamic cellular microenvironment is continuously being remodeled especially during the natural processes of development, differentiation, and wound repair. Understanding the complex dynamic relationship between cells and the ECM is critical to the successful design of synthetic biomaterial scaffolds that promote tissue regeneration.

3. Synthetic PEG-based ECMs for the support of 3D cell and tissue growth

Naturally derived as well as synthetic hydrogels have been extensively investigated in tissue engineering. Natural hydrogels are primarily obtained from ECM proteins (e.g., collagen, fibrin) and polysaccharides (e.g., alginate, chitosan, hyaluronic acid, dextran) (Zhu, 2010). Collagen and fibrin are clinically well-established FDA-approved materials for the healing of burns and chronic wounds, and used as tissue sealants, respectively (Lutolf & Hubbell, 2005). In general, hydrogels derived from natural sources are advantageous over synthetic hydrogel formulations since they possess the structural complexity and functional capacity of a particular tissue of interest. Despite these advantages, their use is often restricted due to complexities associated with purification, immunogenicity, and pathogen transmission (Patterson et al., 2010; Zhu, 2010). In addition, their mechanical properties cannot be readily manipulated, especially independent of alterations in their biochemical composition. These associated disadvantages of naturally-derived ECM components have resulted in a shift over the last several decades towards the use of synthetic biomaterials that offer greater and systematic control of material properties for the design of scaffolds of multiple tissue types.

Among the classes of synthetic biomaterials, PEG hydrogels have been extensively investigated as scaffolds in tissue engineering. Crosslinked PEG hydrogel networks swell extensively in aqueous environments providing a 3D highly swollen network with viscoelastic properties similar to soft tissues enabling diffusive transport and interstitial flow characteristics (Lutolf & Hubbell, 2005). The physical and mechanical properties of these synthetic biomaterials can be readily manipulated via alterations in selected polymerization conditions that directly influence hydrogel crosslink density, swelling, and the elastic modulus resulting in scaffolds with rigidities ranging from those found in soft tissues such as the liver and skin up to rigidity values of articular cartilage and bone (Nemir & West, 2010). PEG scaffolds are also biocompatible and intrinsically resistant to non-specific cell adhesion and protein adsorption. This inherently inert feature offers the advantage for selective incorporation of a variety of identified biofunctional oligopeptide sequences and proteins of the native ECM whose concentration and spatial distribution can be easily modulated within the crosslinked network to provide fundamental insight of signaling events involved in specific cell-matrix interactions. Finally, using a variety of different chemistries, PEG-based hydrogels can be polymerized under mild physiological conditions in the presence of cells and in situ with negligible loss in cell viability and function (Nuttelman et al., 2006; Underhill et al., 2007).

3.1 Mechanisms of biofunctional PEG hydrogel formation

While various forms of gelation exist (i.e., physical, ionic, and covalent interactions), covalent crosslinking leads to more stable PEG hydrogel networks with tunable physical and mechanical properties (Lin & Anseth, 2009). Here we focus our attention on the various mechanisms of covalent crosslinking used for the fabrication of PEG hydrogel ECM mimics as well as the chemistries employed to engineer biofunctional monomers containing key signaling molecules of the native ECM that can be copolymerized and subsequently immobilized within PEG hydrogel networks.

3.1.1 Mechanisms of covalently crosslinked hydrogel formation

The synthesis of covalently crosslinked PEG hydrogels falls into one of three major categories: chain growth or free-radical photopolymerization, step-growth polymerization, and mixed-mode polymerization (a combination of chain and step growth reactions). Chain growth reactions such as those occurring in free-radical photopolymerization are initiated upon the photocleavage of initiator molecules in the presence of UV or visible light to yield primary radical species. These free-radicals propagate through unsaturated vinyl or acrylate bonds on PEG macromers. Covalent crosslinking and subsequent gelation is induced in the presence of PEG macromers containing multiple acrylate groups as in the case of the crosslinking agent PEG diacrylate (PEGDA). In these systems, complete gelation is achieved at relatively short times (on the order of seconds up to a few minutes). Photopolymerization of PEGDA hydrogels has been extensively investigated in tissue engineering and regenerative medicine applications, especially since it has been found as a suitable approach for the in situ encapsulation of proteins and cells.

Step-growth polymerization and gelation occurs when at least two multifunctional macromers that contain complementary reactive chemical groups are reacted under stoichiometric balanced or imbalanced ratios. This polymerization can proceed under physiologic conditions without the use of free-radical initiators and permits more precise control over the network crosslink density (Lin & Anseth, 2009). Hubbell and colleagues have developed a step-growth approach to form crosslinked PEG hydrogel networks that proceed via a Michael-type addition reaction using acrylated star PEG polymers and dithiol monomers (Elbert et al., 2001; Rizzi et al., 2006; Rizzi & Hubbell, 2005). This polymerization mechanism was later extended to fabricate PEG hydrogel ECM mimics using multiarm PEG vinyl sulfone macromers (n-PEG-VS) to allow a Michael-type addition reaction between acrylated PEG and thiol groups presented on free cysteine amino acids on peptides and proteins (Raeber et al., 2005). Michael-type addition reactions are not limited to acrylates as they can also occur between maleimide and thiol groups (Patterson et al., 2010). As compared to free-radical polymerization times, Michael addition reactions take longer to complete (on the order of hours). Another type of step-growth gelation mechanism recently employed for the fabrication of PEG hydrogels is Click chemistry. This approach has elicited tremendous interest in tissue engineering due the well-defined network structures created, improved mechanical properties, its resulting versatility with respect to bioconjugation, and enhanced swelling capacities over other gelation approaches (DeForest et al., 2009; Malkoch et al., 2006; Polizzotti et al., 2008). In this method of gelation, macromers bearing azide and alkyne functional groups are "clicked" together in the presence of copper catalysts to form stable covalent linkages after which biofunctional molecules can be conjugated (Lin & Anseth, 2009). One major drawback with conventional Click PEG hydrogels is that the reactions are catalyzed by copper which is cytotoxic to cells and not suitable for their encapsulation during the gelation process. Recently, Bertozzi and coworkers have developed a copper-free-click chemistry approach which offers great potential for a variety of tissue engineering applications (Laughlin et al., 2008).

PEG hydrogels crosslinked using a mixed-mode approach have been developed by Anseth and colleagues using thiol acrylate photopolymerization (Salinas & Anseth, 2008). In this approach the polymerization is initiated photochemically and proceeds via step growth reactions between thiol and acrylate groups followed by acrylate homopolymerization which proceeds by chain polymerization. The propagation of polymer chains in this mixed mode process of hydrogel formation involves three reactions: the first step involves the step-wise reaction of a thiyl radical with a vinyl group, the second step involves the chain transfer of a radical from a carbon intermediate to a thiol, and the third reaction involves the homopolymerization of the acrylate radical via chain growth propagation. Thiol acrylate polymerization involves the chain transfer of growing polymer chains to thiol monomers which do not occur during the stepwise mechanism of Michael-type addition reactions. Unlike chain growth polymerizations, the use of photoinitiators is not necessarily required in mixed mode photopolymerization. Finally, the competition between the acrylate homopolymerization with the step growth thiol acrylate reaction results in different kinetics with more versatile network structures than those formed with pure step growth or free-radical polymerization gelation chemistries (Lin & Anseth, 2009).

3.1.2 Engineering ECM cues into PEG hydrogel scaffolds

Throughout the last several decades, extensive research has been conducted to systematically incorporate cell signaling molecules of the native ECM within PEG scaffolds which have provided fundamental insight on specific cell-substrate interactions. As previously mentioned, in the absence of cell signaling molecules, PEG hydrogels are inert biocompatible matrices, which do not elicit cell adhesion or protein adsorption. This discovery along with the identification of small peptide sequences of ECM adhesion proteins (Ruoslahti, 1996) has led to their widespread use as matrices upon which to introduce key biofunctional signals of the native ECM in either soluble or covalently immobilized form for modulating specific cellular responses *in vitro* and *in vivo*.

The minimum requirement for anchorage-dependent cells to survive on and within PEG scaffolds is to render them cell-adhesive. Thus, numerous researchers have covalently incorporated cell adhesion sequences, the most common being the RGD sequence, within PEG hydrogels using a variety of conjugation approaches, as described below. However, the incorporation of cell adhesion sequences in PEG hydrogels does not enable three-dimensional (3D) prolonged cell survival when cells are encapsulated within these scaffolds due to the mesh size of the crosslinked PEG network being less than the average diameter of a cell, which poses a physical barrier to the critical process of cell migration (Raeber et al., 2005). Therefore, studies investigating cell behavior on PEG hydrogels modified with cell adhesion ligands have been primarily limited to two dimensions (2D) (at the cell-substrate interface), which is not representative of the actual processes of cell behavior and tissue remodeling in 3D. Nonetheless, these studies have provided significant insight on the effects of scaffold properties on 2D cell behavior which have led to the future design of scaffolds in tissue engineering (DeLong et al., 2005 a, 2005 b; Guarnieri et al., 2010; Hern & Hubbell, 1998).

The applicability of PEG hydrogels in promoting 3D cell behavior and tissue regeneration is to render the scaffolds degradable. This requires that material degradation rate be matched with the rate of tissue regeneration (Papavasiliou et al., 2010). Although various approaches have been used to fabricate degradable PEG hydrogels with pre-engineered degradation rates, as in the case where hydrolytically degradable segments of poly(lactic acid) PLA, poly(glycolic acid) PGA, or poly(lactic-*co*-glycolic acid) PLGA are copolymerized into the crosslinked network, the ability to modify hydrogel degradation rates post-gelation using this approach is limited (Lin & Anseth, 2009). In addition, hydrolytic degradation is not representative of the cellularly mediated and dynamic process of proteolysis that takes place within the native ECM. Therefore, numerous studies have covalently incorporated peptide sequences susceptible to cleavage by cell-secreted proteases (i.e., plasmin-sensitive or MMP-sensitive sequences) into PEG hydrogels thus manipulating gel degradation dynamically in response to cellularly mediated events (Gobin & West, 2002; Lee et al., 2007; Lutolf et al., 2003; Moon et al., 2007; Moon et al., 2010; Patterson & Hubbell 2010; Phelps et al., 2009; Seliktar et al., 2004; West & Hubbell, 1999; Zisch et al., 2003).

In the native ECM, growth factors are stored and released from ECM proteoglycans upon cellular demand, or secreted by cells, thus playing critical roles in controlling cellular function and tissue regeneration (Chen & Mooney, 2003). While one therapeutic approach would be the bolus delivery of growth factors to the damaged and or diseased tissue, growth factors have short half-lives in the circulation and their distribution and rapid degradation may result in undesirable systemic effects and toxicity (Patterson et al., 2010). An alternative approach is to encapsulate growth factors in scaffolds such as those of PEG; however, this has been shown to result in a rapid burst release during the initial hydrogel swelling phase (Zhu, 2010). Alternatively, growth factors can be covalently incorporated into PEG hydrogels in order to prolong their biological activity and to mimic growth factor binding to ECM proteins.

In an effort to covalently immobilize peptides and/or proteins into PEG hydrogels, the pioneering work of Hern et al. showed that the maintenance of biological activity of synthetic peptide sequences upon immobilization requires that the peptide be flexible and experience minimal steric hindrance (Hern & Hubbell, 1998). This study also showed that the tethering of RGD was required for cells to adhere onto crosslinked PEGDA hydrogel surfaces that contained RGD groups modified with a PEG spacer arm (MW=3400) whereas hydrogel surfaces immobilized with RGD in the absence of a spacer exhibited limited cell adhesion. Other studies used different molecular weights of PEG analogues to investigate the effect of linker length on osteoblast adhesion and determined that the effective distance between the peptide and the substrate was 3.5 nm corresponding to an approximate length of a PEG of molecular weight of 3500 Da (Kantlehner et al., 1999). Later studies combined both tethered cell adhesion ligands and growth factors that included a PEG pacer into PEG hydrogels in order to enhance cell proliferation and migration on hydrogel surfaces (DeLong et al., 2005). Early studies by West and Hubbell synthesized PEG hydrogels containing collagenase- and plasmin-sensitive domains between crosslinks that would degrade upon proteolysis to mimic the natural process of ECM degradation (West & Hubbell, 1999). These pioneering studies led to the synthesis of PEG hydrogel networks that incorporated a variety of different pendant cell adhesion sites and growth factors, and crosslinked enzyme-sensitive domains for studying a variety of cell-substrate interactions. The resulting polymer structure of a covalently crosslinked PEG hydrogel network with incorporated biofunctionality for promoting 3D cell proliferation, migration, and tissue regeneration is shown in Figure 2.

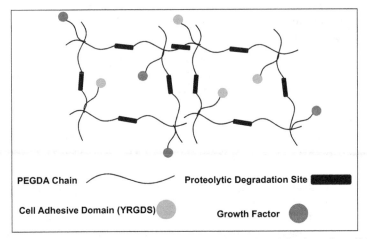

PEGDA Chain   Proteolytic Degradation Site

Cell Adhesive Domain (YRGDS)   Growth Factor

Fig. 2. Schematic of ECM signals immobilized in crosslinked PEG hydrogel scaffolds.

Various conjugation approaches have been used to covalently immobilize cell adhesion ligands, proteolytically-sensitive domains, and growth factors within PEG hydrogels crosslinked via the gelation mechanisms described above. In PEGDA hydrogels formed by free-radical photopolymerization, biofunctionality is incorporated into the hydrogel network primarily through the copolymerization of peptide and/or protein monoacrylate and diacrylate macromers. These biofunctional macromers may be synthesized by reacting acryl-PEG-NHS (NHS = N-hydroxyl succinimide) with peptide resulting in the coupling of the N-terminal α-amine of the peptide to the acrylate PEG macromer. In general the synthesis of biofunctional PEG monoacrylate macromers is achieved by reacting the Acryl-PEG-NHS with peptide in a 1:1 mole ratio so that the resulting macromers are immobilized in a pendant fashion within the hydrogel network upon gelation as in the case where tethered cell adhesion and growth factors are desired. The synthesis of biofunctional PEG diacrylate macromers results via the reaction of the Acryl-PEG-NHS derivative with peptide in a 2:1 mole ratio to produce Acryl-PEG-peptide-PEG-Acryl. This biofunctional PEGDA peptide conjugate therefore becomes crosslinked within the hydrogel network upon polymerization as in the case of crosslinked enzymatically degradable peptide sequences. The physical (swelling ratio), mechanical (elastic modulus), and degradative properties of biofunctional PEG hydrogels can be manipulated by adjusting the polymerization conditions such as the molecular weight of the PEG spacer, or the PEGDA prepolymer weight percentage. Recently, an alternative approach has been used to synthesize crosslinkable PEGDA peptide conjugates by Michael-type addition reaction (Miller et al., 2010). This conjugation involves the step-wise reaction of bis-cysteine peptides with PEGDA to make high molecular weight PEGDA macromer peptide conjugates by controlling the stoichiometric ratio of the functional groups (Tibbitt & Anseth, 2009). These biofunctional PEG diacrylates could then be crosslinked via free-radical photopolymerization to form hydrogels susceptible to cleavage by cell-mediated proteolysis. The techniques mentioned above are not ideal for synthesizing PEG-peptide macromer conjugates since they often result in the formation of multiple types of macromer species of various molecular weights and reactive functionality that alter the kinetics of crosslinking and affect the hydrogel mechanical and physical properties. In order to circumvent these issues more extensive purification methods may be required to separate these undesired products post-conjugation and prior to polymerization.

Step-growth gelation mechanisms such as Michael-type addition allow for facile incorporation of cysteine-containing peptides with various functionalized types of PEG or multiarm PEG macromers (i.e., acrylate, maleimide, and vinyl-sulfone) in a one-step process. For example, Hubbell and colleagues have reacted multi-arm PEG vinyl sulfone with thiol groups of cysteine RGD containing peptides as well as dithiol bearing enzymatically degradable peptide sequences to form crosslinked cell adhesive and proteolytically degradable PEG hydrogels (Lutolf & Hubbell, 2005; Patterson & Hubbell 2010; Raeber et al., 2005; Zisch et al., 2003).

In the mixed mode gelation processes of thiol-acrylate photopolymerization, Anseth and coworkers have synthesized biofunctional PEG hydrogels via the use of thiol-bearing peptides (Lin & Anseth, 2009; Salinas & Anseth, 2008). Since thiol bearing compounds act as chain transfer agents during free-radical polymerization, the use of cysteine-containing peptides enables the peptides to act as chain transfer agents resulting in their incorporation within the hydrogel network. As compared to the bioconjugation approaches of pure free-radical photopolymerization described above, this mixed-mode approach does not rely on the additional step of synthesizing the acrylate and diacrylate PEG-peptide conjugates. Rather, the peptides are directly tethered in a single step fashion within the hydrogel network during photopolymerization as in the case of Michael-type addition.

3.2 Mechanical properties and incorporated biofunctionality of proteolytically-sensitive PEG hydrogels influence 3D cell behavior and tissue regeneration

PEG hydrogels immobilized with cell adhesive peptides, growth factors, and enzyme-sensitive peptides have been widely investigated as ECM mimics for numerous tissue engineering applications. A number of enzyme-sensitive peptides have been incorporated into PEG hydrogels such as collagen-derived GPQG↓IAGQ or MMP-sensitive (Lutolf et al., 2003), fibrin-derived YK↓NRD that are plasmin-sensitive (Pratt et al., 2004; Raeber et al., 2005), and elastase-sensitive peptides such as AAPV↓RGGG (Aimetti et al., 2009), where ↓denotes the peptide cleavage site. Here, we discuss the effects of hydrogel mechanical properties, proteolytically mediated degradation rate, and growth factor presentation on 3D cell behavior and tissue regeneration within PEG hydrogels immobilized with MMP-sensitive domains. We limit our attention to MMP-sensitive PEG hydrogels due to the importance of MMP enzymes in critical processes such as wound healing and vascularization.

Numerous studies have shown that the 3D presentation of specific soluble and immobilized biofunctional cues as well as the mechanical properties (Bott et al. 2010) of biofunctional PEG scaffolds have a profound impact on *in vitro* (Leslie-Barbick et al., 2009; Saik et al., 2011; Seliktar et al., 2004; Zisch et al., 2003) and *in vivo* (Moon et al., 2010; Phelps et al., 2009 ; Zisch et al., 2003) cell behavior. Our recent data (unpublished results) illustrate that the degradation rate of MMP-sensitive PEGDA hydrogels formed by free-radical photopolymerization in the presence of visible light (λ = 514nm) can be controlled through variations in the weight percent of the MMP-sensitive PEG diacrylate macromer when incubated in collagenase enzyme solution. Increases in the weight percentage of the MMP-sensitive diacrylate increases the presentation of MMP-sensitive domains in the network, but also results in a higher crosslink density and in decreased diffusion of enzyme in the gel, thus increasing hydrogel degradation times (Fig. 3).

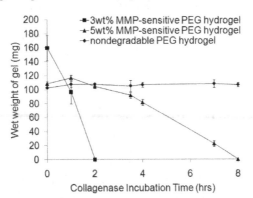

Fig. 3. Degradation profiles of photopolymerized MMP-sensitive PEGDA hydrogels.

For hydrogels that display the fastest degradation, the encapsulation of fibroblasts within PEGDA scaffolds immobilized with MMP-sensitive crosslinks and tethered RGD results in robust invasion of the PEG matrix after 2 weeks in culture (Fig. 4). Other studies using

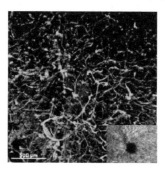

Fig. 4. Fibroblast invasion within an MMP-sensitive PEGDA hydrogel matrix 2 weeks post-encapsulation. Scale bar = 500 μm.

Michael-type addition for the fabrication of PEG hydrogels immobilized with collagen-based MMP-sensitive sequences and the cell adhesion ligand RGD showed that the proliferation of fibroblasts can be manipulated via changes in the hydrogel modulus (ranging from 250 to 1100 Pa) regardless of the sensitivity of the PEG matrix to proteolysis and the presence of cell adhesion motifs (Bott et al., 2010). Studies using UV-based free-radical photopolymerization for the formation of MMP-sensitive, VEGF-bearing PEGDA hydrogels as potential matrices for the induction of vascularization indicated that endothelial cell tubule formation and maintenance was directly affected by gel mechanical properties. Hydrogels formed with an intermediate polymer weight percentage of 10% resulted in the formation of tubules, while less crosslinked PEGDA matrices (7.5% PEG content) caused complete tubule regression, and matrices (15% PEG content) reduced tubule formation *in vitro* (Moon et al., 2010). PEG hydrogels of this intermediate stiffness were implanted into mouse cornea using a micropocket angiogenesis assay and resulted in neovascularization in the presence of vascular endothelial growth factor (VEGF) (Rogers et al., 2007). In this study, the stiffness of the hydrogel and sensitivity to proteolysis were simultaneously altered through variations in the weight percentage of the MMP-sensitive-conjugated PEGDA macromer with compressive moduli ranging from 30 to 110 kPa. This study and others have demonstrated that hydrogel degradation and the formation of functional blood vessels within biofunctional PEG hydrogels can be manipulated by changing the biophysical properties of the hydrogel.

While the physical and mechanical properties of PEG hydrogels play a critical role in dictating 3D cell behavior, extensive research has been conducted on investigating the effects of growth factor identity and delivery on cell behavior and tissue regeneration within PEG scaffolds. The successful application of PEG scaffolds is highly dependent on their ability to promote stable neovascularization (new blood vessel formation) to meet the *in vivo* oxygen and nutrient demands associated with tissues (Papavasiliou et al., 2010). Growth factors, such as VEGF and the fibroblast growth factor (FGF) families, in addition to MMPs, play critical roles during vascularization. Therefore, these biofunctional signals have been incorporated into PEG scaffolds to study their effects on vascularization *in vitro* and *in vivo*.

Growth factors have been incorporated into PEG scaffolds in either soluble or immobilized form or in combinations. For example, MMP-sensitive PEG hydrogels were investigated as a bioactive co-encapsulation system for vascular cells and thymosin β4, a small bioactive molecule. This system was able to induce adhesion, survival, migration and organization of human umbilical vein endothelial cells as well as the controlled release of thymosin β4 triggered by MMP-2 and MMP-9 enzymes (Kraehenbuehl et al., 2009). In a similar study, collagenase-degradable PEG hydrogels modified with covalently immobilized VEGF and RGDS showed a fourteen-fold increase in endothelial cell motility and a three-fold increase in cell-cell connections in the presence of VEGF (Leslie-Barbick et al., 2009). While the above-mentioned studies have focused on the presentation of a single growth factor to cells encapsulated in PEG hydrogels, multiple growth factors are required to drive the process of tissue regeneration to completion *in vivo* (Chen & Mooney, 2003). Therefore, PEGDA hydrogel studies have recently investigated the effects of immobilized combinations of multiple growth factors on the ability of cells to induce key responses involved in angiogenesis (Saik et al., 2011). The immobilization of two key angiogenic proteins, platelet-derived growth factor-BB (PDGF-BB) and fibroblast growth factor-2 (FGF-2), resulted in increased endothelial cell migration *in vitro* compared with the immobilization of each factor alone. Furthermore, the combination of soluble PDGF-BB and immobilized PDGF-BB induced a more robust vascular response compared with soluble PDGF-BB alone when the hydrogels were implanted *in vivo* in a mouse cornea micropocket angiogenesis assay.

In the native ECM, growth factors are released by cell-associated enzymatic activity that degrades the matrix, or that cleaves the matrix binding domain between the growth factor and the ECM component proteins (Zisch et al., 2003). To mimic a delivery system whereby the rate of growth factor release could be controlled by matrix degradation and by cleavage of the matrix-growth factor binding site, PEG hydrogels were engineered with MMP-sensitive domains as well as with immobilized VEGF that was designed to contain an enzymatically degradable plasmin linker. This plasmin linker would allow for VEGF release in response to plasmin secretion as principal forms of VEGF are known to contain a binding site for heparin sulfate proteoglycans which maintain VEGF in its immobilized state until released from local cellular enzymes (Zisch et al., 2003). These matrix associations stabilize the growth factor in its active conformation protecting it from proteolytic inactivation. When these hydrogels were implanted subcutaneously in rats, they were completely remodeled into native, vascularized tissue.

Most of the studies focusing on proteolytically degradable PEG hydrogels for tissue engineering applications have primarily utilized the MMP-sensitive substrate site found within the α-chain of type I collagen (GPQG↓IAGQ or GPQG↓IWGQ). (Seliktar et al., 2004) However, these substrates do not degrade particularly fast which may limit cellular infiltration within the scaffold and these peptides can be also be cleaved by a variety of MMPs. Thus, recent strategies have focused on enhancing proteolytic degradation of PEG hydrogels by targeting peptide substrates with increased catalytic activity (Patterson & Hubbell, 2010, 2011) or by increasing the spatial presentation of these signaling molecules within the hydrogel network (Miller et al., 2010). To increase the spatial presentation, studies have used step-growth polymerization of PEG diacrylate and bis-cysteine MMP-sensitive peptides and via Michael-type addition formed biodegradable ultra-high molecular weight macromers in the form of acrylate-PEG-(peptide-PEG)$_m$-acrylate that when photopolymerized into hydrogels resulted in increased swelling and degradability compared to previously published systems. This system induced angiogenesis in an *ex vivo* aortic arch explant assay and demonstrated that capillary sprouting can be tuned by engineering the susceptibility of the hydrogels to MMP enzymes and the adhesive ligands (Miller et al., 2010). Additionally, using combinatorial methods of peptide libraries, MMP substrate sequences that show increased enzymatic degradation and specificity have been identified (Patterson & Hubbell, 2010, 2011; Turk et al., 2001). In an effort to enhance proteolytic degradation within these synthetic matrices, one study screened these substrates for enhanced degradability by MMP-1 and MMP-2 enzyme solutions in soluble form and as biodegradable crosslinkers within PEG hydrogels (Patterson & Hubbell, 2010). The study showed that increases in catalytic activity resulted in faster degradation times of the hydrogels formed with the different peptides and that certain peptides were more specific to either MMP-1 or MMP-2, while others were susceptible to both enzymes. In another study, these peptides were screened for plasmin sensitivity in addition to MMP-1 and MMP-2 to enhance hydrogel degradation. Hydrogels that displayed increased sensitivity to these enzymes resulted in faster fibroblast proliferation and greater cell invasion from aortic ring segments (Patterson & Hubbell, 2011). These hydrogel systems allow tuning to enzymes that are produced by specific cell types giving a broader *in vivo* application.

4. Biofunctional and mechanical gradients in PEG hydrogels

While the above studies focused on investigating the effects of biofunctional ECM signals and mechanical properties that can be homogeneously distributed within PEG hydrogels on cell behavior, the spatial presentation of these factors plays a critical role in directing cell behavior and tissue regeneration. For example, critical cellular fate processes involved in tissue regeneration, such as cell migration, are directed by the complex spatial and temporal presentation of physical and chemical signals found in the ECM (Barkefors et al., 2008; Lo et al., 2000; Papavasiliou et al., 2010; Smith et al., 2009). Cellular responses to gradients of different stimuli can be separated into durotaxis, the process in which cells respond to changes in matrix rigidity (Lo et al., 2000), haptotaxis, a response to matrix bound chemoattractants (Lo et al., 2000; Smith et al., 2004; Kim et al., 2009), and chemotaxis, a response to a soluble factor that diffuses freely from its source (Barkefors et al., 2008; Lo et al., 2000). *In vivo*, studies of the retina have shown that gradients of secreted VEGF help shape vascular patterns during angiogenic sprouting by guiding endothelial tip cell formation and subsequent migration (Gerhardt et al., 2003). Furthermore, computational models predict that VEGF gradients in hypoxic muscle tissue are capable of directing the formation of new vessels (Mac Gabhann et al., 2007). Other studies have demonstrated that gradients of platelet derived growth factor (PDGF) guide fibroblasts towards wounded regions (Schneider et al., 2010) and that leukocytes migrate to areas of inflammation, infection or injury in response to secreted chemokines and bacterial byproducts (Moissoglu & Schwartz, 2006). To date, a limited number of studies have sought to recreate both haptotactic and durotactic gradients within synthetic PEG scaffolds. PEG hydrogels with gradients of biofunctional ECM signals and/or stiffness have been generated using gradient makers (DeLong et al., 2005a, 2005b; Nemir & West, 2010), microfluidic techniques (Burdick et al., 2004; Guarnieri et al., 2008, 2010) and perfusion-based frontal polymerization (Turturro & Papavasiliou, 2011). In other instances, soluble growth factors have been included in PEG hydrogels to stimulate a chemotactic response as they diffuse from the implanted hydrogels into surrounding tissue (Papavasiliou et al., 2010).

4.1 Gradient makers

The use of gradient makers to generate spatial variation in either matrix stiffness or covalently immobilized biofunctional moieties is attractive since these devices are easy to use and commercially available. Gradient makers were originally developed for the separation of proteins via gel electrophoresis but have been miniaturized to form biofunctional hydrogel gradients. This approach of gradient generation uses two feed streams containing different prepolymer compositions that are combined in varying proportions as they pass through a control valve. The mixture is gently added to a mold such that layers of varying prepolymer composition are retained (Fig. 5). The prepolymer solution is then converted into a hydrogel by photopolymerization resulting in a uniform gradient that transitions from one of the prepolymer compositions to the other. In one study, a gradient maker was employed to generate a gradient of covalently immobilized basic fibroblast growth factor (bFGF) by adding bFGF to only one of the feed streams (DeLong et al., 2005). Using this technique, it was shown that smooth muscle cells migrated towards regions of increasing concentration of bFGF. In another study, PEGDA hydrogels with gradients in elastic modulus ranging from 100 kPa to 5 kPa were formed with a gradient maker using feed streams containing PEGDA macromers of two molecular weights, 3.4 kDa and 20 kDa, respectively (Nemir et al., 2010). This study showed that that macrophages seeded on PEG hydrogel surfaces primarily adhered to regions of greater stiffness.

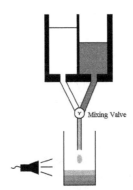

Fig. 5. Schematic illustration of gradient generation in photopolymerized PEG hydrogels using gradient makers.

4.2 Microfluidics

Similar to gradient makers, microfluidic techniques are based on the mixing of two prepolymer feed streams at varying ratios. The difference is that microfluidics uses a series of interconnected microchannels to intermix the feed streams and simultaneously create multiple layers each with a unique composition (Fig. 6). The resultant layers are then converted to a hydrogel by bulk photopolymerization. Using this approach, Burdick et al. combined two monomer solutions (with and without RGDS) to create gradients of RGDS (Burdick et al., 2004). They demonstrated that endothelial cell surface attachment increased with increasing immobilized RGDS content in PEGDA hydrogels formed by free-radical photopolymerization. Furthermore, when two different solutions of varying PEGDA

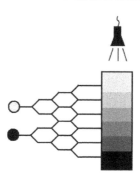

Fig. 6. Illustration of a microfluidic device for the generation of gradients in photopolymerized PEG hydrogel scaffolds.

molecular weight and concentration were injected into the microfluidic chamber, hydrogels with gradients of thickness were produced, suggestive of a gradient in crosslink density Other studies using microfluidic gradient makers have demonstrated directional fibroblast alignment (Guarnieri et al., 2008) and migration (Guarnieri et al., 2010) on PEGDA hydrogel surfaces with gradients of immobilized YRGDS. In the latter study, 2D cell migration was found to increase with increases in the slope of the gradient of immobilized RGD (Guarnieri et al., 2010).

4.3 Perfusion based frontal polymerization

In an effort to generate hydrogel scaffolds with controllable gradients of mechanical properties and incorporated biofunctionality, Turturro et al. have developed a novel free-radical photopolymerization technique, perfusion based frontal photopolymerization (PBFP) (Turturro & Papavasiliou, 2011), which is a variant of a previously developed frontal polymerization process (Chechilo et al., 1972). Originally reported in the 1970's for the synthesis of poly(methyl methacrylate) under high pressure, frontal polymerization is based on the generation of a localized reaction zone that propagates through a monomeric mixture in a fashion similar to a reaction wave (Alzari et al., 2009; Chekanov & Pojman, 2000). The reaction front is created by an external energy source, such as heat or light, and propagates through the system via consumption of the monomeric species. The PBFB technique is distinct from the previously published methods of frontal polymerization (Chechilo et al., 1972; Chekanov & Pojman, 2000; Pojman et al., 1996) in that it is based on the perfusion of a photoinitiator resulting in localized initiation and subsequent formation of crosslinked PEGDA hydrogels with gradients of mechanical properties and of immobilized YRGDS. In PBFP, gradients are generated by the controlled delivery of eosin Y, a photoinitiator, to the base of the monomeric solution via a perfusion pump and a glass frit filter disk. Due to density differences between the PEGDA precursor solution and the eosin Y photoinitiator, buoyant eosin Y rises to the surface and initiates propagation upon exposure to visible light (λ = 514 nm). Additional photoinitiator is trapped by the polymer layer causing a descending polymer front that is characteristic of frontal polymerization (Fig. 7). Through controlled delivery of the photoinitiator and alteration of the polymerization conditions, this technique allows for manipulation of the magnitude of the gradients generated and is capable of producing multiple gradients in photopolymerized PEGDA scaffolds. As shown

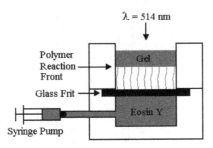

Fig. 7. Schematic of perfusion-based frontal polymerization.

in Figure 8, simultaneous gradients of elastic modulus and immobilized concentration of the YRGDS adhesion ligand are achieved using PBFP. *In vitro* cell studies in which aggregates of 3T3 fibroblasts are seeded on the surface of these hydrogels indicated that cells are capable of detecting and responding to the gradients generated via directional outgrowth. As shown in Figure 9, cells seeded on PBFP hydrogels spread roughly twice as far in the direction parallel to the gradient as in the perpendicular direction.

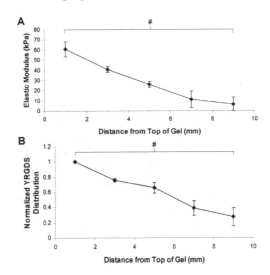

Fig. 8. Distribution of (A) elastic modulus and (B) immobilized YRGDS in PBFP PEGDA hydrogels (n=3; # = p < 0.001).

4.4 Microsphere packing

In addition to the above mentioned techniques, gradients in PEG hydrogels have also been formed by microsphere packing (Roam et al., 2010). This technique relies upon the formation of PEG microspheres of varying density by thermally induced phase separation. The microspheres of varying density are combined in a serum rich solution and centrifuged to form a density gradient. The presence of the serum proteins allows the microspheres to

Fig. 9. Aggregate of 3T3 fibroblasts 13 days post-seeding on a PFBP PEGDA hydrogel. The arrow indicates the direction of increasing elastic modulus and immobilized YRGDS concentration. Scale bar = 100 μm.

crosslink together and form solid hydrogels. Using this technique, PEG hydrogels containing gradients with either sharp interfaces or gradual transitions with up to 5 distinct layers were formed. This approach has been also used to fabricate macroporous PEG hydrogels with gradients in porosity and homogeneously distributed PEG microspheres loaded with RGD and sphingosine 1-phosphate that promoted endothelial cell infiltration through the scaffolds, but the effects of gradients on cell behavior have yet to be determined (Scott et al., 2010). This technique offers the potential for incorporating gradients of biofunctionality or even regions of different proteins/drugs that are capable of diffusing from a microsphere of a particular density and stimulating a desired cellular response.

The above mentioned techniques have proven successful in generating gradients of mechanical properties and biofunctional moieties that are capable of recreating the natural microenvironment of cells. However, the effects of PEG hydrogel gradients on cell behavior have been primarily limited to the cell-substrate interface and have not been quantified in 3D. Therefore the future of gradients will rest on the ability to combine gradient techniques with existing methods of incorporating degradable domains and in particular those susceptible to cell-mediated proteolysis within PEG hydrogels. Success in doing so will allow researchers to gain a better understanding of how the 3D spatial presentation of matrix signals influence cell behavior and tissue regeneration.

5. Prediction of scaffold properties via computational hydrogel models

In order to design a functional synthetic scaffold that supports 3D cell behavior, the mechanical properties, degradation kinetics, and spatial presentation of biochemical signals must be appropriately tuned prior to *in vitro* and/or *in vivo* cell studies (Tibbitt & Anseth, 2009). Polymerization and hence gelation processes are kinetically controlled and consequently the resultant polymer properties are highly dependent on polymerization conditions. In systems where multiple biofunctional signals are to be incorporated into crosslinked PEG hydrogel networks, the number of reactions between functional monomeric species increases, and optimization of the physical, mechanical, and biofunctional spatial and temporal presentation of these matrix cues can become challenging for engineering specific tissues. The future design of PEG scaffolds would greatly benefit from computational models of biofunctional hydrogel formation that accurately predict the final concentrations of immobilized ECM peptide signals as well as the mechanical properties as a function of space and time due to alterations in polymerization conditions (e.g., initial concentrations of biofunctional monomers, polymerization time, polymer content). The predictive capability of these models would provide alternative approaches to detailed experimental sensitivity analysis for optimization of cell-substrate interactions.

A computational model of hydrogel formation of multiarm comb PEG crosslinked with transglutaminase enzyme has been developed on the basis of the Flory-Stockmyer theory (Sperinde & Griffith, 2000). Gelation kinetics were predicted in terms of variations in the macromer structure and composition, stoichiometric ratios of reactants, and crosslinking enzyme concentration. Model predictions correlated well with gelation times with alterations induced via variations in enzyme concentration. These models, however, are based on statistical approaches that make it difficult to characterize the dynamics of the complex process of gel formation.

Other studies of computational modeling of free-radical photopolymerization of crosslinked PEGDA hydrogels formed via interfacial photopolymerization have predicted the formation of crosslink density gradients within planar hydrogel surfaces as a function of space and time (Kizilel et al., 2006). These models were based on previous models of gel formation occurring via free-radical crosslinking of vinyl and divinyl monomers developed by Gossage (Gossage, 1997). The models utilized the Numerical Fractionation (NF) technique which avoids the numerical divergence of the second and higher molecular weight moments at the gel point (Teymour & Campbell, 1992). Utilization of this technique in polymerization models of gelation has provided extensive insight of the resulting polymer molecular weight distribution, the dynamics of gelation, and the evolution of the crosslink density of the gel (Gossage 1997; Kizilel et al., 2007; Papavasiliou et al., 2002; Papavasiliou & Teymour, 2003; Teymour & Campbell, 1992). Computational models of interfacial polymerization of PEG hydrogels using the NF technique were later extended to account for the formation of crosslinked PEG hydrogel films immobilized with glucagon-like peptide-1 for the prediction of film thickness and crosslink density in designing membranes for the encapsulation of islet cells. While these hydrogel models have provided insight into the effects of polymerization conditions on hydrogel crosslinking, they have yet to be validated in terms of incorporated biofunctionality, and hydrogel crosslink density.

The future use and implementation of computational models of hydrogel formation for the design and optimization of scaffolds in tissue engineering will require that they be validated in terms of their mechanical and physical properties as well as the immobilized concentrations of multiple ECM signals. A critical parameter in validating these models will be the experimental determination of unknown kinetic rate constants that play a critical role in dictating the kinetics of polymerization. For hydrogels formed via free-radical photopolymerization, the incorporation of biofunctional-PEG macromer conjugates into PEG scaffolds formed using either Acryl-PEG-NHS chemistry or Michael-type addition reaction between bis-cysteine and PEGDA requires further optimization. These biofunctional macromer conjugation schemes can often result in multimodal molecular weight distributions (Miller et al., 2010) which can directly affect hydrogel crosslinking, gel degradation rate, and the resulting biomaterial properties. Therefore, computational modeling at the macromer level would provide significant insight in defining the synthesis conditions required to eliminate the need of undesired species that may compromise biomaterial properties.

6. Conclusion

Significant progress has been made using PEG hydrogel scaffolds as ECM mimics for supporting and directing cell behavior and tissue regeneration. Ongoing challenges however remain in PEG-based biomaterial strategies for the engineering of new tissues. Although numerous efforts have focused on investigating the effects of biological signal identity, gel degradation rate, and mechanical properties on cell behavior, little work has been done to independently tune these properties in order to isolate and quantify the individual effects of these factors on cell behavior. Coupling computational models with experimental data to identify conditions that result in independent tuning of these properties would provide significant insight on how PEG scaffolds could be better designed to enhance tissue regeneration prior to complete material degradation. The ability to stimulate neovascularization within PEG hydrogels is also critical for the clinical success of designing tissues that require the formation of a stable and functional vasculature. Research in this regard has focused on vascularization of small polymer scaffolds which do not represent the clinical volumes needed for the generation of vessels in large tissues and therefore, strategies for the design of scaffolds that promote vascularization of larger tissues are needed to address this limitation. The modulation of the spatial arrangement of ECM signals and scaffold mechanical and physical properties in PEG hydrogels in directing cell behavior in response to gradients has been primarily limited to the cell-substrate interface. Future studies quantifying the effects of 3D gradients of multiple matrix cues and how this influences cellular fate processes will provide significant insight in effectively designing PEG-based scaffolds with gradients for processes that require more rapid and guided cell migration and tissue regeneration. Finally, alternative polymer strategies where the delivery of growth factors and other biofunctional signals can be dynamically regulated by the scaffolds themselves and in combination with cellularly-mediated events are critical to the success of PEG hydrogels in tissue engineering. These design criteria need to be considered for the continued enhancement of these scaffolds for regenerative medicine applications.

7. Acknowledgement

The authors would like to acknowledge funding from the Pritzker Institute of Biomedical Science and Engineering at the Illinois Institute of Technology and the National Institutes of Health (Grant No. R21HL094916).

8. References

Aimetti, A.A.; Tibbitt, M.W., et al. (2009). Human neutrophil elastase responsive delivery from poly(ethylene glycol) hydrogels. *Biomacromolecules*, Vol. 10, No.6, pp. 1484-1489

Alzari, V.; Monticelli, O., et al. (2009). Stimuli responsive hydrogels prepared by frontal polymerization. *Biomacromolecules*, Vol. 10, No.9, pp. 2672-2677

Barkefors I.; Le Jan S., et al. (2008). Endothelial cell migration in stable gradients of vascular endothelial growth factor A and fibroblast growth factor 2 - Effects on chemotaxis and chemokinesis. *Journal of Biological Chemistry*, Vol. 283, No.20, pp. 13905-13912

Bott, K.; Upton, Z., et al. (2010). The Effect of Matrix Characteristics on Fibroblast Proliferation in 3D Gels. *Biomaterials*, Vol. 31, No.32, pp. 8454-8464

Alberts,B.; Johnson A.; Lewis, J.; Raff, M.; Roberts, K.; Walter, P. (2008). *Molecular Biology of the Cell* (5th Edition), Galrland Science, ISBN, 978-0-8153-4105-5, New York, New York

Burdick J.A.; Khademhosseini A., et al. (2004). Fabrication of Gradient Hydrogels Using a Microfluidics/Photopolymerization Process. *Langmuir*, Vol. 20, No.13, pp. 5136-5156

Chechilo, N.M.; Khvilivitskii, R.J., et al. (1972). On the Phenomenon of Polymerization Reaction Spreading. *Doklady Akademii Nauk SSSR*, Vol. 204, pp. 1180-1181

Chekanov, Y.A.&Pojman, J.A. (2000). Preparation of Functionally Gradient Materials via Frontal Polymerization. *J. Appl. Poly. Sci.*, Vol. 78, No.13, pp. 2398-2404

Chen R.R. & Mooney D.J. (2003). Polymeric Growth Factor Delivery Strategies for Tissue Engineering. *Pharmaceutical Research*, Vol. 20, No.8, pp. 1103-1112

Daley W.P.; Peters S.B., et al. (2008). Extracellular Matrix Dynamics in Development and Regenerative Medicine. *Journal of Cell Science*, Vol. 121, No.3, pp. 255-264

DeForest C.A.; Polizzotti B.D., et al. (2009). Sequential Click Reactions for Synthesizing and Patterning Three-Dimensional Cell Microenvironments. *Nature Materials*, Vol. 8, pp. 659-654

DeLong S.A.; Gobin A.S., et al. (2005). Covalent Immobilization of RGDS on Hydrogel Surfaces to Direct Cell Alignment and Migration. *Journal of Controlled Release*, Vol. 109, No.1-3, pp. 139-148

DeLong S.A., Moon J.J., et al. (2005). Covalently Immobilized Gradients of bFGF on Hydrogel Scaffolds for Directed Cell Migration. *Biomaterials*, Vol. 26, No.16, pp. 3327-3234

Elbert D.L.; Pratt A.B., et al. (2001). Protein Delivery from Materials formed by Self-Selective Conjugate Addition Reactions. *Journal of Controlled Release*, Vol. 76, No.1-2, pp. 11-25

Gerhardt H.; Golding M., et al. (2003). VEFG Guides Angiogenic Sprouting Utilizing Endothelial Tip Cell Filopodia. *The Journal of Cell Biology*, Vol. 161, No.6, pp. 1163-1177

Gobin A.S. &West J.L. (2002). Cell Migration Through Defined, Synthetic Extracellular Matrix Analogs. *The FASEB Journal*, Vol. 16, No.3, pp. 751-753

Gossage J.L.(1997). Numerical Fractionation Modeling of Nonlinear Polymerization, *PhD Thesis*, Department of Chemical Engineering, Insitute of Technology, Chicago, IL, USA

Guarnieri D.; Capua A.D., et al. (2010). Covalently Immobilized RGD Gradient on PEG Hydrogel Scaffold Influences Cell Migration Parameters *Acta Biomaterialia*, Vol. 6, No.7, pp. 2532-2539

Guarnieri, D.; Borzacchiello, A., et al. (2008). Engineering of Covalently Immobilized Gradients of RGD Peptides on Hydrogel Scaffolds: Effect on Cell Behavior. *Macromololecular Symposia*, Vol. 266, No., pp. 36-40

Hern, D.L. & Hubbell, J.A. (1998). Incorporation of Adhesion Peptides into Nonadhesive Hydrogels Useful for Tissue Resurfacing. *Journal of Biomedical Materials Research*, Vol. 39, No.2, pp. 266-276

Kantlehner, M.; Finsinger, D., et al. (1999). Selective RGD-Mediated Adhesion of Osteoblasts at Surfaces of Implants. *Angewandte Chemie*, Vol. 38, No.4, pp. 560-562

Kizilel. S.; Papavasiliou, G., et al. (2007). Mathematical Model for Vinyl-divinyl Polymerization. *Macromolecular Reaction Engineering*, Vol. 1, No.6, pp. 587-603

Kizilel, S.; Pérez-Luna, V.H., et al. (2006). Mathematical Model for Surface-initiated Photopolymerization of Poly(ethylene glycol) Diacrylate *Macromolecular Theory and Simulations*, Vol. 15, No.9, pp. 686-700

Kraehenbuehl, T.P.; Ferreira, L.S., et al. (2009). Cell-responsive Hydrogel for Encapsulation of Vascular Cells. *Biomaterials*, Vol. 30, No.26, pp. 4318-4324

Laughlin, S.T.; Baskin, J.M., et al. (2008). In Vivo Imaging of Membrane-Associated Glycans in Developing Zebrafish. *Science*, Vol. 320, No.5876, pp. 664-667

Lee, S.H.; Moon, J.J., et al. (2007). Poly(ethylene glycol) Hydrogels Conjugated with a Collagenase-sensitive Fluorogenic Substrate to Visualize Collagenase Activity During Three-dimensional Cell Migration. *Biomaterials*, Vol. 28, No.20, pp. 3163-3170

Leslie-Barbick, J.E.; Moon, J.J., et al. (2009). Covalently-immobilized Vascular Endothelial Growth Factor Promotes Endothelial Cell Tubulogenesis in Poly(ethylene glycol) Diacrylate Hydrogels. *Journal of Biomaterial Science Polymer Edition*, Vol. 20, No.12, pp. 1763-1779

Levi, E.; Miao, H.Q., et al. (1996). Matrix Metalloproteinase 2 Releases Active Soluble Ectodomain of Fibroblast Growth Factor Receptor 1. *Proceedings of the National Academy of Sciences of the United States of America*, Vol. 93, No.14, pp. 7069-7074

Lin, C.C. & Anseth, K.S. (2009). PEG Hydrogels for the Controlled Release of Biomolecules in Regenerative Medicine. *Pharmaceutical Research*, Vol. 26, No.3, pp. 631-643

Lo, C.M.; Wang, H.B., et al. (2000). Cell Movement is Guided by the Rigidity of the Substrate. *Biophysical Journal*, Vol. 79, No.1, pp. 144-152

Lutolf, M.P. & Hubbell, J.A. (2005). Synthetic Biomaterials as Instructive Extracellular Microenvironments for Morphogenesis in Tissue Engineering. *Nature Biotechnology*, Vol. 23, No.1, pp. 47-55

Lutolf, M.P.; Lauer-Fields, J.L., et al. (2003). Synthetic Matrix Metalloproteinase-sensitive Hydrogels for the Conduction of Tissue Regeneration: Engineering Cell-invasion Characteristics. *Proceedings of the National Academy of Sciences of the United States of America*, Vol. 100, No.9, pp. 5413-5418

Mac Gabhann, F.; Ji, J.W., et al. (2007). VEGF Gradients, Receptor Activation, and Sprout Guidance in Resting and Exercising Skeletal Muscle. *Journal of Applied Physiology*, Vol. 102, No.2, pp. 722-734

Malkoch, M.; Vestberg, R., et al. (2006). Synthesis of Well-defined Hydrogel Networks Using Click Chemistry. *Chemical Communications*, No.26, 2006, pp. 2774-2776

Miller, J.S.; Shen, C.J., et al. (2010). Bioactive Hydrogels Made from Step-growth Derived PEG-peptide Macromers. *Biomaterials*, Vol. 31, No.13, pp. 3736-3743

Moissoglu, K. & Schwartz, M.A. (2006). Integrin Signalling in Directed Cell Migration. *Biology of the Cell*, Vol. 98, No.9, pp. 547-555

Moon, J.J.; Lee, S.H., et al. (2007). Synthetic Biomimetic Hydrogels Incorporated with Ephrin-A1 for Therapeutic Angiogenesis. *Biomacromolecules*, Vol. 8, No.1, pp. 42-49

Moon, J.J.; Saik, J.E., et al. (2010). Biomimetic Hydrogels with Pro-angiogenic Properties *Biomaterials*, Vol. 31, No.14, pp. 3840-3847

Nemir, S.; Hayenga, H.N., et al. (2010). PEGDA Hydrogels with Patterned Elasticity: Novel Tools for the Study of Cell Response to Substrate Rigidity. *Biotechnolgy and Bioengineering*, Vol. 105, No.3, pp. 636-644

Nemir, S. & West, J.L. (2010). Synthetic Materials in the Study of Cell Response to Substrate Rigidity *Annals of Biomedical Engineering*, Vol. 38, No.1, pp. 2-20

Nuttelman, C.R.; Benoit, D.S.W., et al. (2006). The Effect of Ethylene Glycol Methacrylate Phosphate in PEG hydrogels on Mineralization and Viability of Encapsulated hMSCs. *Biomaterials*, Vol. 27, No.8, pp.

Papavasiliou, G.; Birol, I., et al. (2002). Calculation of Molecular Weight Distributions in Non-linear Free-radical Polymerization Using the Numerical Fractionation Technique. *Macromolecular Theory and Simulations*, Vol. 11, No.5, pp. 533-548

Papavasiliou, G.; Cheng, M., et al. (2010). Strategies for Vascularization of Polymer Scaffolds. *Journal of Investigative Medicine*, Vol. 58, No.7, pp. 838-844

Papavasiliou, G. &Teymour, F. (2003). Reconstruction of the Chain Length Distribution for Vinyl-divinyl Copolymerization using the Numerical Fractionation Technique. *Macromolecular Theory and Simulations*, Vol. 12, No.8, pp. 543-548

Papavasiliou, G.; Cheng, M.H., et al. (2010). Strategies for Vascularization of Polymer Scaffolds. *Journal of Investigative Medicine*, Vol. 58, No.7, pp. 838-844

Patino, M.G.; Neiders, M.E., et al. (2002). Collagen as an Implantable Material in Medicine and Dentistry. *Oral Implantology*, Vol. 28, pp. 220-225

Patterson, J. & Hubbell, JA (2010). Enhanced Proteolytic Degradation of Molecularly Engineered PEG hydrogels in Response to MMP-1 and MMP-2. *Biomaterials*, Vol. 31, No.30, pp. 7836-7845

Patterson J.; Martino, M.M., et al. (2010). Biomimetic Materials in Tissue Engineering. *Materials Today*, Vol. 13, No.1-2, pp. 14-22

Patterson, J. & Hubbell, J.A. (2011). SPARC-derived Protease Substrates to Enhance the Plasmin Sensitivity of Molecularly Engineered PEG hydrogels. *Biomaterials*, Vol. 32, No.5, (February, 2011), pp. 1301-1310

Phelps, E.A.; Landázuric N., et al. (2009). Bioartificial Matrices for Therapeutic Vascularization. *Proceedings of The National Academy of Sciences of the United States of America*, Vol. 107, pp. 3323-3328

Pojman, J.A.; Ilyashenko, V.M., et al. (1996). Free-radical Frontal Polymerization: Self-Propagating Thermal Reaction Waves. *Journal of the Chemical Society, Faraday Transactions* Vol. 92, No.16, pp. 2825-2837

Polizzotti, B.D.; Fairbanks, B.D., et al. (2008). Three-Dimensional Biochemical Patterning of Click-Based Composite Hydrogels via Thiolene Photopolymerization. *Biomacromolecules*, Vol. 2008, No.9, pp. 1084-1087

Pratt, A.B.; Weber, F.E., et al. (2004). Synthetic Extracellular Matrices for In Situ Tissue Engineering. *Biotechnology and Bioengineering*, Vol. 86, No.1, pp. 27-36

Raeber G.P.; Lutolf M.P., et al. (2005). Molecularly Engineered PEG hydrogels: A Novel Model System for Proteolytically Mediated Cell Migration. *Biophysical Journal*, Vol. 89, No.2, pp. 1374-1388

Rizzi, S.C. & Hubbell, J.A. (2005). Recombinant Protein-co-PEG Networks as Cell-Adhesive and Proteolytically Degradable Hydrogel Matrixes. Part I: Development and Physicochemical Characteristics. *Biomacromolecules*, Vol. 6, No.3, pp. 1226-1238

Rizzi, S.C.; Ehrbar, M., et al. (2006). Recombinant Protein-co-PEG Networks as Cell-Adhesive and Proteolytically Degradable Hydrogel Matrixes. Part II: Biofunctional Characteristics. *Biomacromolecules*, Vol. 7 No.11, pp. 3019-3029

Roam, J.L.; Xu, H. et al. (2010). The Formation of Protein Concentration Gradients Mediated by Density Differences of Poly(ethylene glycol) Microspheres. *Biomaterials*, Vol. 31, No.33, pp. 8642-8650

Rogers, M.S.; Birsner, A.E., et al. (2007). The Mouse Cornea Micropocket Angiogenesis Assay. *Nature Protocols*, Vol. 2, No.10, pp. 2545-2550

Ruoslahti, E. (1996). RGD and Other Regognition Sequences for Integrins. *Annual Review of Cell and Developmental Biology*, Vol. 12, pp. 697-715

Saik, J.E.; Gould, D.J., et al. (2011). Covalently Immobilized Platelet-derived Growth Factor-BB Promotes Angiogenesis in Biomimetic Poly(ethylene glycol) Hydrogels. *Acta Biomaterialia*, Vol. 7, No.1, pp. 133-143

Salinas, C.N;. & Anseth K.S. (2008). Mixed Mode Thiol-Acrylate Photopolymerizations for the Synthesis of PEG-Peptide Hydrogels. *Macromolecules*, Vol. 41, No.16, pp. 6019-6026

Saltzman, W.M. (2004). *Tissue Engineering. Principles for the Design of Replacement Organs and Tissues*. Oxford University Press, ISBN, 978-0-19-514130-6, New York, New York.

Schneider, L.; Cammer, M., et al. (2010). Directional Cell Migration and Chemotaxis in Wound Healing Response to PDGF-AA are Coordinated by the Primary Cilium in Fibroblasts. *Cellular Physiology and Biochemistry*, Vol. 25, No.2-3, pp. 279-292

Scott ,E.A.; Nichols, M.D., et al. (2010). Modular Scaffolds Assembled Around Living Cells Using Poly(ethylene glycol) Microspheres with Macroporation via a Non-cytotoxic Porogen. *Acta Biomaterialia*, Vol. 6, No.1, pp. 29-38

Seliktar, D.; Zisch, A.H., et al. (2004). MMP-2 sensitive, VEGF-bearing Bioactive Hydrogels for Promotion of Vascular Healing. *Journal of Biomedical Materials Research Part A*, Vol. 68A, No.4, pp. 704-716

Singer, A.J. & Clark, R.A. (1999). Cutaneous Wound Healing. *New England Journal of Medicine*, Vol. 341, No.10, pp. 738-746

Smith, J.T.; Kim, D.H., et al. (2009). Haptotactic Gradients for Directed Cell Migration: Stimulation and Inhibition Using Soluble Factors. *Combinatorial Chemistry & High Throughput Screening*, Vol. 12, No.6, pp. 598-603

Somerville, R.; Oblaner S., et al. (2003). Matrix Metalloproteinases: Old Dogs with New Tricks. *Genome Biology*, Vol. 4, No.216, pp. 1-11

Sperinde, J.J. & Griffith, L.G. (2000). Control and Prediction of Gelation Kinetics in Enzymatically Cross-Linked Poly(ethylene glycol) Hydrogels. *Macromolecules*, Vol. 33, No.15, pp. 5476–5480

Steffensen, B.; Häkkinen, L., et al. (2011). Proteolytic Events of Wound-Healing – Coordinated Interactions Among Matrix Metalloproteinases (MMPs), Integrins, and Extracellular Matrix Molecules. *Critical Reviews in Oral Biology and Medicine*, Vol. 12, No.5, pp. 373-398

Teymour, F. & Campbell, J.D. (1992). Analysis of the Dynamics of Gelation in Polymerization Reactors Using the Numerical Fractionation Technique. *Macromolecules*, Vol. 27, No.9, pp. 2460-2469

Tibbitt, M.W. & Anseth, K.S. (2009). Hydrogels as Extracellular Matrix Mimics for 3D cell Culture. *Biotechnology and Bioengineering*, Vol. 103, No.4, pp. 655-663

Turk, B.E.; Huang; L.L., et al. (2001). Determination of Protease Cleavage Site Motifs Using Mixture-Based Oriented Peptide Libraries. *Nature Biotechnology*, Vol. 19, No.7, pp. 661-667

Turturro, M.V. & Papavasiliou, G. (2011) Generation of Mechanical and Biofunctional Gradients in PEG Diacrylate Hydrogels by Perfusion-Based Frontal Photopolymerization. *Journal of Biomaterials Science Polymer Edition*, http://www.ingentaconnect.com/content/vsp/bsp/pre-prints/jbs3298)

Underhill, G.H.; Chen, A.A., et al. (2007). Assessment of Hepatocellular Function within PEG hydrogels. *Biomaterials 28, 256, 2007*, Vol. 28, No.2, pp. 256-270

Vu, T.H. & Werb, Z. (2000). Matrix metalloproteinases: Effectors of Development and Normal physiology. *Genes and Development*, Vol. 14, No.17, pp. 2123-2133

West, J.L. & Hubbell, J.A. (1999). Polymeric Biomaterials with Degradation Sites for Proteases involved in Cell Migration. *Macromolecules*, Vol. 32, No.1, pp. 241-244

Zhu, J. (2010). Bioactive Modification of Poly(ethylene glycol) Hydrogels for Tissue Engineering. *Biomaterials*, Vol. 31, No.17, 2010), pp. 4369-4656

Zisch, A.H.; Lutolf, M.P., et al. (2003). Cell-demanded Release of VEGF from Synthetic, Biointeractive Cell-ingrowth Matrices for Vascularized Tissue Growth. *The Journal of the Federation of Americal Societies of Experimental Biology*, Vol. 17, No.13, 2003), pp. 2260-2262

11

In Vitro Selection of Salt Tolerant Calli Lines and Regeneration of Salt Tolerant Plantlets in Mung Bean (*Vigna radiata* L. Wilczek)

Srinath Rao* and Prabhavathi Patil

Department of Botany, Gulbarga University, Gulbarga Karnataka,
India

1. Introduction

Salinity is one of the important abiotic factors in limiting plant productivity (Munns, 2002). 19.5% of the irrigated agricultural land is considered saline (Flowers & Yeo, 1995). Although minerals are essential for plants, their excess quantity in the soil is injurious to plants. Plants exposed to saline environment suffer from ion excess or water deficit and oxidative stress linked to the production of reactive oxygen species (ROS), which cause damage to lipids, proteins and nucleic acids (Hernandez, *et al.,* 2000). Oxidative stress is considered to be one of the major damaging factors in plant cells exposed to salinity (Gossette, *et al.,* 1994; Hernandez, *et al.,* 1995; Khan & Panda, 2002; Queiros *et al.,* 2007). The process of salt response and tolerance has been studied at the whole plant level (Hasegawa *et al.,* 2000; Jogeshwar, *et al.,* 2006). However the structural complexity of the whole plant makes it difficult to separate systemic from cellular salinity tolerance mechanism (Hawkins & Lips, 1997). The importance of plant tissue culture in the improvement of salt tolerance in plants has been pointed long back (Dix, 1993; Hasegawa, *et al.,* 1994; Nabors *et al.,* 1980; Tal 1994). In recent years tissue culture techniques are being used as a useful tool to elucidate the mechanism involved in salt tolerance by using *in vitro* selected salt tolerant cell lines (Davenport, *et al.,* 2003; Gu, *et al.* 2004; Lutts *et al.,* 2004 ; Naik & Harinath, 1998; Rao & Patil. 1999; Venkataiah. *et al.,* 2004). Besides, these lines have been used to regenerate salt tolerant plants (Chen *et al.,* 200; Jaiwal & Singh, 2001; Miki *et al.* 2001; Ochatt *et al.,* 1999; Rao & Krupanidhi, 1996; Shankhdhar, *et al.,* 2000). The selection of crop varieties for greater tolerance to saline environment will allow greater productivity from large saline lands.

2. Materials and methods

2.1 Callus induction and culture

Seeds of mung bean cultivar (cv) PS-16 were obtained from Agriculture Research Station Gulbarga, India. Seeds were washed thoroughly and imbibed in sterile water for six hours and then soaked in 70% (v/v) ethanol for 4 min and then surface sterilized for 1 min by 0.1%

---

* Corresponding Author

$HgCl_2$ (m/v). The seeds were germinated on sterile filter paper soaked in sterile water in culture tubes and allowed to germinate for 5 days at 16/8h (light/dark) photoperiod at 42 μ mol m^{-2} s^{-1} irradiance provided by cool fluorescent tubes and incubated at 26 ±1 ^{0}C. Cotyledons from 5 days old seedlings were used as source of explants for callus induction. The cotyledons were cut along the four side (i.e. leaf apex, petiole region and leaf margin of both the sides) with a sterile scalpel and cultured on Murashige and Skkog's (MS) medium supplemented with 9.09μM 2, 4-Dichlorophenoxy acetic acid (2,4-D) and 5.76μM kinetin(Kn) solidified with 8% (m/v) agar, before solidifying the pH of the media was adjusted to 5.6, the cultures were incubated under 16/8h (light/dark) photoperiod and allowed to grow for four weeks.

2.2 Selection of salt tolerant lines

Callus tissue grown on MS medium as described above was directly exposed to different concentrations of NaCl (50, 100, 150 and 200 mM) and the concentration at which growth was completely inhibited was determined. Cell survival was nonexistent at 200 mM. Consequently a concentration of 150 mM was used for selection and a portion of calli developed at this concentration was thereafter sub- cultivated on fresh MS medium containing same concentration (150mM) of NaCl and allowed to grow for 8 weeks. After this, calli were then transferred on MS medium without NaCl for 4 weeks. To test the stability of salt tolerant trait. Callus line growing healthily, on medium without NaCl were again transferred to MS medium containing 150 mM NaCl for 4 weeks, surviving callus lines showing good growth at this stage was considered to be NaCl tolerant. The salt tolerant calli was picked up   and were further grown for a period of 4 weeks and used for enzyme, protein and proline estimation.

2.3 Assessment of growth

Callus tissue (~250 mg) of both selected and non-selected were grown MS medium supplemented with 9.09 μM 2, 4-D and 5.76μM Kin containing 0-150 mM NaCl. The increase in fresh weight of the callus was determined after four weeks. Three replicates each consisting of 10 culture tubes were maintained per treatment, the relative growth rate of callus was calculated as the (FM_f - FM_i) /FM_i, where FM_f and FM_i were the final and initial fresh weight respectively. Samples of callus tissues were stored in – 80 °C for further analysis.

2.4 Lipid peroxidation

Lipid peroxidation was determined in terms of Malondialdehyde (MDA) content using the Thiobarbutric acid reaction (TBARS) according to Heath and Packer (1968). Frozen callus tissue was ground to a fine powder in liquid nitrogen with a mortar and pestle. Subsequently about l00 mg of powder was homogenized in 5 cm^3 of TCA (0.1% w/v) and centrifuged at 10,000g for l0min at 5 °C. 1 cm^3 of supernatant was mixed with 4 cm^3 of 0.5% TBA reagent in 20% TCA. The mixture was then heated at 95 °C for 30 min, cooled over ice and centrifuged at 10,000g for 10 min. The absorbance of the supernatant was recorded at 532 nm and corrected for non specific absorbance at 600 nm. MDA content was calculated using an extinction coefficient (ϵ) of 155 mM^{-1} $cm^{-1}$and and expressed as nmol g^{-1} FW.

2.5 Determination of ascorbate

Ascorbic acid (As A) was extracted in 5% (m/v) methophosphoric acid with sand at 4 °C. The homogenate was then centrifuged at 3000 g for 20 min at 4 °C Ascorbate content was quantified in the supernatant as described by Shukla, et al., (1979). An aliquot of 1cm³ of the supernatant was mixed with 2.5 cm³ of 1% (v/v) freshly diluted Folin - Ciocalteu reagent. The reaction mixture was allowed to stand for 40 min at room temperature, then the absorbance was recorded at 730 nm, using ascorbic acid as standard.

2.6 Superoxide dismutase activity (SOD) (EC 1.15.11)

The activity of SOD was determined based on its ability to inhibit the auto oxidation of pyrogallol. The measurement was based on the modified method of Markland and Markland, (1947). The reaction mixture consisted of 0.25 M pyrogallol in 0.1 M sodium phosphate buffer (pH 7.4) and 0.1 cm³ of enzyme extract. The reaction was initiated by light illumination and the rate of oxidation was measured spectrophotometrically at 429 nm as per the procedure described by Nakono & Asada, (1981). SOD activity is expressed in units/mg protein. One unit is defined as the amount of the enzyme which caused 50% inhibition of pyrogallol oxidation.

2.7 Ascorbate peroxidase activity (APX)

The activity of APX was determined by measuring the decrease in absorbance at 290 nm and the amount of Ascorbate oxidized to dehydroascorbate was calculated from the extinction coefficient 2.8 (Nakano and Asada, 1981). The reaction mixture (2 cm³) consisted of 25 mM phosphate buffer (pH 7.0), 0.1 mM EDTA, 0.25 mM ascorbic acid, 1.0 mM H_2O_2 and 0.2 cm³ enzyme extract.

2.8 Catalase activity (CAT; EC 1.11.1.6)

Catalase activity was measured according to method of Chandlee and Scandalios (1984). One gram of frozen callus was homogenized in a pre chilled pestle & mortar with 5 ml of ice cold 50 mM phosphate buffer. The extract was centrifuged at 4°C for 20 Min. at 12,500 X g. The supernatant was used for enzyme assay. The assay mixture contained 2.6 ml of 50 mM potassium phosphate buffer (pH 7.0) 0.4 cm² 15 mM H_2O_2 and 0.1 cm² of enzyme extract. The decomposition of H_2O_2 was followed by decline in absorbance at 240 nm.

2.9 Peroxidase activity (POX; EC 1.11.1.7)

Peroxidase activity was assayed by the method of Kumar and Khanna (1982). Assay mixture contained 2 ml of 0.1 M Phosphate buffer (pH 6.8) 1ml of 0.01M pyrogallol, 1 ml of 0.005 M H_2O_2 and 0.5 ml of enzyme extract. The solution was incubated for 5 min at 25 °C after which the reaction was terminated by adding 1 ml of 2.5N H_2SO_4. The purpurogallin formed was determined by measuring the absorbance at 420 nm against a blank prepared by adding the extract after the addition of 2.5N H_2SO_4 at zero times. The activity was expressed in unit mg⁻¹ protein. One unit (U) is defined as the change in the observance by 0.1 min⁻¹ mg⁻¹ protein.

2.10 Protein estimation

Samples of frozen callus tissue were ground in Tris-HCl (60 mM, pH 6.8) at 4°C, using prechilled mortar and pestle. The extract was centrifuged (4000g for 15 min at 4°C) and the supernatant was used for protein estimation according to Bradford (1976).

2.11 Proline estimation

Proline was extracted and determined calorimetrically by the method of Bates, *et al.,* (1973). Callus tissue (~200 mg) was homogenized with 4 cm³of 50 mM phosphate buffer (pH 7.8) containing 1% (w/v) polyvinyl pyrrolidone (PVP) 0.01% (w/v) Triton X-100, and centrifuged at 800 rpm for 15 min. Proline was determined in the supernatant by measuring the absorbance of proline-ninhydrin product formed, at 520 nm in a spectrophotometer using toluene as a solvent.

2.12 Regeneration of plantlets from NaCl tolerant calli

Plantlets were regenerated from salt tolerant and non-tolerant (control callus) callus. For regeneration of plantlets, callus was transferred in fragments of about 300 mg directly to MS medium supplemented with 0.5-2 mg/l BAP, alone or in combination with 0.25 and 0.5 mg/l NAA. After the shoots reached a height of 3-4 cm individual shoots were excised and transferred to rooting media which consisted of MS medium supplemented with 0.5 and 1 mg/l IBA.

2.13 NaCl tolerance test of regenerated plantlets

After the regenerated shoots produced roots, they were transferred to Erlenmeyer flasks containing the same rooting medium supplemented with 150 mM NaCl (Chen, et al., 2001). The survival rate of plantlets regenerated from non selected callus and NaCl selected callus was determined by using the formula

$$\frac{\text{The number of surviving plantlets} \times 100.}{\text{Total number of tested plantlets}}$$

2.14 Statistical analysis

The results presented are an average over three independent experiments. Data were processed with analysis of variance (ANOVA) and the means were compared using students test at a significance level of = 0.05.

3. Results and discussion

3.1 Establishment of salt tolerant lines

The NaCl tolerant mungbean cell lines obtained in this study were selected from callus cultures initiated from cotyledon explants. The direct recurrent selection procedures were successful in selection of mungbean cell lines exhibiting tolerance to 150 mM NaCl.

Callus tissue cultured on medium containing 50 mM NaCl showed good cell proliferation and appeared morphologically similar to control, where as callus tissue cultured on medium containing 100 mM NaCl showed moderate cell proliferation and small portions showing brownish color indicating early stages of necrosis. When callus tissue was cultured on medium supplemented with 150 mM NaCl, a great portion of the cells did not show any sign of growth and turned brown within 2 weeks of culture. However, small clusters of cells survived on 150 mM NaCl containing medium. Subsequently the cell clumps that proliferated in the presence of the salt were picked up and sub cultivated on NaCl free medium for 4 weeks. The salt tolerant cell lines were again transferred on to a fresh medium supplemented with 150 mM NaCl. On this medium cell clumps proliferated in to a mass of callus exhibiting friable nature. The selection processes resulted in establishment of NaCl-tolerant callus lines showing good proliferation of callus.

Cell lines exhibiting tolerance to NaCl have been previously selected from a range of legumes Gosal & Bajaj,1984) pasture legumes, (Smith & MeComb, 1981) chick pea, (Pandey & Ganapathy,1984,1985) Alfalfa (Wincicov, 1991) red gram (Rao & Krupanidhi,1996) pea (Ochatt, et al., 1999) soybean (Liu & Van Staden,2000) ground nut (Jain, et al., 2001). And in several other plant species such as, canola (Tyagi & Rangaswamy, 1997, Jain, et al., 1991a, b) finger millet, (Pius, et al. 1993) Cotton (Gosset,t et al., 1994) Potato (Olmos & Hellin, 1996; Queiros, et al., 2007) rice (Dang & Nuguyen, 2003; Shankdhar et al., 2000) Sunflower (Davenport, et al., 2003) chilli (Venkataiah, et al., 2004) sugarcane (Gandonou, et al., 2006).

3.2 Growth of callus

Growth rate of adapted callus line in NaCl supplemented media was steady and sustainable but it was lower than the growth of the control callus grown on stress-free medium.

Reduction in the growth is a common phenomenon in cultured cells grown on medium supplemented with NaCl (Cushman et al., 1990; Greenway & Munns, 1980; Jain et al., 1991 a, b; Luts et al., 2004; Rus et al. 1999; Shankdhar et al., 2000; Thomas et al., 1992; Venkataiah et al., 2004) and it has been interpreted that a certain amount of the total energy available for tissue metabolism is channeled to resist the stress (Cushman, et al., 1990). However 200 mM NaCl tolerant calli obtained either by direct or indirect selection process did not sustained a regular growth on salt supplemented media, these calli suffered with time a decrease in growth and all callus tissues died when the culture period was prolonged beyond one month. A similar response was reported in salt tolerant potato calli lines (Queiros, et al., 2007) and in Chrysanthemum sp. (Hossain, et al., 2007).

However, at a given concentration of NaCl, selected calli maintained higher growth rate than non-selected calli when grown on NaCl supplemented medium (Table-1). Similar to our findings, Kumar and Sharma (1989), Patil & Rao (1999) & Gulati & Jaiwal (2010) reported that NaCl tolerant lines maintained higher growth rate when compared to non selected calli in this species. Most of the previously published reports in legumes indicated that selected cell lines maintained higher rate of growth on saline media when compared to their wild type for example, Pea (Hassan & Wilkins, 1988; Olms, et al., 1995) red gram (Krupanidhi & Rao,1996) and in other plants like tobacco (Hascgawa et al., 1980) colt cherry (Ochatt & Power, 1989) rice (Basu, et al., 2002; Kishor & Reddy, 1986) finger millet (Pius, et al., 1993) sugar cane (Gandonou, et al., 2006).

Serial No	NaCl (µm)	non- selected callus (mg)	selected callus (mg)	MDA (nmol g^{-1}freshweight)	
				Non-selected callus (mg)	Selected callus (mg)
1	00 (Control)	778.0 ± 0.67[b]	780.0 ± 0.45[b]		
2	50	806.6± 0.34[a]	865.3 ±0.83[a]	84.0±0.75[c]	82.6±0.13[c]
3	100	647.6± 0.97[c]	740.0 ±0.56[c]	90.0 ± 0.86[c]	82.3±0.21[c]
4	150	449.3 ± 0.53[d]	614.6±0.36[d]	140.6 ±0.54[b]	88.0±0.43[b]

Data represents average of three replicates; each replicate consists of 10 cultures.
Mean ± Standard error.
Mean followed by the same superscript in a column is not significantly different at
P = 0.05 levels.

Table 1. Effect of NaCl on fresh weight and MDA content of selected and non-selected callus of *Vigna radiata* L.

3.3 Analysis of proline and protein contents

Proline and protein content in the selected and non-selected callus line is presented in table-2. Both protein and proline contents were significantly (p= < 0.05) affected with an increase in the concentration of NaCl. NaCl tolerant callus line maintained increased protein levels than control callus when exposed to 150 mM NaCl. Increased protein content in NaCl tolerant callus has been reported earlier (Chen, *et al.*, 2001; Queiros, *et al.*, 2007). Extra protein bands in *In Vitro* raised NaCl tolerant *V. radiata* plants have been reported by (Hassan, *et al.*, 2004). Similarly a gradual increase in the Proline content of tolerant callus line was noticed with an increase in the concentration of NaCl from 100 to 150 mM, maximum being at 150 mM NaCl level. (Kumar & Sharma, 1989; Patil & Rao, 1999) reported higher proline in selected calli when compared to non selected calli in this species. Accumulation of proline has been widely advocated for use as parameter of selection of cell lines for salt tolerance and it has been shown that proline over producing lines are more tolerant than their parent cultivars (Kavi Kishor *et al.*, 1995). Increased proline levels in NaCl tolerant callus has been earlier reported in many legumes like, soybean (Liu, & Van Staden, 2000) ground nut (Jain, *et al.*, 2001) and in other plants like tobacco (Wattad, *et al.*, 1983) canola (Jain, *et al.*, 1991 a,b; SashiMadan, *et al.*, 1995) tobacco (Gangopadhyay, *et al.*, 1997b) barley (Chaudhuri, *et al.*, 1997) *Citrus* (Piqueras, *et al.*, 1996) alfalfa (Petrusa, *et al.*, 1997) basmati rice (Basu, *et al.*, 2002) bamboo (Singh, *et al.*, 2003) chili (Venkataiah, *et al.*, 2004) sugarcane (Gandonou,, *et al.*, 2006). On the contrary some workers did not observe any appreciable increase in free proline content (Dix & Pearce, 1981, Errabii, *et al.*, 2006; Jain, *et al.*, 1987). Higher proline levels in salt tolerant callus may be due to an increased rate of synthesis or a decreased rate of oxidation of this compound as suggested by Wyn-Jones & Gorham (1984). Patnaik & Dabata (1997) suggested that salinity induced proline accumulation could be due to putrescence oxidation. There are divergent reports concerning the role of proline in salt tolerance. High levels of proline in NaCl selected cells have been suggested as a factor conferring salt tolerance (Kumar & Sharma, 1989; Pandey & Ganapathy, 1985; Wattad *et al.*, 1983). In contrast proline accumulation has also been excluded as the mechanism of tolerance in some cases (Dix & Pearce, 1981; Errabii *et al.*, 2007; Tal *et al.* 1979). The significant levels of proline observed in our study supports the former. It is well known fact that proline stabilizes the structure and function of various macromolecules (Kavi Kishor, *et al.*, 2005; Rhodes, 1987;Smirnoff & Cumbes, 1989).

In Vitro Selection of Salt Tolerant Calli Lines and Regeneration of Salt Tolerant Plantlets in Mung Bean (Vigna radiata L. Wilczek)

179

Seria l No.	NaCl (μm)	Protein mg g^{-1} FW		Proline mg g^{-1} FW		Ascorbic acid mg g^{-1} FW	
		Non-selected callus	Selected callus (mg)	Non-selected callus	Selected callus (mg)	Non-selected callus	Selected callus (mg)
1	00 (Control)	22.6 ±0.43c	27.2 ± 0.76c	7.6±0.53c	8.8 ± 0.54d	8.4 ± 0.44a	22.9 ± 0.45c
2	50	28.2 ±0.87b	42,0±0.56b	10.5 ±0.19b	18.2 ±0.76b	8.6±0.35a	26.5 ± 0.45b
3	100	32.2 ± 0.56a	47.0 ± 0.23a	16.2±0.65a	28.2±0.55a	6.4±0.87b	42.7±0.50a
4	150	16.4±0.32d	19.2 ±0.65d	5.9 ±0.25d	12.2±0.22d	5.5 ± 0.43c	27.1±0.25b

Data represents average of three experiments; Mean± standard error
Mean followed by the same super script in a column is not significantly different at P= 0.05 level.

Table 2. Effect of NaCl on protein, proline, and Ascorbic acid content in non-selected and selected callus on *Vigna radiate* L.

3.4 Lipid peroxidation

Oxidative damage to lipids was determined as lipid peroxidation in terms of amount of malondialdehyde (MDA). MDA content of tolerant callus was less when compared to that of sensitive callus line (Table-1). Sumithra, *et al.*,2006) reported low levels of lipid peroxidation in salt tolerant *V. radiata* cv Pusa bold than in salt sensitive cv CO 4. Jain, *et al.*, (2001) reported 4 fold increases in MDA content of sensitive callus of ground nut when grown on saline medium where as in the tolerant callus line the increase was only 1.1 fold. The results indicate that NaCl tolerant callus maintained membrane integrity when growing in saline environment however sensitive calls line were unable to maintain membrane integrity under salinity stress resulting in decreased growth and metabolic imbalance. Changes in the cell wall have been shown to be important for salt adaptation (Binzel, *et al.*, 1985; Curz, *et al.*, 1992). Salt stress is known to result in extensive lipid per oxidation (Davenport, *et al.*, 2003; Hernandez, *et al.*, 2000; Khan & Panda, 200;, Kholova *et al.*, 2009; Queiros *et al.*, 2007). In the present investigations it was noticed that the extent of lipid peroxidation was higher in NaCl sensitive cell lines than NaCl adopted cell lines, such observations were also made in ground nut (Jain *et al.*, 2001) wheat (Sarin, *et al.*, 2005) eggplant (Yaser, *et al.*, 2006). It seems that due to low peroxidation in NaCl tolerant callus line, membrane integrity might have been maintained, thus preventing protein denaturation which is supported by the fact that NaCl tolerant callus lines had higher protein content. Lipid membranes are vulnerable targets for stress induced cellular damage and the extent of damage is commonly used as a measure of stress (Gadallah, 1999; Zhou, *et al.*, 1992).

3.5 Ascorbic acid content

Ascorbic acid content increased gradually in NaCl tolerant calli line than in the control line. An increase of about 18% was found in callus tissue grown at 50 mM NaCl, while in callus grown on medium supplemented with 100 and 150 mM NaCl ascorbic acid content significantly increased to 30% and 59% respectively (Table-2). It was previously reported that the response of many plants cells to salinity is the increased synthesis of ascorbic acid. A positive correlation between ascorbic acid content and salinity was also reported in NaCl tolerant cell lines in several species (Gosset, *et al.*, 1996; Sarin, *et al.*, 2005; Qlmos & Hellin, 1996; Queiros, *et al.*, 2007). This antioxidant compound is one of the most effective free radical scavengers implicated in the adaptations of plants to stress (Shigeoka, *et al,.* 2002; Smirnoff, *et al.*, 2000; Queiros, *et al.*, 2007).It is reported that, high .levels of endogenous ascorbate is essential to effectively maintain the anti-oxidant system that protects plants from oxidative damage due to abiotic stress (Shigeoka, *et al.*, 2002).

3.6 Antioxidant enzymes

NaCl stress affect plant processes that lead to the formation of reactive oxygen species (ROS) super oxide radical (O_2) Hydrogen peroxide (H_2O_2) and hydroxide radicals (OH). The ROS cause oxidative damage to membrane lipids, proteins and nucleic acid. Accelerated detoxification is fundamental in development of tolerance to NaCl. To control the levels of ROS and protect the cells from injury under stress conditions, it is important that ROS should be scavenged.

Plants protect cells and sub cellular systems from the effects of reactive oxygen species (ROS) by enzymes such as Super oxide dismutase (SOD) Ascorbate peroxidase (APX) and Catalase (CAT). In the present investigation it was noticed that NaCl significantly altered the activities of SOD, APX & CAT with an increase in the levels of NaCl from 50 to 150 mM NaCl level. Maximum activity of these enzymes (Table-3) was noticed in callus tissue tolerant to 150 mM NaCl, where as in sensitive callus the activity of these enzymes was highly inhibited.

Serial No.	NaCl (μm)	SOD (EU mg^{-1} protein)		APX (EU mg^{-1} protein		CAT (EU mg^{-1} protein)	
		Non-selected callus (mg)	Selected callus (mg)	Non-selected callus (mg)	Selected callus (mg)	Non-selected callus (mg)	Selected callus (mg)
1	00 (Control)	40.0 ± 0.76^b	40.6±0.34^c	0.95 ±0.23^b	0.88 ± 0.67^d	22.3 ± 0.56^a	22.9±0.41.6^c
2	50	42.8 ± 0.45^a	50.5 ± 0.56^b	1.40±0.12^a	2.10 ±0.56^b	20.9 ± 0.75^b	26.5 ± 0.45^b
3	100	38.2 ± 0.36^c	75.8 ± 0.65^a	0.71 ±0.34^c	2.94 ± 0.67^a	16.4 ± 0.67^c	42.7± 0.50^a
4	150	32.6±0.65^d	50. 6 ± 0.48^b	0.58 ± 0.56^b	1.80 ± 0.56	12.6 ± 0.45^d	27.1± 0.25^b

Data represents average of three experiments. Mean ± standard error.
Mean followed by the same superscript in a column is not significantly different at
P= 0.05 levels.

Table 3. Effect of NaCl on antioxidant enzymes in non-selected and selected callus in *Vigna radiata* L.

In Vitro Selection of Salt Tolerant Calli Lines and Regeneration of Salt Tolerant Plantlets in Mung Bean (Vigna radiata L. Wilczek)

181

Serial No.	Growth regulators (mg/l)	Frequency of shoot induction (%)	Number of shoot buds per explants.	Average length of plant let (cm)
1	BAP 0.5	0.0	00	00
2	BAP 1.0	21.66 ± 1.6 a	2.90 ± 0.16 a	3.8 ±0.25 a
3	BAP 2.0	26.00 ± 1.4 b	8.20 ± 0.16 b	5.8 ± 0.34 b
4	BAP 2+NAA0.25	30.10 ± 1.8 c	12.80 ± 0.34 c	7.2 ± 0.55 c
5	BAP 2 +NAA 0.5	40.13 ± 1.8 d	14.80± 0.64 d	8.6 ± 0.68 c

Data represents average of three experiments. Mean ± standard error.
Mean followed by the same superscript in a column is not significantly different at P= 0.05 levels.

Table 4. Effect of growth regulators on Frequency, Number of shoots, and mean shoot length from cotyledon derived callus in *Vigna radiata*.

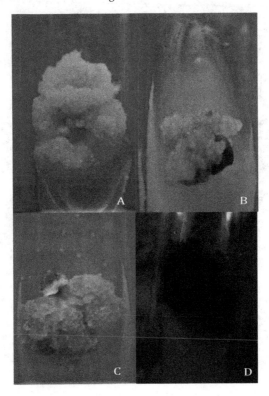

A) Callus on NaCl-free medium.

B) Surviving callus on 150mM NaCl.

C) NaCl-tolerant callus growing on 150mM NaCl.

D) NaCl sensitive callus on 150mM NaCl supplemented medium (Note complete darkening and death of Callus.

Bar ▭ =1 cm

Fig. 1. Photographs showing effect of NaCl on callus cultures of *Vigna radiata*

SOD, APX and CAT are considered as useful enzymes to help plants defend salt stress (Koca, *et al.* 2006,Shen, *et al.*, 2002). Sumithra *et al.*, (2006) have reported increased SOD and CAT activity in NaCl tolerant cultivars in this species. There are previous reports indicating increased activity of SOD in pea (Hernandez, *et al.*, 1993) wheat (Sairam, *et al.*, 2003) cotton (Gossett, *et al.*, 2004) potato (Queiros, *et al.*, 2007) sweet potato (He, *et al.*, 2008). Apart from SOD activity considerable increase in the activity of the enzymes APX and CAT was noticed in tolerant callus line than in sensitive callus when grown on different concentrations of NaCl. increase in the activities of SOD, CAT, APX in callus tissue raised from NaCl tolerant callus in cotton than in plants obtained from sensitive callus was reported by Gossett, *et al.*, (2004). EIkahoui, *et al.*, (2004) reported increased APX activities in NaCl adopted cell lines in periwinkle and Mittova, *et al.*, (2002) in salt tolerant wild tomatoes.

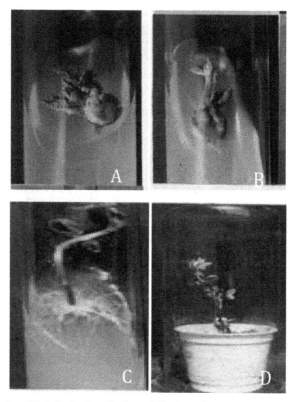

A)   Initiation of multiple leafy shoot buds on MS +2 mg/l BAP + 0.5 mg/ NAA

B)   Further elongation of shoot after 25 days of culture

C)   Rooting of the shoot on MS + 1 mg/l IBA.

D)   A potted plant in a polycup containing 3 : 1 garden soil and sand.

Fig. 2. Photographs showing organogenesis from NaCl tolerant callus

Catalase is a common enzyme found in all living organisms which are exposed to oxygen, where it functions to catalyze the decomposition of H_2O_2 to water and oxygen (Chelikani, *et al.*, 2004). H_2O_2 is a harmful by product formed as a result of stress conditions to prevent damage it must be quickly converted into less dangerous substance like gaseous oxygen and water molecules. Hence the role of catalase is important in reducing stress related damages. Willekens, *et al.*, (1997) suggested that the function of catalase in the cell is lo remove the bulk of H_2O_2 where as peroxidase would be involved mainly in scavenging H_2O_2 that is not taken by catalase. In the present investigations it was noticed that both catalase and peroxidase activity increased in NaCl tolerant callus. From this observation it can be inferred that catalase/peroxidase might have acted co-operatively to remove H_2O_2 at a minimum expense of reducing power.

3.7 Regeneration of plant lets from NaCl selected and non selected callus

For regeneration of plant lets calli pieces approximately 250±10 mg were transferred to MS medium supplemented with 0.5-2 mg/l BAP alone or in combination with 0.25 and 0.5 mg/l NAA. On MS medium without growth regulators and 0.5 mg/l shoot bud initiation was not observed; however on MS medium supplemented with 1 & 2 mg/l BAP shoots were initiated and the frequency of shoot formation was 21.66 and 26.00% respectively. Supplementing NAA at0.25-0.5 mg/l further enhanced the frequency of shoot formation increased to 30.10 & 40.13%respectively.

Regeneration from various seedling explants viz shoot tip, cotyledons, cotyledonary node of mung on BAP supplemented medium has been reported earlier (Mathews, 1987; Gulati & Jaiwal, 1992; Kaviraj & Rao, 2009; Mendoza, et al., 1999).there is only one report of regeneration from callus cultures in mungbean (Rao, *et al.*, 2005), similarly plantlets have been regenerated from NaCl tolerant callus in many species like *Citrus sinensis*(Ben-Hyyim & Goffer,1989) *Brassica juncea* (Jain, *et al.*,1991,Kerthi, et al., 1989; Sashimadan, *et el.*, 1995) *Triticum durum*(Zair, *et al.*, 2003) *Ipomea batats*(He, *et al.*, 2009) In the present study however, it was noticed that the formation of shoots from NaCl tolerant callai was lower (14.00% than that from NaCl-non selected calli (40.13%) and the number of multiple shoots per callai piece was reduced to 3.33 from 14.80, a phenomenon reported earlier by few workers(Basu, et al., 1997;Chen *et al.*, 2001; Ochatt, *et al.*,1999; Queiros, *et al.*, 2007; Zair *et al.*, 2003).

4. Conclusions

In conclusion NaCl stress induces sever oxidative stress in mungbean callus where the antioxidant defense system seemingly fails to combat with the stress induced oxidative damage. However it is possible to select callus line tolerant to elevated levels of NaCl stress by sudden exposure to high concentration of NaCl, accordingly a NaCl tolerant cell line was selected from cotyledon derived callus of mung bean which proved to be a true cell line variant. It can also be concluded that, the salt tolerant cell line could overcome the adverse effect of NaCl and maintained better growth on NaCl supplemented medium when compared to salt sensitive cell lines (controls). This conclusion is based on the following observations (a) cells which have been removed from the selection pressure for at least four passages retained tolerance to NaCl after transferring to NaCl (150 mM) medium (b) High accumulation of proline was noticed in salt tolerant cell lines compared to salt sensitive cell lines. (c)The activity of anti-oxidant enzymes was high. The results presented in this paper suggest that the mechanism of enhanced tolerance in mungbean callus is via improved synergistic and protective effects of antioxidant enzyme like SOD, APX and CAT. It is also concluded that the callus lines were able to grow in elevated concentration (150mM NaCl) by maintaining membrane integrity, high levels of protein and proline ,which supports the hypothesis that proline plays a significant role in protecting the cells from oxidative stress as reported in other grain legumes like chickpea and pigeonpea.

5. References

Al-Naggar, AMM. Saker, MM. Shabana, R. Ghanem, SA. Reda, AH. & Eid, S. 2008 *In vitro* selection and molecular characterization of salt tolerant canola plantlets *Arab.J Biotech* 11(2): 207-218,

Basu, S., Gangopadhyay, G., Gupta, S. & Mukherjee, BB. 1996. Screening for cross tolerance against related osmotic stress in adapted calli of salt sensitive and tolerance varieties of rice. *Phytomorphology.* 46(4): 357-364

Basu, S.,gangopadhyay, G.,Mukherjee, BB. & Gupta, S. Plant regeneration of salt adapted callus of indica rice (var.Basmati370) in saline conditions. *Plant Tissue & Organ Culture* 50: 153-159.

Ben-Hayyim, G. &. Kochba, J. 1983 Aspects of salts Tolerance in a NaCl-Selected Stable Cell Line of *Citrus sinensis. Plant Physiol.* 72: 685-690

Bhaskaran, S., Smith, RH. & Schertz, K. 1983 Sodium chloride tolerant callus of *Sorghum bicolor* (L.). Z *Pflanzenphysiol.* 112: 459-463

Binzel, ML. Hasegawa, PM. Rhodes, RH. Handa, D. Handa, AK. & Raym, AB. 1987 Solute Accumulation in Tobacco Cells adapted to NaCl. *Plant Physiol.* 84: 1408-1415

Binzel, MC. Hasegawa, PM. Handa, AK. & Bressan, RA. 1985 Adaptation of tobacco cells to NaCl. *Plant Physiol.* 79: 118-125

Borochov, NH. & Borochov, A. 1991. Response of melon plants to salt 1: Growth. Morphology and root membrane properties", *J. Plant Physiol.* Vol. 139: Pp. 100-105

Bradford, M.M.: A rapid and sensitive method for quantification of microgram quantities of protein utilizing the principle of protein dye binding. *Anal Biochem* 72: 248-254 1976

Chaudhuri P Sengupta RK Ghosh P 1997 *In Vitro* assessment of salinity tolerance in *Hordeum vulgare. Indian J Plant Physiol* 2 (2): 123-126

Chen, R., Gyokusen, K. & Saito, A. 2001. Selection, regeneration and protein profile characteristics of NaCl-Tolerant callus of *Robinia pseudoacaia* L. *J For Res* 6: 43-48

Croughan, TP., Stavarek, SJ., Rains, DW. 1978. Selection of a NaCl tolerant line of cultured alfalfa cells. *Crop Sci* 18: 959-963

Dix, PJ. 1993. The role of mutant cell lines in studies on environmental stress tolerance an assessment. Plant J 3: 309-313

Dix, PJ. & Pearce, RS. 1981 Proline accumulation in NaCl-resistent and sensitive lines of *Nicotiana sylvestris. Zeitschrift fur Planzen-schuts Physiologie* 102: 243-248

Datta, KS., Kumar, A., Varma, SK. & Angrish, R. 1996 Effects of salinity on water relations and ion uptake in three tropical forage crops. *Indian J Plant Physiol* 1: 102-108

Errabii, T., Gandonou, CB., Eassalmani, H. Abrini, J. Idaomar, M. & Skali-Senhaji, N. 2007. Effect of NaCl and mannitol induced stress on sugarcane (Saccharum Sp.) callus cultures. *Acta Physiol Plant* 29: 95-102

In Vitro Selection of Salt Tolerant Calli Lines and Regeneration of Salt Tolerant Plantlets in Mung Bean
(Vigna radiata L. Wilczek)

185

Gadallah MAA 1999 Effect of proline and glycine-betaine on Visia faba responses tosalt stress. *Biol Plant* 42: 247-249,

Gandonou CB Errabii T Abrani J Idamora M Skali-senhaji 2006 N Selection of callus cultures of sugarcane (Saccharum Sp.) tolerant to NaCl and their response to salt stress. *Plant Cell Tissue Organ Cult.* 87: 9-16

Gangopadhyay G Basu S Mukherjee BB and Gupta S 1997 Effects of Salt and osmotic Shocks on unadapted and adapted callus lines of tobacco. *Plant Cell, Tissue Organ Cult.* 49: 45-52

Gosal SS and Bajaj YPS 1984. Isolation of sodium chloride resistant cell lines in some grain legumes. *Indian Journal of Experimental Biology* 22: 209-214

Gosset DR Millhollon EP Cranlucs M Banks SW and Marney M 1994 The effects of NaCl on antioxidant enzyme activities in callus tissue of salt tolerant and salt sensitive cotton cultivars. *Plant Cell Rep.* 13: 498-503

Gossett DR Millhollon EP Lucas MC 1994 Antioxidant response to NaCl stress in salt-tolerant and salt-sensitive cultivars of cotton. *Crop Sci* 34: 706-714

Greenway H Munns R 1980 Mechanism of salt tolerance in non- halophytes. *Ibid.* 31: 149-190

Gu, R., Liu,Q., Pie, D. & Jiang, X. 2004 Understanding saline and osmotic tolerance of *Populus euphratica* suspended cells. *Plant Cell Tissue & Organ Culture* 78: 261-265

Gulati, A. & Jaiwal,PK. 1992 *In Vitro* induction of multiple shoots and regeneration from shoot tips of mung bean (*Vigna radiata* L. Wilczke). *Plant Cell Tissue & Organ Culture* 29: 199-205

Hasegawa, PM., Bressan. RA., Nelson, DE., Samaras, Y & Rhodes D. 1994. Tissue culture in the improvement of salt tolerance in plants. In: Monographs on Theoretical and Applied Genetics: Breeding plants with resistance to problem soils

Hasegawa, PM., Bressan, RA. & Handa, AK. 1980 Growth characteristics of NaCl selected and non-selected cells of *Nicotiana tabacum* L. *Plant and Cell Physiol.* 21 (8): 1347-1355

Hassan NS and Willkins DA 1988 *In vitro* selection for salt tolerant lines in *Lycopersicon peruvianum. Plant Cell Reports.* 7: 463-466

Hassan NM Serag MS El-Feky FM2004 Changes in nitrogen content and protein profiles following In vitro selection of NaCl resistant mung bean and tomato. *Acta Physiologiae Plantarum.* 26(2): 165-175

Hawkins HJ Lips SH1997 Cell suspension cultures of *Solanum tuberosum* L. as a model system for N and salinity response. Effect of salinity on NO_3^- uptake and PM-ATPase activity. *J Plant Physiol* 150: 103-109

Heath RL Packer L1968 Photo peroxidation in isolated chloroplasts 1. Kinetics and stoichiometry of fatty acid peroxidation. *Arch Biochem Bio Phys* 125: 189-198

Hernandez, J.A., Olmos, E., Corpas,F.J., Sevilla,F.,Del Rio,LA Salt induced oxidative stress in chloroplasts of pea plants *Plant Sci.* 105: 151-167,1995.

Hernandez JA Jimenez A Mullineaux P Sevilla F 2000 Tolerance of pea (*Pisumsativum* L.) to long term salt stress is associated with induction of antioxidant defenses. *Plant Cell Environ.* 23: 853-862

He, S., Han, Y. Zhai, H. & Liu, Q. 2007 *In vitro* selection and identification of sweetpotato (*Ipomoea batatas* (L.) Lam.) Plants tolerant to NaCl. *Plant Cell Tissue Orga Cult.* 96:69-74

Hossain, Z., Azad, AK., Datta, M SK. & Biswas, AK. 2009 Development of NaCl-tolerant line in *Chrysanthemum morifolium* Ramat. Through shoot organogenesis of selected callus line. *Biotechnology.* 129(4): 658-667

Jain, RK., Dhawan, RS., Sharma, DR. & Chowdhury, JB. 1987 salt tolerance and proline accumulation: a comparative study in salt tolerant and wild type cultured cells of egg plant. *Plant Cell Reports* 6: 382-384

Jain, M., Mathur, G., Koul, S. & Sarin,NB. 2001Ameliorative effects of proline on salt stress-induced lipid peroxidation in cell lines of groundnut (*Arachis hypogaea* L.) *Plant Cell Rep* 20: 463-468

Jaiswal R Singh NP 2001 Plant Regeneration from NaCl Tolerant Callus/Cell Lines of Chickpea. *ICPN* 8: 21-23

Karadimova M Djambova G 1993 Increased NaCl - tolerance in wheat (*Triticum aestivm* L. and *T. durum desf.*) through *in vitro* selection. *In vitro Cell Dev Biol* 29: 180-182.

Kavi Kishor, PB & Reddy, GM. 1985. Resistance of rice Callus Tissues to Sodium Chloride and Polyethylene Glycol. 59:1129-1131

Kavi Kishor, PB. ; Sangam. S.; Amrutha, RN.; Sri Laxmi, P. ; Naidu, KR.; Rao, KRSS., Rao,S. Reddy, KJ., Theriappan & Sreenivasulu, N. 2005 Regulation of proline biosynthesis, degradation, uptake and transport in higher plants: Its implications in plant growth and abiotic stress tolerance. 88(3): 424-43.

Kaviraj, CP. & Rao, S. 2009. High frequency of plant regeneration from cotyledonary node explants of green gram (*Vigna radiata* L.)-A recalcitrant grain legume. *The Bioscan* 4(2): 267-271.

Kumar V Sharma DR Isolation and characterization of sodium chloride resistant callus cultures of *Vigna radiata* (L.) Wilczek. var. radiata. *J Expt Botany* 40: 143-147

Kumar KB Khan PA Peroxidase and Polyphenol oxidase in excised ragi (*Elaucine coracona* cv. 2020 leaves during senescence Indian *J Expt Botany.* 20: 412-416 1982

Liu, T & Vanstadn, J. 2000 Selection and characterization of sodium chloride-tolerant callus of *Glycine max* (L.) Merr cv. Acme. Plant Growth Regulation 31: 195-207

Lutts, S., Majerus, V. & Kinet, JM. 1999. NaCl effects on proline metabolism in rice (*Oryza sativa*) seedlings. *Physiol. Plant.* 105: 450-458.

Miki Y Hashiba M Hisajima S 2001 Establishment of salt stress tolerant rice plants through step-up NaCl treatment *in vitro.* Biol. Plant. 44: 391-395

Mathews, H., 1987 Morphogenetic responses from *In Vitro* cultured seedling explants of mungbean (*Vigna radiata* L. Wilczek) plant cell Tissue & organ Culture. 11, 233-240.

Murashige, T. & Skoog, F. 1962 A revised medium for rapid growth and bio assays with tobacco tissue cultures. *Plant Physiol.* 15: 473-497,

Nabors, MW., Gibbs, SE., Bernstein, CS. & Meis ME. 1980. NaCl tolerant plants from cultured cells. Z. *Pflanzenphysiol.* 97: 13-17.

Naik, GR. & Harinath Babu, K. 1987 *In vitro* response of sugarcane to salinity tolerance. 37th convention D.S.T.A. 187-194.

Nakona, Y., Asada, K.: Hydrogen peroxide is scavenged by ascorbic specific peroxidase in spinach chloroplasts. *Plant Cell Physiol.* 22: 867-880, 1981.

Ochatt, SJ ; Marconi, P.L., Radice, S., Arnozis, P.A., Caso, O.H.: *In vitro* recurrent selection of potato: production and characterization of salt tolerant cell lines and plants. *Plant Cell, Tissue and Organ Culture.* 55: 1-8, 1999.

Pandey, R., Ganapathy, P.S.: Effect of Sodium Chloride stress on callus cultures of *Cicer arietinum* L. Cv. BG. 203: Growth and Ion accumulation. *Journal. Exp. Botany.* 157 (35): 1194-1197, 1984

Patil, P. & Rao S. 1999 Selection and characterization of NaCl tolerant callus cultures of *Vigna radiate* (L.) Wilczek. *Plant Tissue Culture and Biotechnology Emerging Trends.*

Petrusa, LM. & Winicov, I. 1997 Proline status in salt tolerant and salt sensitive alfalfa cell line and adapted plants in response to NaCl. *Plant Physiol Biochem* 35: 303-310.

Pius, J., Eapen, S., George, L. & Rao, P.S. Isolation of sodium chloride tolerant cell lines and plants in finger millet. *Biologia Plantarum.* 35(2): 267-271 1993.

Queiros, F., Fidalgo, F., Santos, I., Salema, R. *In vitro* selection of salt tolerant cell lines in *Solanum tuberosum* L. *Biologia Plantarum.* 51(4): 728-734, 2007.

Rao Srinath, Patil, P. 1998 Effect of sodium chloride on callus culture of *Cicer arietinum* L. *Bulletin of pure and Applied Sciences.* 17B (2): 81-87.

Rao, Srinath, Patil, P. & Kaviraj, CP. 2005 Callus induction and organogenesis from various explants in *Vigna radiata* (L.) Wilczek. *Indian Journal of Biotechnology.* 4: 556-560.

Rout, G.R., Senapatil, S.K., Panda, J.J.: Selection of Salt tolerant Plants of *Nicotiana tabacum* L. Through *In vitro* and its biochemical characterization. *Acta Biologica Hungarica.* 59(1): 77-92, 2008.

Rhodes, D., Verslues, P.E., Sharp, R.E.: Role of amino acids in abiotic stress resistance. In: Singh, B.K. (Ed.) Plant amino acids: Biochemistry and Biotechnology. Arcel Dekker, N.Y. pp 319-356.

Rus, A.M., Panoff, M., Perez-Alfocea, F., Bolarin, M.C.: NaCl response in tomato calli and whole plants. *J. Plant Physiol.* 155: 727-733, 1999.

Sairam, R.K., Srivastava, G.C., Agarawal, S., Meena, R.C. Differences in antioxidant activity in response to salinity stress in tolerant and susceptible wheat genotypes. *Biologia Plantarum* 49 (1): 85-91, 2005.

Shigeoka, S., Ishikawa, T., Tamoi, M., Miagawa, Y., Takeda,T., Yabuta, Y., Yoshimora, K.: Regulation and function of ascorbate peroxidase isoenzymes. *J. Exp.Bot.* 53: 1305-1319, 2000.

Shashi Madan, Nainawatee, HS., Jain, RK. & Chowdhury, JB. 1995. Proline and Proline metabolizing enzymes *in-vitro* Selected NaCl-tolerant *Brassica juncea* L. under Salt Stress. *Annals of Botany.* 76:51-57

Singh, M., Jaiswal, U., Jaiswal, V.S.: *In Vitro* selection of NaCl-Tolerant callus lines and regeneration of plantlets in a bamboo (*Dendrocalamus strictus* Ness.) *Biol-Plant.* 39: 229-233, 2003.

Subhashini, K. & Reddy, GM. In vitro selection for salinity and regeneration of plants in rice. *Current Science.* 58: 584-586.

Sumithra, K., Jutur, K., Dalton-Caramel, B. & Reddy, A.R. 2006. Salinity induced changes in two cultivars of *Vigna radiata*: response of antioxidative and proline metabolism. *Plant Growth Regul.* 50: 11-22.

Tal, M., Katz, A., Heikin, M. & Dechan, K. 1979 Salt tolerance in the wild relatives of the cultivated tomato: Proline accumulation in *Lycopersicon esculentum* Mill. *L.peruvianum* Mill., and *Solanum pennelli* Cor. Treated with NaCl and polyethylene glycol. *New Phytol.* 82: 349-355

Tal, M. 1994. *In vitro* selection for salt tolerance in crop plants: Theoretical and practical considerations. *In vitro Cell Dev. Biol.* 30: 175-180.

Thomas, JC., Armond, RL. and Bonnert, HJ. 1992. Influence of NaCl on growth, proline and phosphenol pyruvate carboxylase levels in *Mesembryanthemem crytalinum* suspension cultures. *Plant Physiol* 98: 626-631.

Tyagi, RK. & Rangaswamy, NS.1997 *In vitro* induction and selection for salt tolerance in oilseed brassicas and regeneration of the salt tolerant soma clones. *Phytomorphology.* 47(2): 209-220

Venkataiah, P., Christopher, T. & Subhash K. 2004 Selection and characterization of sodium chloride and mannitol tolerant callus lines of red pepper (*Capsicum annuum L.*) *Plant Physiol.* 9(2): 158-163

Watad, AA., Reinhold, L., Lerner, HR. 1983 Comparison between a Stable NaCl-Selected *Nicotiana* Cell Line and the Wild Type. *Plant Physiol* 73: 624-629

Winicov, I. 1991. Characterization of salt tolerant alfalfa (*Medicago sativa* L.) plants regenerated from salt tolerant cell lines. *Plant Cell Reports* 10: 561-564

Winicov, I. & Bastola, DR. 1997 Salt tolerance in crop plants: new approaches through tissue culture and gene regulation. *Acta Physiol Plant* 19(4): 435-449

Yasar, Ellialyioglu, S. and Kusvuran,S. 2006 Ion and lipid peroxide content in sensitive and tolerant eggplant callus cultured under salt stress. *Europ J Hort Sci* 71(4): 169-172

Zair,I., Chlyah, A., Sabounji,K., Tittahsen,M & Chlyah, H. 2003 Salt tolerance improvement in some wheat cultivars after application of in vitro selection pressure. *Plant Cell Tissue & Organ Culture* 73: 237-244

Zhao, Y., Aspinall, D. & Palge, LG. 1992 Protection of membrane integrity in *Medicago sativa* (L.) by glycinebetaine against the effects of freezing. *J Plant Physiol* 140: 541-543

12

Environmental Friendly Sanitation to Improve Quality and Microbial Safety of Fresh-Cut Vegetables

Ji Gang Kim

National Institute of Horticultural and Herbal Science,
Rural Development Administration, Suwon,
Korea

1. Introduction

Fresh-cut products have a limited shelf-life due to rapid deterioration caused by microbial growth as well as physiological disorder. Cutting of fruits and vegetables increases microbial spoilage of fresh-cut produce through transfer of microflora on the outer surface to the interior tissue where microorganisms have access to nutrient-laden juice (Das & Kim, 2010a). Washing with sanitizer is an important step in reducing the microbial population and quality deterioration. The use of chemical compounds to extend postharvest life of fruit and vegetables has become lesser accepted by consumers since these compounds may be contaminant of the environment or harmful to human health. Therefore, the proper application of different sanitizing agents should be highly optimized to guarantee a minimal number of spoilage microorganisms.

Chlorine has been widely used in produce washes in order to inactivate microorganisms and ensure quality and safety. However, increasing public health concerns about the possible formation of chlorinated organic compounds and the emergence of new more tolerant pathogens, have raised doubts in relation to the use of chlorine by the fresh-cut industry (Kim, 2007). The use of chlorine as a sanitizing agent is prohibited in some countries due to the hazardous byproducts formed by chlorine with process water and other organic matters. Consequently, sanitization of fresh vegetables with chlorine in the industry renders a negative impact to the environment and human health as a whole. Recently, to avoid chlorine the use of non-chemical sanitizers in fresh-cut industries is becoming the more popular trend universally. Therefore, the industry is searching for alternative environment-friendly sanitizing methods to maintain the quality of fresh-cut produce at the best level.

Ozone is a strong antimicrobial agent with high reactivity, penetrability and spontaneous decomposition to a non-toxic product (Khadre et al., 2001; Kim et al., 2007a). Ozone is found in natural form in the atmosphere or it can be produced by generators. Ozone as an aqueous disinfectant was declared to be generally recognized as safe (GRAS) for food contact applications. Ozone's primary advantages include fast decomposition in water to oxygen, no residue, and improved microbial reduction efficacy against bacteria and fungal spores than hypochlorite. Ozone forms oxidated radicals in the presence of water that penetrate and act on cell membranes. The use of ozonated water has been applied to fresh-cut vegetables for sanitation purposes reducing microbial populations and extending the shelf-life of some of these products (Beltran et al., 2005; Hassenberg et al., 2007; Kim et al., 2007a). Ozone has been declared in many countries to have potential use for food processing including sanitation of fresh and fresh-cut vegetables. When compared to chlorine, ozone treated with optimum conditions has a greater effect against certain microorganisms and rapidly decomposes to oxygen, leaving no residues (Kim, 2007; Rico et al., 2007). Inactivation of microorganism by ozone is a complex process that attacks various cell membrane and wall constituents (e.g. unsaturated fats) and cell content constituents (e.g. enzymes and nucleic acids). The micro-organism is killed by cell envelope disruption or disintegration leading to leakage of the cell contents. Disruption or lysis is a faster inactivation mechanism than that of other disinfectants which require the disinfectant agent to permeate through the cell membrane in order to be effective (Kim et al., 1999). Bacteria are more sensitive than yeasts and fungi. Gram-positive bacteria are more sensitive to ozone than Gram-negative organisms and spores are more resistant than vegetative cells (Das & Kim, 2010b; Pascual, 2007).

Electrolyzed water (EW), the second most popular sanitizer in Korea (Kim, 2008) is considered as an environment-friendly sanitizer compared to chlorine. However, there are different opinions on the effect of EW among industrial users. EW generated by adding NaCl to pure water from non-diaphragm system is one of the major EW systems in Korea. This electrolytic process facilitates the conversion of chlorine oxidants (Cl_2, HClO/ OCl-) which are effective for inactivating a variety of microorganisms (Kim, 2007; Yang et al., 2003). The bactericidal effect of EW have been evaluated on several fresh-cut vegetables such as lettuce, carrots, spinach, and cucumber (Izumi, 1999; Nimitkeatkai & Kim, 2009). Acidic EW has a strong bactericidal effect against pathogens and spoilage microorganisms due to its low pH, high oxidation reduction potential (ORP) and the presence of residual chlorine (Kim, 2007). Using acidic EW resulted in moderate control of aerobic bacterial growth during storage of fresh-cut cilantro (Wang et al., 2004). Acidic EW was also tested for its efficacy in inactivating Salmonella on fresh-cut produce. However, the concentration of acid used can influence the organoleptic quality of vegetables, i.e., loss of texture (Kim, 2007; Zhang & Farber, 1996). Electrolyzed water at high pH (pH 6.8, 20 mg/L available chlorine) was tested as a disinfectant and the research found that it did not affect tissue pH, surface color, or general appearance of fresh-cut vegetables (Izumi, 1999).

Heat treatments such as hot water and hot air are non-chemical methods that have been used to control microorganism and senescence-related symptoms of fresh produce. Recently, mild heat treatment as physical technology to extend shelf-life of fresh and fresh-cut produce have become of interest. Heat treatment using hot water was used for fungal and insect control, but has been extended to improve the storage quality of fresh-cut produce. Hot water dips to control both decay and quality change of fresh and fresh-cut produce are often applied for 30 seconds to a few minutes at temperatures of 40-60°C (Kim, 2007; Kim et al., 2011). Heat treatments combined with other agents have also been used to prevent the microbial quality, browning, and maintaining texture in various vegetables (Das & Kim, 2010a). A combination of heat treatment followed by calcium dip has also been applied for the primary purpose of controlling postharvest pests and/or diseases and has been found to have very good results in maintaining or improving the texture of various horticultural products. Combined heat treatments with UV-C were applied to fresh-cut processed broccoli (*Brassica oleracea* L.) florets to investigate their effects on several quality and senescence parameters. Heat treatments and UV-C radiation have been utilized to extend storage quality of fresh-cut products (Lemoine et al., 2008).

Acidity is a commonly used factor to control the growth of microorganisms in foods. Organic acids such as citric acid and ascorbic acid have been applied for preserving physicochemical qualities (Rosen & Kader, 1989) and preventing microbial growth at levels that did not adversely affect taste and flavor (Yildiz, 1994). Therefore, organic acids could be a potential sanitizer for fresh-cut vegetables. However, different studies have shown that the inhibitory or bactericidal effect depends on the characteristics of the acid used to adjust the medium pH (Buchanan et al., 1993; Eswaranandam et al., 2004; Parish & Higgins, 1989). Application of organic acids as sanitizers at higher concentration can reduce the overall quality and produce off-flavor in leafy vegetables after few days of storage (Kim 2007; Chandra & Kim, 2011). The acid tolerance of fresh-cut vegetables varies among different microorganisms and products. Organic acid alone treatment is not successful sanitation in controlling pathogens and maintaining food quality during storage. Hurdle technology or combined technology, which involves simultaneous multiple preservation approaches, is generally better to control both microbial safety and food quality of fresh-cut produce.

Natural antimicrobial agents derived from fruit, herb, and shell have been investigated as preservatives. The interest in the possible use of natural alternatives to food additives to prevent bacterial and fungal growth has notably increased. Edible coatings containing natural antimicrobial agents are gaining importance as potential treatments to reduce the deleterious effects imposed by fresh-cut processing on fresh-cut fruits. However, application of natural edible coatings for fresh-cut vegetables has not received interest as much as fresh-cut fruits. Essential (volatile) plant oils occur in edible, medicinal and herbal plants which minimizes questions regarding their safe use in food products. Essential oils and their constituents have been widely used as flavouring agents in foods since the earliest recorded history and it is well established that many have wide spectra of antimicrobial action (Holley & Patel, 2005). No fresh-cut company used natural antimicrobial agent to wash or preserve fresh-cut produce in Korea due to less effects than chemical sanitizers and non-economic efficiency. However, these days natural agents are good candidate to replace tap-water washing because these agents have more potency against microorganisms. Natural agents can be used for washing microgreens and organic fresh-cut produce which are sold at a higher price. The growing demand for fresh and fresh-cut produce by consumers had led to the need for natural food preservation methods such as the use of natural antimicrobials and their combination with other hurdles, without adverse effects on the consumer or the food itself (Tiwari et al., 2009).

Although a wide range of different microbial agents are available for sanitizing fresh-cut produce, their efficacies vary and none are able to ensure elimination of pathogen completely without compromising sensory quality. In addition, recent studies have shown that chlorine lacks efficacy on pathogen reduction; the formation of the chlorine by-products are also deleterious to human health . Thus, there is much interest in developing a safer and more environmental friendly antimicrobial alternative to chlorine. Ozone, electrolyzed water, mild heat treatment, organic acid, and natural antimicrobial agents, or the combination of those sanitizing methods have been applied to various fresh-cut vegetables. It is generally accepted that an ideal sanitizing agent should have two important properties: a sufficient level of antimicrobial activity and a negligible effect on the sensory quality of the product.

2. Ozonated water washing

Ozonated water washing is getting more popular now-a -days due to its high biocidal efficacy, wide antimicrobial spectrum and environment friendly. Research has shown that treatment with ozone appears to have a beneficial effect in extending the storage life of fresh produce such as cucumber, apples, grapes, oranges, pears, raspberries and strawberries by reducing microbial populations and by oxidation of ethylene (Kim, 2007). The effect of ozonated waters with different concentrations and contact times on the quality attributes and microbial population of fresh-cut produce were studied. Two types of ozone generators were used to investigate the efficacy of microbial reduction and quality maintenance of fresh-cut lettuce, cilantro, carrot, broccoli, and paprika. One aqueous ozone solution was prepared by continuously circulating the water through an ozone generator and a stainless steel water tank. The circulating type ozone generator was equipped with a vortexer to facilitate dissolving of gaseous ozone in the water, and a de-gassing system to remove the undissolved ozone. The other ozone solution was prepared by flowing ozonated water into plastic bucket through an ozone generator. The flowing type ozone generator was equipped with a cylinder used as compression tank to facilitate high concentration of ozone. Both circulating type and flowing type ozone solutions were used immediately after the required ozone concentration were reached.

2.1 Sanitation with low ozone concentration (≤1 ppm)

Ozonated water using circulating type with low ozone concentration (less than 1 ppm) and insufficient contact time was not much effective compared to 100 ppm chlorine in reducing microbial population and maintaining quality of fresh-cut cilantro, iceberg lettuce, romaine lettuce, and baby leaves. Fresh-cut cilantro was washed in tap water, 100 ppm chlorine solution (pH 7), and 0.7 ppm ozonated water for 1 minute separately. The initial total aerobic plate count (APC) on the unwashed cilantro leaves was 6.45 log CFU/g. There was a significant decrease in APC between washed cilantro and unwashed sample after washing. However, no significant difference was found in microbial reduction of fresh-cut cilantro between tap water and ozonated water throughout 6 days storage at 5°C. The chlorine treatment maintained a low level of microbial count compared to other treatments. Fresh-cut romaine lettuce, spinach, microgreens, and baby leaves sanitized with 0.5-0.8 ppm of ozonated water had higher microbial population compared to samples washed in 50-100 ppm of chlorine. Ozonated water sanitation with low ozone concentration (less than 1 ppm) is not inadequate to be used in fresh-cut industry practically because the ozonated water was not effective in microbial decontamination and maintaining storage quality of fresh-cut vegetables (Kim, 2007).

Continuous flowing type ozonated water containing 1 ppm of ozone concentration was also used to sanitize fresh-cut iceberg lettuce if the sanitation could get much effect in reducing microbial population. The ozonated water was compared with tap water and 100 ppm chlorine solution (pH 6.5). During storage at 5 °C, there was a significant increase in APC among all treatments. The highest numbers of APC and coliform plate count (CPC) were observed in tap water washing followed by ozone washing. Chlorine treatment had the most reduction on microbial population on fresh-cut iceberg lettuce throughout storage. In fresh-cut lettuce, cut edge browning commonly occurs during storage making unsuitable for consumers. Discoloration occurred in all fresh-cut iceberg lettuces on day 6. Samples washed in ozonated water had lower degree of cut edge browning index score than samples washed in tap water. However, chlorine solution showed the lowest degree of cut edge browning. Though ozonated water containing 1 ppm of ozone concentration was effective in delaying discoloration and reducing microbial population the effectiveness was lower compared to 100 ppm chlorine solution (Kim, 2007).

2.2 Sanitation with high ozone concentration (>1 ppm)

2.2.1 Washing fresh-cut vegetables with 1.5 ppm of ozone

Flowing ozonated water washing was also used to sanitize 'Tah Tasai' Chinese cabbage baby leaves and fresh-cut romaine lettuce. Those fresh-cut products were washed in tap water, 100 ppm chlorine (pH 7.0), and continuous flow of 1.5 ppm ozonated water for 2 minutes separately. Samples treated with the ozonated water had lower APC compared to those washed in tap water. Ozonated water containing 1.5 ppm of ozone concentration reduced APC on fresh-cut baby leaves and romaine lettuce by 0.3 and 0.6 log CFU/g, respectively, on day 0. On the other hand, 100 ppm chlorine solution treatment reduced APC on fresh-cut baby leaves and romaine lettuce by 0.5 and 0.9 log CFU/g on day 0. Fresh-cut produce washed in ozonated water had lower microbial population than samples washed in tap water until middle period of 9 days-storage at 5 °C. Ozonated water washing containing 1.5 ppm of ozone concentration and washed for 2 minutes was not sufficient to decontaminate microorganism of fresh-cut romaine lettuce and baby leaves as much as effectiveness of 100 ppm chlorine solution washing. Therefore, it has been required to find optimum ozonated water washing conditions for improving storage quality and microbial food safety of each fresh-cut product to apply to fresh-cut industry as a chlorine alternative.

2.2.2 Washing fresh-cut carrots with initial 2 ppm ozone

Higher ozone concentration with longer contact time has been applied to get much effect in reducing microbial population and maintaining quality of fresh-cut produce. Fresh-cut carrot shreds were washed in tap water for 1 minute, 50 ppm chlorinated water (pH 6.3) once or two times for each 1 minute, or initial 2 ppm ozonated water using circulating type at varying times (1, 5, and 20 minutes). The samples were then centrifuged to remove excess water, packaged in 50 μm PE film bags, and stored at 5°C. Different ozonated water washing time affected microbial growth, off-odor development, color, and overall quality of carrot shreds. A single chlorine wash and 20 minutes ozonated water wash treatments had lower APC and lactic acid bacteria compared to other washing treatments until 2 weeks storage (Fig. 1). The 20 minutes ozone treatment reduced APC on carrot shreds by1.4 and 1.1 log CFU/g on week 1 and week 2, respectively.

Ozonated water washing for 20 minutes maintained quality by inhibiting off-odor and high overall quality score due to less whiteness development (Fig. 2). The single chlorine water wash was effective and resulted in better quality compared to two time chlorine water wash. However, samples washed for 20 min in ozonated water had better quality with less off-odor and higher overall visual quality scores than samples washed in chlorine water washed once. The efficacy of optimum ozonated water washing on microbial reduction and quality of those fresh-cut produce was similar to chlorine or better than chlorine. Ozonated water containing initial 2 ppm ozone concentration with sufficient washing time would be effective in reducing microbial population and maintaining quality of fresh-cut carrot and could be an alternative method to maintain quality and shelf-life of fresh-cut carrot shreds (Kim et al., 2007a).

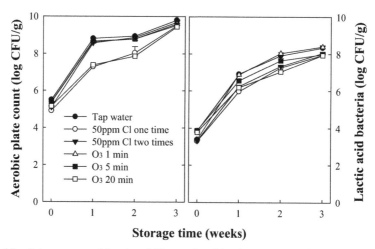

Fig. 1. Aerobic plate count and lactic acid bacteria of fresh-cut carrot shreds washed in different sanitizers and stored at 5°C for up to 3 weeks.

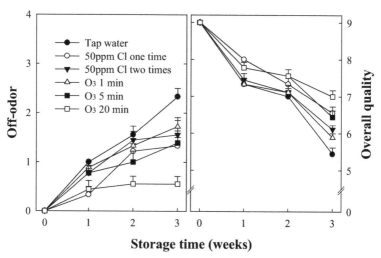

Fig. 2. Off-odor development and Overall quality of fresh-cut carrot shreds washed in different sanitizers and stored at 5°C for up to 3 weeks.

2.2.3 Washing fresh-cut broccoli with 2 ppm ozone

Ozonated water washing using flow type with different contact times on storage quality and microbial growth in fresh-cut broccoli was conducted to compare ozone with chlorine.

Fresh-cut broccoli samples were washed each for 90 and 180 seconds in normal tap water, 100 ppm chlorinated water (pH 7), and 2 ppm ozonated water separately and respectively. Then, samples were packaged in 30 μm polyethylene bags and stored at 5°C for 9 days. No significant differences were observed in gas composition and color among different sanitizers with contact times. No off-odor was detected during the 9 days storage. Sanitizers affected microbial population of fresh-cut broccoli. In the color characteristics no difference were marked in L* and a* value and hue angle of the samples among different washing solutions and contact times during the storage period. It was found that electrical conductivity increased with the longer contact time in all washing solutions compared to shorter contact time. Electrolyte leakage is generally considered as an indirect measure of plant cell membrane damage. Ozonated water washing for 180 seconds contact time initially showed highest electrical conductivity probably due to its highly oxidizing nature than chlorine and tap water washing, but at the end of the storage the value is low and nearly equal to the above washings (Fig.3, left). It may be due to quality maintenance without texture damage or decay. Electrical conductivity is relatively high immediately after fresh-cut processing and decrease rapidly, then either decrease gradually or remain relatively stable until the samples have good quality in many fresh-cut produce (Kim et al., 2005a; Kim, 2007). This typical response pattern to processing and storage is similar to the result of fresh-cut broccoli sanitized in ozonated water and stored for 9 days at 5°C. No color difference was found among the treatments during 9 days storage (Fig. 3, right).

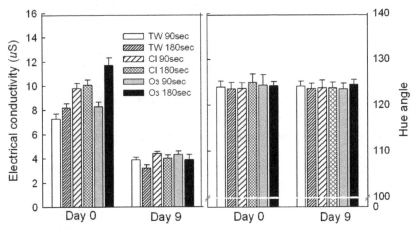

Fig. 3. Electrical conductivity and hue angle of fresh-cut broccoli washed in different sanitizers and stored for 9 days at 5°C. In figure, samples washed for 90 and 180 seconds in tap water (TW), 100 ppm chlorinated water (Cl), and 2 ppm ozonated water (O₃).

Among the sanitizers, ozonated water with 180 seconds maintained the lowest numbers of aerobic plate count throughout the storage days in comparison with others (Fig. 4). Ozonated water with 90 seconds was not much effective in reducing microbial population compared to chlorine. However, samples washed with ozonated water for 180 seconds showed the lowest coliform count. Absolutely no coliform were observed in ozonated water with 180 seconds washing treatment on day 0. The result reveals that longer contact time of ozone affects positively whereas other sanitizers don't affect on the microbial quality and safety aspects of fresh-cut broccoli. The difference of microbial population is may be due to the following causes; the surface wash off might be attached to the samples again during washing time, higher electrical conductivity observed during storage and the reactivity against pathogen may be less effective in comparison with 2 ppm of ozone. Ozone effectiveness against microorganisms depends not only on the amount applied, but also on the effectiveness of ozone delivery method, type of material, the target microorganisms, physiological state of the bacteria cells at the time of treatment (Das & Kim, 2010b).

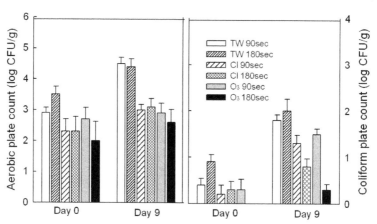

Fig. 4. Aerobic plate count and coliform plate count of fresh-cut broccoli washed in different sanitizers and stored for 9 days at 5°C. In figure, samples washed for 90 and 180 seconds in tap water (TW), 100 ppm chlorinated water (Cl), and 2 ppm ozonated water (O₃).

2.2.4 Washing fresh-cut vegetables with 3 to 4 ppm ozone

Ozonated water treatment containing 3 ppm of ozone with 3 minutes washing reduced APC and coliform/E. Coli count in both fresh-cut paprika and iceberg lettuce, similar to 100 ppm chlorine at day 0 (Table 1). Treatment with the ozonated water showed the lowest numbers of aerobic plate count and coliform/E. Coli count on day 6 in fresh-cut paprika. There was no difference in quality parameters such as color, off-odor, and visual quality among treatments. The highest numbers of aerobic and coliform/E. Coli were observed in tap water washing. Fresh-cut broccoli washed in 4ppm ozonated water for 5 minutes had lower microbial populations than samples washed in 100 ppm chlorine. No quality deterioration

Sanitizer	Fresh-cut paprika				Fresh-cut iceberg lettuce			
	APC		Coliform/E. Coli		APC		Coliform/E. Coli	
	Day 0	Day 6	Day 0	Day 6	Day 0	Day 6	Day 0	Day 6
Tap water	2.3	5.3	1.5	1.2	3.6	5.3	0.8	0.9
100ppm Chlorine	2.0	4.3	0.5	1.0	3.1	4.7	0.2	0.7
3ppm Ozone	2.1	3.5	0.7	0.7	3.0	4.7	0.2	0.7

Table 1. Microbial population (log CFU/g) of fresh-cut paprika and iceberg lettuce washed in different sanitizers for 3 minutes, packaged, and stored for 6 days at 5°C.

or side effects of higher ozone concentration were found in fresh-cut broccoli. These results showed that ozonated water reduced microbial growth more effectively than 100 ppm chlorine solution. Therefore, ozonated water washing with optimum ozone concentration and sufficient contact time could be a favorite alternative sanitation to chlorine.

3. Electrolyzed water

Strong acidic EW (pH2.7) and weak acidic EW (pH 5-6.5), which were generated by electrolysis of NaCl solution and HCl, respectively have been used as disinfectant in fresh-cut industry. Weak alkaline EW (pH 7.5-8) from non-diaphragm EW generator, recently developed is getting popular among three types of EW. In general, strong acidic EW had stronger bactericidal effect compared to alkaline EW which has high pH levels. However, strong acidic EW can cause tissue damage in some fresh-cut produce, especially leafy vegetables during storage or distribution. Little information exists on the efficacy of weak alkaline or weak acidic EW on quality and microbial reduction in fresh-cut produce.

3.1 Washing fresh-cut lettuce, sesame leaf, and strawberry

The effect of strong acidic and weak alkaline EW containing 80ppm available chlorine concentration as well as general chlorine solution on storage quality and microbial growth of fresh-cut iceberg lettuce has been studied. The effectiveness of strong acidic EW on microbial reduction was greater than weak alkaline EW at initial storage. However, strong acidic EW affected quality deterioration due to texture damage after 6 days at 10°C (Table 2). Weak alkaline EW reduced off-odor development and was as effective as chlorine in inhibiting total aerobic bacterial and coliform group on fresh-cut iceberg lettuces (cultivar; *U-lake*). Fresh-cut sesame leaves washed in weak alkaline and neutral EW had less total plate counts and better sensorial quality compared to samples washed in chlorine and strong acidic EW. The result reveals that, weak acidic EW affects positively whereas strong acidic EW was not effective in maintaining quality. Strong acidic EW may be used for sanitation of fruits which has firm texture like apple. Nimitkeatkai & Kim (2009) reported that apples washed in strong acidic EW for 5 minutes had less hue angle value throughout storage period and lower sensory evaluation score at the end of storage. However, apples washed in strong acidic EW for 2 min was effective in reducing microbial growth and maintaining sensorial quality of apples. Different result was found that microbial quality of strawberry (cultivar; *Maehyang*) washed in weak alkaline EW (pH 8.0, 60ppm HClO) was inferior compared to samples washed in chlorine (pH 6.5, 60ppm HClO). These results indicated that EW does not affect quality maintenance and microbial safety for all fresh and fresh-cut products. Though weak alkaline EW did not show the effectiveness in strawberry weak alkaline EW could be an effective alternative to chlorine for washing fresh-cut leafy vegetables.

Treatment	Gas composition (%)		Total plate count (log CFU/g)	Off-odorz	Discolorationz
	O_2	CO_2			
Chlorine	0	25.8 b	7.0 a	3.3 b	0.5 b
Weak alkaline EW	0	31.2 a	7.3 a	3.8 a	1.8 a
Strong acidic EW	0	23.8 b	7.1 a	2.7 c	0.7 b

z 0 = none, 1 = slight, 2 = moderate, 3 = severe, 4 = strong

Table 2. Gas composition, aerobic plate count, and quality of fresh-cut iceberg lettuce washed in different sanitizers and stored at 10°C for 6 days

3.2 Washing fresh-cut broccoli

Weak alkaline EW was also used to sanitize fresh-cut broccoli. The fresh-cut samples were washed for 90 seconds in tap water, 80 ppm chlorinated water (pH 7), and EW (pH 7.2) containing 80 ppm free chlorine separately and respectively. Then, samples were packaged in 30 μm polyethylene bags and stored at 5°C for 9 days. No significant differences were observed in gas composition and color among different sanitizers. No off-odor was detected during the storage. Samples washed with EW showed the lowest total aerobic bacterial population and coliform count. The result reveals that weak alkaline EW affects microbial population of fresh-cut broccoli positively.

	Electrical conductivity		Aerobic plate count		Coliform plate count	
	Day 0	Day 9	Day 0	Day 9	Day 0	Day 9
Tap water	6.1	4.2	4.28	5.52	1.39	2.20
Chlorine	8.9	4.5	3.62	4.61	-	1.89
Electrolyzed water	8.2	4.7	3.53	4.24	0.22	1.02

Table 3. Electrical conductivity and microbial population of fresh-cut broccoli washed in different sanitizers and stored at 5°C for 9 days

3.3 Combined EW washing with MA packaging

Study on effect of combined EW washing with modified atmosphere (MA) packaging was carried out to investigate the influence of the combined treatment on quality maintenance and microbial food safety of fresh-cut iceberg lettuce. Fresh-cut iceberg lettuce were washed in alkaline EW (free chlorine 80 ppm), dried, and packaged with 35 μm P-Plus film to compare with conventional technology using combination of chlorine sanitation and vacuum packaging. Samples for control treatment were prepared following industrial practices; 100 ppm chlorine (100ppm, pH 7.5) wash and vacuum packaging with 80μm Ny/PE film. Combined EW and MA technology reduced off-odor development of packaged fresh-cut iceberg lettuce during storage (Table 4). The combined technology using EW washing and MA packaging was as effective as control using chlorine in inhibiting total aerobic bacterial counts on fresh-cut iceberg lettuces. The combined EW washing and MA packaging extended two more days of shelf-life of fresh-cut iceberg lettuce compared to control treatment.

Treatment	Gas composition		Aerobic plate count (log cfu/g)	Off-odor*	Dis-coloration*
	$O_2(\%)$	$CO_2(\%)$			
Chlorine + Vacuum pack.	0	19.0	6.4	2.3	0.6
EW + MA packaging	1.8	11.5	6.2	0.7	0.9

* 0 = none, 1 = slight, 2 = moderate, 3 = severe, 4 = strong

Table 4. Gas composition, aerobic plate count, and quality of fresh-cut iceberg lettuce treated with hurdle technology and stored at 5°C for 12 days

4. Heat treatment

Heat treatments is one of postharvest treatments that has been used to control postharvest decay and; or to improve the storage quality of fresh-cut produce. Heat treatments alone or combined with other agents have also been used to prevent the microbial quality, browning, and maintaining texture in various fresh-cut vegetables.

4.1 Heat treatment for fresh-cut winter squash and lotus roots

The effectiveness of heat treatment for fresh-cut winter squash and lotus roots has been applied. Winter squash which had hard rinds can be treated with hot water at 60-65 °C for 2-3 minutes to reduce microbial contamination before it is fresh-cut processed (Arvayo-Ortiz et al., 1994; Hawthorne, 1989). However, other vegetables which have soft texture are not recommended to treat at high temperature like winter squash. Winter squash (cultivar; *Bouzang*) harvested in summer season was immersed in 65°C hot water for 2 minutes before it is fresh-cut processed. The heat treatment reduced microbial population and maintained good quality with bright yellow or orange with fine, moist texture and high solids, and sugar contents. The changes in the quality of fresh-cut lotus roots treated with hot water were investigated. Lotus roots washed, peeled, and sliced with 1cm-thickness was dipped in water at 30, 55, and 80°C for 45 seconds. Then, samples were air-dried at room temperature, packed in polyethylene films, and stored at 4°C for 12 days. Generally, the weight loss of the lotus roots that were treated with hot water slightly increased. The application of the heat treatment delayed the browning of the lotus roots, especially the treatment with 55°C hot water. The Hunter color 'L' and 'a' values of the lotus roots treated with 80°C hot water significantly increased during their storage. The heat treatment effectively inhibited the growth of mesophilic microorganisms. Therefore, the organoleptic quality of the lotus roots that were treated with 55°C hot water was the best among those temperatures (Chang et al., 2011).

4.2 Mild heat treatment for peeled potato

For peeled potato, the most popular method of retarding surface browning used in the Korean industry is vacuum packaging that induces high CO_2 and low O_2 levels. However, the presence of high CO_2 and low O_2 concentrations may cause off-odor development due to anaerobic respiration. Hence, heat treatment methods that reduce browning and off-odor development were investigated. Potatoes (var. '*Jopung*') kept at 5°C after harvest were heat treated (30°C for 24 hours, 45°C for 3hours, or non-heated), peeled, and immersed in tap

Treatment	Gas composition (%)		Color		Off-odor[z]
	O_2	CO_2	Lightness (L)	Redness(a)	
Control	0.67 b	34.7 a	69.8 b	-1.61 a	3.2 a
Heat treatment at 30°C for 24 hours	2.74 a	16.8 b	71.2 a	-2.47 c	2.5 b
Heat treatment at 45°C for 3 hours	0.56 b	32.4 ab	70.3 ab	-2.15 b	1.7 c

[z] 0 = none, 1 = slight, 2 = moderate, 3 = severe, 4 = strong

Table 5. Effect of heat treatment before peeling on gas composition and quality of peeled potato fresh-cut processed and vacuum packaged, and stored at 5°C for 5 days.

water at 5°C for 3hours. Samples were then vacuum-packaged with 80μm Ny/PE film and stored at 10°C for up to 5 days. Mild heat treatment (30°C) was effective in reducing CO_2 concentrations and off-odor development in the package of samples throughout storage. The 30°C mild heat treatment also delayed browning of peeled potatoes and maintained the highest overall quality score. The mild heat treatment at 30°C before peeling can be a practical method to delay browning and off-odor development of 'Jopung' potato (Kim et. al, 2009).

4.3 Combined heat treatment for fresh-cut paprika

The combined effect of washing solutions and heat treatment were investigated as potential sanitizers for maintaining the quality and microbial safety of fresh-cut paprika. Fresh paprika shreds were washed in tap water and 1% calcium chloride combined with 19°C (normal tap water) and 50°C water temperature (heat treatment) for 2 minutes. Then, samples were packaged in 30 μm polypropylene bags and stored at 5°C for 12 days. No significant differences were observed in color and gas composition of the package among treatments and no off-odor was detected until the end of 12 days storage. However, 50°C water temperature with calcium chloride had lower microbial numbers upto the storage period in comparison with tap water (Fig. 5). The result reveals that 50 °C water temperature with calcium chloride can be used as an washing solution to maintain the microbial quality in fresh-cut Paprika (Das & Kim, 2010a).

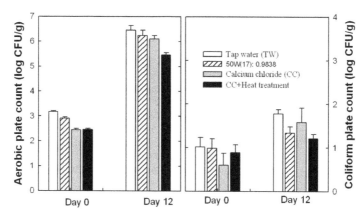

Fig. 5. Aerobic plate count and coliform plate count of fresh-cut paprika during the storage period. In figure, samples washed for 90 seconds in tap water (TW) and calcium chloride solution at 5°C.

Softening, textural changes are one of the main causes of quality losses in case of fresh-cut products. In general, fresh-cut vegetables that maintain firm and crunchy textures are highly desirable. Though there is no significant difference in gas composition and color among treatments calcium chloride and heat treatment tended to increase firmness of fresh-cut paprika during the beginning of the storage compared to tap water washing treatment. It is well known that calcium plays a major role in maintaining the quality of fruit and vegetables. Increasing the calcium content in the cell wall of fruit tissue can help delay softening of fresh-cut produce. The beneficial effects obtained with heat treatments have generally been explained in terms of pectin esterase activation. Calcium dips have been employed to improve firmness and extend the postharvest shelf-life of a wide range of fruit and vegetables.

Similarly, certain commercial additives can maintain the quality of fresh-cut products (Encarna et al., 2008; Luna-Guzman & Barrett, 2000). The changes in electrical conductivity of fresh-cut paprika depend upon the type of washing solution and heat treatment (Fig. 6). Tap water washing treatments (both TW and TW + Heat treatment) showed lower electrolyte leakage compared to calcium chlorine washing during the entire storage period. A combination of heat treatment followed by calcium dip may need more research if it can be applied for the purpose of controlling postharvest pests and/or diseases and have very good results in maintaining or improving the texture of on various fresh-cut produce.

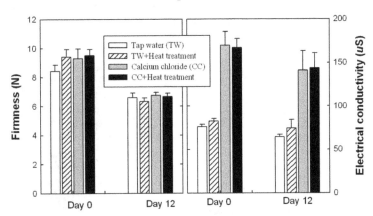

Fig. 6. Firmness and electrical conductivity of fresh-cut paprika heat treated and stored at 5°C for 12 days. In figure, samples washed for 90 seconds in tap water and calcium chloride solution at 5°C.

5. Organic acid and acid compounds

Organic acid and acid compound sanitizers have been used to sanitize fresh and fresh-cut produce. Organic acids are one of the important sanitizers that have been applied largely for preserving physicochemical qualities and for preventing microbial growth in many fresh-cut products. Organic acid with optimum condition did not adversely affect taste and flavor, but leaving no effect on environment. Citric acid can be used to extend the shelf life of fresh-cut produce by reducing the loss of eating quality and disease development. Ibrahim et al. (2009) reported that leaves of some selected vegetables decontaminated with 5% citric acid showed a considerable decrease in microbial count compared to water washing. However, the application of these acids at higher concentration may cause quality deterioration due to off-odor and texture damage in some fresh-cut leafy vegetables. Sequential treatment of citric acid and ethanol on the quality and microbial reduction of organic vegetables has also been examined. Hence, organic acid and the combined technology using have been carried out to find an alternative sanitizer to chlorine.

5.1 Washing fresh-cut lettuce with organic acid and combined acid

Fresh iceberg lettuce leaves were sanitized separately with tap water, 100 μL L^{-1} chlorine, 0.2% citric acid, 50% ethanol, and the combination of citric acid solution and 50% ethanol spray. Samples were then dried with centrifugal dryer, packaged in 80μm Ny/PE films, and stored for 6 days at 5°C. The 50% ethanol solution dipping was the most effective treatment to reduce microbial population of fresh-cut iceberg lettuce (Fig. 7). However, fresh-cut iceberg lettuce sanitized in ethanol solution had severe injury with lowest visual quality score and highest electrical conductivity among treatments after 6 days storage (Fig. 8). The decline of overall visual quality in ethanol treated sample might be a consequence of tissue damage as reflected from the electrical conductivity data. Citric acid alone was not effective in reducing microbial population, similar to 100ppm chlorine solution treatment (Fig. 7). The combination of citric acid and ethanol spray reduced aerobic microbial population by 1.1 log CFU/g as compared to tap water. The combination with citric acid and ethanol spray also maintained good quality with high overall quality score at the end of 6 days storage (Fig. 8). Therefore, the combination of citric acid and ethanol spray could be an alternative to chlorine as an environment-friendly sanitizer for washing fresh-cut leafy vegetables (Kim et al., 2011).

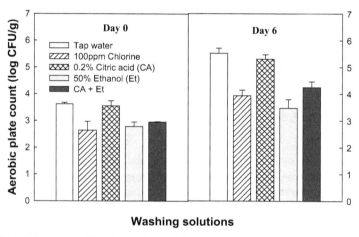

Fig. 7. Aerobic plate count of fresh-cut iceberg lettuce treated with different sanitizers and stored at 5 °C.

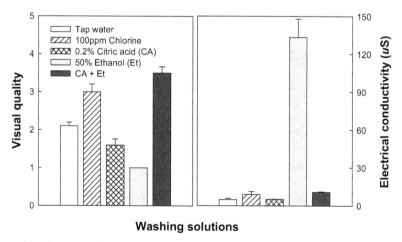

Fig. 8. Visual quality and electrical conductivity of fresh-cut iceberg lettuce treated with different sanitizers and stored for 6 days at 5 °C.

5.2 Washing fresh-cut spinach, baby leaves, and microgreens with the combined organic acid

Fresh organic vegetables such as spinach, 'Tah Tasai' Chinese cabbage baby leaves, and microgreens were also sanitized separately with tap water, 100 ppm chlorine, 0.2% citric acid (CA), and the sequential treatment of 0.2% CA solution and 50% ethanol spray (CA+Et). In case of spinach, chlorine and CA+Et increased CO_2 partial pressures in the headspace of sample packages and generally had higher electrical conductivity compared to tap water. No significant differences were observed in color among different sanitizers during storage at 5°C in fresh-cut spinach samples. The chlorine and CA+Et treatments were effective in reducing microbial population of fresh-cut spinach. However, CA+Et treatment induced off-odor of microgreens resulting more aerobic plate count compared to chlorine treatment. In case of microgreens, samples treated with CA+Et did not have good quality score, worse than score of chlorine at the end of storage probably due to severe texture damage. In 'Tah Tasai' Chinese cabbage baby leaves, sanitizer chlorine treatment showed lower number of total aerobic count immediately after washing. However, citric acid in combination with ethanol spray treatment showed the lowest number until day 7. No significant difference was found in microbial number among the treatments at the end of 10 days storage. Wang et al. (2004) also reported that no significant difference in total aerobic plate count at the end of 14 days storage of cilantro leaves. The possible reason might be due to the baby leave samples became softer with the progress in storage which caused damage in texture for all samples. Citric acid and ethanol, on the other hand, both are used as anti-microbial agents leaving no effect to the environment. Their combined use was almost similarly effective as of chlorine possibly due to the dual sanitization effects on the sample used.

5.3 Acid compound sanitizers

5.3.1 Use of acidified sodium chlorite and peroxyacetic acid-based sanitizer

Acid compound sanitizers such as acidified sodium chlorite (ASC) and peroxyacetic acid-based sanitizer (PA) have been used for food safety of fresh and fresh-cut produce. Peroxyacetic acid is a strong oxidizing agent that has been used extensively to disinfect food processing equipment and has been approved by the U.S. Food and Drug Administration (FDA) as a disinfectant for fruits and vegetables (Gonzalez et al., 2004). Recent studies undertaken to determine the suitability of PA (Tsunami, Ecolab, USA) for washing fresh-cut vegetables showed it to be effective against *Listeria monocytogens*, *Salmonella* spp., and *E. coli* O157:H7 (Beuchat et al., 2004; Gonzalez et al., 2004). Acidified sodium chlorite has also been approved by the FDA for spray or dip application on various food products, including fresh and fresh-cut produce (Kim et al., 2007b). Studies have shown that ASC and PA have a strong antimicrobial efficacy against various human pathogens inoculated onto cantaloupes and asparagus (Kim 2007; Park & Beuchat, 1999). However, the effect of both ASC and PA on quality attributes such as texture and color of fresh-cut produce was not investigated with an in-depth study. It was found that carrot shreds washed in higher PA or ASC concentrations (200 and 500 ppm, respectively) lost firmness and looked melted during storage. There was a significant difference in firmness between 200 and 30 ppm PA concentration or 500 and 30 ppm ASC concentration. Therefore, high PA and ASC concentration may cause faster deterioration of carrot shreds because of inferior texture.

5.3.2 Washing fresh-cut carrot shreds

Fresh-cut shredded carrots were washed in tap water, 100 ppm chlorinated water, 30 ppm ASC, or 30 ppm PA. Samples were then packaged in 35 um polyethylene film and stored for 14 days at 5°C. Sanitizers affected off-odor, skin whitening, and microbial population of carrot shreds. Chlorine and PA wash maintained a lower level of microbial count compared to water or ASC treatments. Though no significant difference was found in total aerobic count between chlorine and PA until day 11, samples treated with PA had a lower APC than samples washed in chlorine at the end of storage. Chlorine caused the highest reduction (0.75 log), followed by PA (0.65 log) and ASC (0.56 log) at day 0. No significant difference in total aerobic count was found between ASC and water treatment after day 4. Therefore, 30 ppm ASC treatment was not effective in reducing microbial organism of fresh-cut carrot shreds during storage. The reason may be due to low ASC concentration (30 ppm). It was also found that 500 ppm ASC concentration was more effective in reducing microbial numbers than 50 ppm chlorine wash water. However, 500 ppm ASC concentration resulted in visual deterioration of the carrot shreds. Kim et al. (2006) reported that 30 ppm PA was effective in reducing microbial contamination as well as 100 ppm chlorine concentration. This would suggest that the residual effect of acetic acid released when PA is degraded causes reduced growth in microflora (Gonzalez et al. 2004).

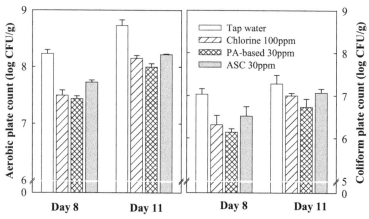

Fig. 9. Aerobic plate count and coliform plate count of fresh-cut carrot shreds. In figure, samples washed for 2 minutes in tap water, chlorinated water, peroxyacetic acid-based sanitizer (PA), and acidified sodium chlorite (ASC).

Peroxyacetic acid-based sanitizer also maintained initial color values, inhibited off-odor development and skin whitening of samples, and achieved the highest overall quality score among those sanitizer treatments. Off-odor was detected in all samples on day 8 and day 11 (Fig. 10, left). Off-odor was lower in samples treated with PA than in any other treated samples. Fresh-cut produce is known to develop undesirable off-odors under low O_2 and elevated CO_2 atmospheres (Kim et al., 2005a). Carrot shreds treated with water, chlorine, or ASC reached score 2.0 (the limit of marketability) at the end of storage, whereas samples PA-treated had score 1.9 after 11 days of storage. In fresh-cut produce, patterns of off-odor development correlate with ethanol and acetaldehyde and there is also a strong relationship between package atmospheric conditions and off-odor development (Kim et al., 2005b). The degree of off-odor of fresh-cut carrot shreds may be influenced by fermentation due to anaerobic microorganism. Surface whitening which is one of major postharvest quality problems in carrot was significantly retarded by applying sanitizer PA. Samples treated with PA maintained inherent color and exhibited the lowest rate of increase in whitening scores (Fig. 10, right). Surface discoloration during storage is most detrimental to the quality of shredded carrots. The possible reasons for whiteness development on carrot surface are dehydration and lignification (Kim et al., 2006). Visual observation of carrot shreds treated with PA showed a moister surface compared to other sanitizer treatments. The PA-treated samples had the highest overall quality score, with relatively low levels of whitening. Therefore, sanitizer PA treatment significantly affected quality and shelf-life of fresh-cut shredded carrots. At present, chlorine is the most practical, efficient, and low cost disinfectant available. Due to concerns about the formation of by-products, however, a safer alternative is needed. Peroxyacetic acid-based sanitizer treatment resulted in comparable antimicrobial effectiveness and sensory quality of carrot shreds throughout storage.

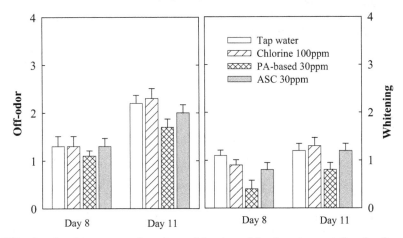

Fig. 10. Off-odor development and surface whitening of fresh-cut carrot shreds after storage at 5°C for 8 and 11 days. In figure, samples washed for 2 minutes in tap water, chlorinated water, peroxyacetic acid-based sanitizer (PA), and acidified sodium chlorite (ASC).

6. Natural antimicrobial agents

Natural compounds can serve as carriers for a wide range of food additives, including anti-browning agents, colorants, and antimicrobials that can extend product shelf-life and reduce the risk of pathogen growth on fresh-cut produce surface. In recent years there has been a considerable pressure by consumers to reduce or eliminate chemically synthesized additives in foods. Plants and plant products can represent a source of natural alternatives to improve the shelf-life and the safety of food. In fact, they are characterised by a wide range of volatile compounds, some of which are important flavour quality factors (Patrignani et al., 2008; Utama et al., 2002). A key role in the defence systems of fresh produce against decay microorganisms has been attributed to the presence of some of these volatile compounds (Patrignani et al., 2008). However, no plant volatiles have been used as natural antimicrobial agent for fresh-cut produce practically. Recently developed natural antimicrobial agents from marine resource product have been applied to fresh-cut vegetables.

6.1 Sanitation fresh-cut lettuce with calcinated calcium

The heated scallop shell powder; calcinated calcium (CC) was investigated as potential sanitizers for maintaining storage quality and microbial safety of fresh-cut iceberg lettuce. Samples were washed in normal tap water, 50 ppm chlorinated water (pH 6.5), 1.5 g ·L⁻¹ CC for 2 minutes separately. Samples were then packaged in 80 μm nylon/polyethylene bags and stored at 5°C. The initial aerobic plate count of unwashed iceberg lettuce was 6.5 log CFU ·g⁻¹. The aerobic plate count on fresh-cut lettuce increased with storing time, reaching 6.05 to 7.05 log CFU ·g⁻¹ on 12 days-storage. Washing in CC was effective in reducing aerobic plate count of fresh-cut lettuce samples by 0.4 to 1.0 log CFU ·g⁻¹ as well as chlorine treatment throughout storage as compared to tap water (Fig. 11, left). Electrical conductivity of all samples decreased during the initial period of storage, remained stable thereafter or increased slightly at the end of storage (Fig. 11, right). Electrical conductivity of fresh-cut lettuces increased after 8 days. Electrical conductivity is generally considered as an indirect measure of plant cell membrane damage and deterioration of fresh-cut vegetables (Jiang et al., 2001; Kim et al., 2005b). Increased electrical conductivity after fresh-cut processing is a common phenomenon due to the leakage from cut ends of the samples or otherwise wounded tissues.

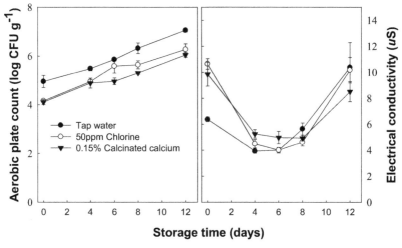

Fig. 11. Aerobic plate count and electrical conductivity of fresh-cut iceberg lettuce sanitized with different washing solutions (tap water, chlorinated water, and calcinated calcium) and stored at 5°C for 12days.

Samples treated with CC had good quality with low off-odor at the end of storage. Visual quality score of all fresh-cut iceberg lettuce samples was lower than score 3, which was considered the limit of marketability at the end of 12 days-storage (Fig. 12, left). The visual quality related to browning was probably induced by relatively high O_2 and low CO_2 concentration in sample packages. Off-odor was first detected in fresh-cut lettuce samples treated with TW after 6 days and increased relatively until the end of storage. Off-odor of fresh-cut lettuce samples washed in chlorine and CC was lower than tap water (Fig. 12, right). Samples sanitized with chlorine or CC reached score 1.3 which was lower score than the limit of marketability on the 12 days-storage. Fresh-cut produce is known to develop undesirable off-odors under low O_2 and elevated CO_2 atmospheres (Kim, 2007). In fresh-cut lettuce, patterns of off-odor development correlated with ethanol and acetaldehyde and there was also a strong relationship between package atmospheric conditions and off-odor development (Kim et al., 2005a; Kim et al 2005b). The degree of off-odor in the packaged fresh-cut lettuce was influenced by subsequently fermentation due to anaerobic microorganism.

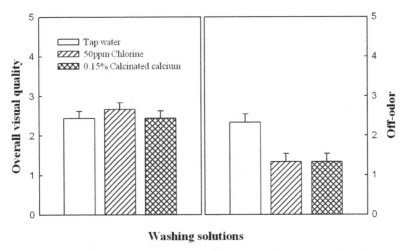

Fig. 12. Overall visual quality and off-odor development of fresh-cut iceberg lettuce sanitized with different washing solutions (tap water, chlorinated water, and calcinated calcium) after 12 days storage at 5°C.

6.2 Sanitation fresh-cut Bok choi and broccoli with calcinated calcium

Natural materials, calcinated calcium and fruit extract compound from Japanese apricot were used as sanitizer to maintain quality and reduce microbial population of bok choi with different maturities. Microgreen, baby leaf, and mature bok choies were washed in tap water, 50 ppm chlorine, and 500 fruit extract compound, for 2 min separately. Those samples were then packaged in 50 μm PE film and stored at 5°C for 6 days. One of natural compounds, the fruit extract compound was not effective significantly in reducing microbial population and quality such as off-odor. However, samples treated with CC had better quality with less off-odor until 4 to 6 days-storage in baby leaf and mature Bok choi. Calcinated calcium affected in reducing microbial population of microgreen, baby leaf, and mature Bok choi for 2, 4, and 6 days, respectively (Fig. 13). Bok choi micrggreens had highest microbial population, followed baby leaves, and mature samples in terms with maturity. Therefore, mature and baby leaf Bok choi samples can have 6 and 4 days of shelf-life with CC sanitation, respectively. Fresh-cut broccoli was also washed in CC at normal tap water temperature. Broccoli samples sanitized in CC solution had good quality with lower off-odor and microbial count at the end of 9 days-storage (Kim et al., 2010). To avoid chlorine which may lead to the formation of carcinogenic compounds, CC can be used as environmental friendly sanitizer and an alternative to chlorine washing for fresh-cut broccoli without affecting microbial and sensorial quality.

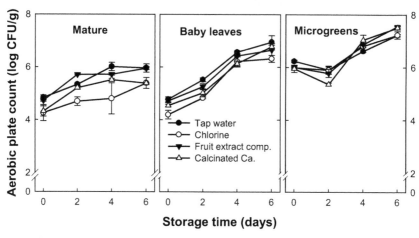

Fig. 13. Aerobic plate count of mature, baby leaf, and microgreen fresh-cut bok choi samples sanitized with different washing solutions (tap water, chlorine, fruit extract compound, and calcinated calcium).

7. Conclusion

Use of chlorine to reduce microbial populations of fresh-cut vegetables has faced with challenges to find alternatives which are more environmental friendly and not harmful to human health. Wide ranges of different agents are available for sanitizing fresh-cut produce, their efficacies vary and none are able to ensure elimination of pathogen completely without compromising sensory quality. Therefore, application of different environment-friendly sanitizing agents has been conducted to investigate highly optimized condition to guarantee a minimal number of spoilage microorganisms in many fresh-cut vegetables. As alternatives to chlorine, ozone, electrolyzed water, mild heat treatment, organic acid with ethanol, and natural antimicrobial agents, or the combination of those sanitizing methods can be used for fresh-cut vegetables. But, selection of washing solution and use of optimum condition to meet each fresh-cut vegetable should be performed to get similar or better efficacy to chlorine. Ideal sanitizing agent should have effectiveness in two important properties: antimicrobial activity and sensory quality of the fresh-cut product.

For organic fresh-cut leafy vegetables which are facing challenges to find the means to extend shelf-life and to enhance microbial food safety, combined citric acid with ethanol spray or calcinated calcium alone solution can be used in fresh-cut industry. Those environment friendly agents are good candidate to replace tap water washing or organic acid solution because these agents have more potency against microorganisms. Heat treatment without chemical use can be used for washing of fresh-cut produce which have firm texture such as winter squash, potato, lotus roots, and paprika. To get much effectiveness in reducing microbial population of fresh-cut vegetables constant 1~4 ppm of ozonated water and week alkaline or weak acidic electrolyzed water can be used practically. However, the concentration of ozone and contact time is very important for microbial safety of fresh-cut vegetables such as broccoli, iceberg lettuce, carrot, etc. Calcinated calcium, a natural and an environment-friendly sanitizer can be an alternative to mild heat treatment for washing of fresh-cut vegetables without affecting sensorial quality.

8. References

Arvayo-Ortiz, R.M., Garza-Ortega, S. & Yahia, E.M. (1994). Postharvest response of Winter squash to hot-water treatment, temperature and length of storage. *HortTechnology* 4:253-255.

Beltran, D., Selma, M.V., Marın, A. & Gil, M.I. (2005). Ozonated water extends the shelf life of fresh-cut lettuce, *Journal of Agricultural and Food Chemistry* 53, 5654–5663.

Beuchat, L.R., Alder, B.B. & Lang, M.M. (2004). Efficacy of chlorine and a peroxyacetic acid sanitizer in killing *Listeria monocytogenes* on on iceberg and romaine lettuce using simulated commercial processing conditions. *Journal of Food Protection.* 67(6): 1238-1242.

Buchanan, R. L., Golden, M. H., & Whiting, R. C. (1993). Differentiation of the effects of pH and lactic or acetic acid concentration on the kinetics of Listeria monocytogenes inactivation. *Journal of Food Protection,* 56, 474e478.

Chang, M.S., Kim, J.G., & Kim, G.H., (2010) Quality characteristics of fresh-cut lotus roots according to the temperature of the wash water. *Korean Journal of Food Preservation.* 18(3): 288-293

Chandra, D & Kim, J.G. (2011). Effects of different sanitizers on the quality of 'Tah Tasai' Chinese cabbage (*Brassica campestris* var. narinosa) baby leaves. *Korean Journal of Food Preservation* . In press

Das, B.K & Kim, J.G.(2010a) Combined effect of heat treatment and washing solutions on the quality and microbial reduction of fresh-cut paprika. *Horticulture, Environment, and Biotechnology.* 51(4): 257-261

Das, B.K & Kim, J.G.(2010b) Microbial quality and safety of fresh-cut broccoli with different sanitizers and contact times. *Journal of Microbiology and Biotechnology.* 20(2): 363-369

Encarna, A., Victor, H.E. & Francisco, A. (2008). Effect of hot water treatment and various calcium salts on quality of fresh-cut 'Amarillo' melon. *Postharvest Biology and Technology.* 47:397–406.

Eswaranandam, S., Hettiarachchy, N. S., & Johnson, M. G. (2004). Antimicrobial activity of citric, lactic, malic or tartaric acids and nisin-incorporated soy protein film against Listeria monocytogenes, Escherichia coli O157:H7 and Salmonella gaminara. *Journal of Food Science,* 69, 79e84.

Gonzalez, R.J., Luo, Y. Ruiz-Cruz, S. & McEvoy, J.L. (2004). Efficacy of sanitizers to inactivate Escherichia coli O157:H7 on fresh-cut carrot shreds under simulated process water conditions. *Journal of Food Protection.* 67(11):2375-2380

Hassenberg, K., Christine Idler, Eleanor Molloy, Martin Geyer, Matthias Plöchl, Jeremy Barnes, (2007). Use of ozone in a lettuce-washing process: an industrial trial. *Journal of the Science of Food and Agriculture.* 87 (5), P914 – 919.

Hawthorne, B.T. (1989). Effects of cultural practices on the inidence of storage rots in Cucurbita spp. New Zealand *Journal of Crop Horticultural Science.* 17:49-54.

Holley, R.A. & Patel, D. (2005) Improvement in shelf-life and safety of perishable foods by plant essential oils and smoke antimicrobials. Food Microbiology 22 (2005) 273–292

Ibrahim, T.A., Jude-ojei, B.S., Giwa, E.O. & Adebote, V.T. (2009) Microbiological analysis and effect of selected antibacterial agents on microbial load of fluted pumpkin, cabbage and bitter leaves. *Research Journal of Agriculture and Biological Sciences.*, 3: 1143-1145

Izumi, H. (1999). Electrolyzed water as a disinfectant for fresh-cut vegetables. *Journal of Food Science.* 64: 536-539.

Jiang, Y., Pen, L. & Li, J. (2004) Use of citric acid for shelf life and quality maintenance of fresh-cut Chinese water chestnut. *Journal of Food Engineering.* 63: 325-328

Khadre, M.A., Yousef, A.E. & Kim, J.G. (2001). Microbiological aspects of ozone applications in food: a review, *Journal of Food Science* 66 (9), 1242–1252.

Kim, J.G., Yousef, A.E. & Chism, G.W. (1999). Applications of ozone for enhancing the microbiological safety and quality of foods: a review, *Journal of Food Protection*, 62 (9), 1071–1087.

Kim, J.G. (2007). *Fresh-cut produce industry and quality management.* Semyeong Press, Suwon, Republic of Korea. ISBN 978-89-957887-1-4

Kim, J.G. (2008). Quality maintenance and food safety of fresh-cut produce in Korea. *The Asian and Australasian Journal of Plant Science Biotechnology.* 2:1-6.

Kim, J.G., Choi,S.T. & Pae, D.H. (2009). Effect of heat treatment and dipping solution combination on the quality of peeled potato 'Jopung'. *Korean Journal of Horticultural Science Technology.* 27(2):256-262, 2009

Kim, J.G., Luo, Y., Saftner, R.A. & Gross, K.C. (2005a), Delayed modified atmosphere packaging of fresh-cut Romaine lettuce: Effects on quality maintenance and shelf-life. *Journal of American Society for Horticultural Science.*130(1):116-123

Kim, J.G., Luo, Y., Tao, Y., Saftner, R.A. & Gross, K.C., (2005b), Effect of initial oxygen concentration and film oxygen transmission rate on the quality of fresh-cut Romaine lettuce, *Journal of the Science of Food and Agriculture.* 85(10): 1622-1630

Kim, J.G., Luo, Y. & Lim, C.H. (2007a). Effect of ozonated water and chlorine water wash on the quality and microbial de-contamination of fresh-cut carrot shreds. *Korean Journal of Food Preservation.* 14(1):54-60

Kim, J.G., Luo, Y. & Tao, Y., (2007b), Effect of the sequential treatment of 1-MCP and acdified sodium chlorite on microbial growth and quality of fresh-cut cilantro. *Postharvest Biology and Technology.*, 46:144-149

Kim, J.G., Nimitkeatkai, H., Choi, J.W. & Cheong, S.R. (2011), Calcinated calcium and mild heat treatment on storage quality and microbial populations of fresh-cut iceberg lettuce, *Horticulture, Environment, and Biotechnology.* 52(4): 408-412

Kim, J.G., Nimitkeatkai, H. & Das,B.K. (2010), Effect of calcinated calcium washing solution and heat treatment on quality and microbial reduction of fresh-cut lettuce and broccoli. *ACTA Horticulturae.*875:237-242

Kim, J.G. , Yaptenco, K.F. & Lim, C.H. (2006). Effect of sanitizers on microbial growth and quality of fresh-cut carrot Shreds. *Horticulture, Environment, and Biotechnology.* 47(6): 313-318

Lemoine, M.L., Civello, P.M., Chaves, A.R. & Mart´ınez, G.A. (2008) Effect of combined treatment with hot air and UV-C on senescence and quality parameters of minimally processed broccoli (*Brassica oleracea* L. var. *Italica*). *Postharvest Biology and Technology* 48: 15–21

Luna-Guzman, I. & Barrett, D.M. (2000). Comparison of calcium chloride and calcium lactate effectiveness in maintaining shelf stability and quality of fresh-cut cantaloupes. *Postharvest Biology and Technology.* 19: 61–72.

Nimitkeatkai, H. & Kim, J.G. (2009). Washing Efficiency of Acidic Electrolyzed Water on Microbial Reduction and Quality of 'Fuji' Apples. *Korean Journal of Horticultural Science and Technology.* 27(2):250-255

Parish, M.E., & Higgins, D. P. (1989). Survival of Listeria monocytogenes in low Ph model broth systems. *Journal of Food Protection*, 52, 144e147.

Park, C.M. & Beuchat, L.R.. (1999). Evaluation of sanitizers for killing *Escherichia coli* O157:H7, *Salmonella* and naturally occurring microorganisms on cantaloupes, honeydew melons, and asparagus. *Dairy Food Environment and Sanitation*.19: 842-847.

Pascual, A., Llorca, I. & Canut, A. (2007). Use of ozone in food industries for reducing the environmental impact of cleaning and disinfection activities. *Trends in Food Science & Technology*. 18 (1): S29-S35.

Patrignani, F., Iucci, L., Belletti, N., Gardini, F., Guerzoni, M.E., & Lanciotti, R. (2008). Effects of sub-lethal concentrations of hexanal and 2-(E)-hexenal on membrane fatty acid composition and volatile compounds of Listeria monocytogenes, Staphylococcus aureus, Salmonella enteritidis and Escherichia coli. *International Journal of Food Microbiology*. 123:1-8

Rico, D., Martín-Diana, A.B., Barat, J.M. & Barry-Ryan, C. (2007). Extending and measuring the quality of fresh-cut fruit and vegetables: a review. *Trends in Food Science & Technology*. 18 (7): 373-386.

Rosen, J. & Kader, A. (1989) Postharvest physiology and quality maintenance of sliced pear and strawberry fruits. *Journal of Food Science*. 54: 656–659

Tiwari, B. K., Valdramidis, V. P., O'Donnell, C. P., Muthukumarappan, K., Bourke, P. & Cullen, P. J. (2009). Application of natural antimicrobials for food preservation. *Journal of Agricultural and Food Chemistry*, 57: 5987-6000.

Utama, I.M.S., Willis, R.B.H., Ben-Yehoshua, S. & Kuek, C. (2002). In vitro efficacy of plant volatiles for inhibiting the growth of fruit and vegetable decay microorganisms. *Journal of Agricultural and Food Chemistry* 50, 6371–6377.

Wang, H., Feng, H. & Luo, Y. (2004). Microbial reduction and storage quality of fresh-cut cilantro washed with acidic electrolyzed water and aqueous ozone. *Food Research International*. 37: 949-956.

Yang, H., Swem, B.L. & Li, Y. (2003). The effect of pH on inactivation of pathogenic bacteria on fresh-cut lettuce by dipping treatment with electrolyzed water. *Food Microbiology and Safety* 68: 1013-1017.

Yildiz, F. (1994) Initial preparation, handling and distribution of minimally processed refrigerated fruits and vegetables. In: *Minimally Processed Refrigerated Fruits and Vegetables*, Wiley, R.C. (ed), Chapman & Hall, New York, USA, p. 15–49

Zhang, S. & Farber, J.M. (1996). The effects of various disinfectants against *Listeria monocytogenes* on fresh-cut vegetables. *Food Microbiology*. 13: 311-321.

13

Molecular Structure of Natural Rubber and Its Characteristics Based on Recent Evidence

Jitladda T. Sakdapipanich[1] and Porntip Rojruthai[2]

[1]*Mahidol University,*

[2]*King Mongkut's University of Technology North Bangkok,*

Thailand

1. Introduction

Natural rubber (NR) latex collected from the *Hevea* trees exists as a colloidal suspension (Verhaar, 1959). The amount of latex obtained on each tapping is about 300 mL. The tapping is usually done once every 2–3 days for 9 months each year. Usually, the collected latex is treated with formic acid to coagulate the suspended rubber particles within the latex (Gazeley et al., 1988). After having been pressed between rollers to consolidate the rubber into 0.6-micrometre in thickness slabs or thin crepe sheets, the rubber is air-dried or smoke-dried for shipment. These rubbers are known as air-dried sheet (ADS) and ribbed smoked sheet (RSS), respectively. The treatment of latex with $NaHSO_3$ is used to produce rubber of pale color, termed pale crepe. The other forms of rubber as block rubber are known as Standard Malaysian Rubber (SMR) or Standard Thai Rubber (STR), which are mainly graded by dirt content.

Fresh latex from the *Hevea* tree contains 30–35% rubber. After collection, the latex is stabilized with NH_3 and transported from the plantation to a factory where it undergoes continuous centrifugation to produce a concentrated NR latex containing ~60% rubber. For long-term preservation of concentrated latex, the NH_3 content is usually raised to 0.6–0.7%; this is referred to as high-ammonia preserved concentrated latex. Low-ammonia preserved concentrated latex contains only 0.2–0.3% NH_3 plus tetramethyl thiuram disulfide (TMTD), as a bactericide.

Rubber particles in latex show a wide range of diameter, from 0.01 to 5 μm, with the majority being 0.1–2 μm diameter (Pendle & Swinyard, 1991). The latex is composed of the rubber phase, Frey–Wyssling particles, serum, and the bottom fraction (Archer et al., 1969; Jacob et al., 1993). Protein has been considered to be an essential component of NR for its characteristic properties. Recently, several proteins in NR were found to cause type I allergic responses that led to life-threatening anaphylactic reactions. In 1991, the FDA stipulated that rubber products made from NR latex (e.g., gloves and condoms) should be treated to remove extractable proteins. Deproteinization of latex was carried out by using a proteolytic enzyme in the presence of surfactants to reduce the extractable proteins to less than their detection limit (Sakaki, et al., 1996; Nakade, et al., 1997). It is remarkable that the physical properties of deproteinized NR (DPNR) are almost equivalent to those of ordinary natural rubber; moreover, the dynamic properties such as resilience and rebound are improved as a result of increasing the content of rubber hydrocarbon after deproteinization. These findings support the idea that the protein component of NR is not essential to produce its outstanding and characteristic properties.

Commercially available solid NR contains neutral lipids (2.4%), glycolipids and phospholipids (1.0%), proteins (2.2%), carbohydrates (0.4%), ash (0.2%), and other compounds (0.1%) (Sentheshanmuganathan, 1975; Nair, 1987). Free fatty acids in solid rubber are composed of mainly long-chain saturated and unsaturated fatty acids such as stearic, oleic, and linoleic acids (Crafts et al., 1990; Arnold & Evans, 1991). Recently, it was presumed that rubber chains also contain long-chain fatty acids, presumably occurring as phospholipids covalently linked to the chain-end (Eng et al., 1994a). While saturated fatty acids induce the crystallization of rubber chain, unsaturated fatty acids – which are present in NR as a mixture – act as a plasticizer of rubber and accelerate synergistically with saturated fatty acids on the crystallization of rubber chains (Kawahara et al., 1996; Nishiyama et al., 1996). These linked and mixed fatty acids were presumed to bring about outstanding mechanical properties through rapid crystallization of natural rubber. Highly purified NR can be obtained by removing proteins, blended fatty acids, and linked fatty acids (lipids). This can be achieved by the deproteinization of the latex, acetone extraction of the resultant deproteinized solid rubber, and transesterification of acetone-extracted rubber with sodium methoxide in toluene solution (Eng et al., 1994a; Tangpakdee & Tanaka, 1997a). It is remarkable that transesterified NR showed stress-strain properties of unvulcanized rubber (termed 'green strength') similar to synthetic cis-1,4-polyisoprene (Figure 1).

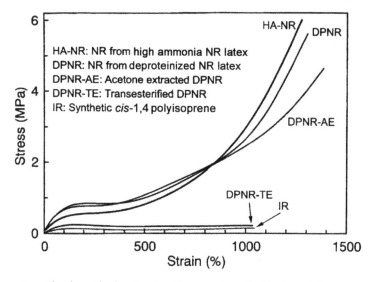

Fig. 1. Green strength of purified natural rubbers and synthetic cis-polyisoprene

The effect of linked and blended fatty acids in NR was confirmed by the preparation of a model cis-1,4-polyisoprene grafted with small amounts of stearic acid at the 3,4 unit after introducing hydroxyl group selectively at 3,4 units by a hydroboration reaction. The C_{18} grafted cis-1,4-polyisoprene mixed with linoleic acid showed a crystallizability similar to that of NR (Kawahara et al., 2000). These findings suggest that the structure of chain-end groups in NR confers major characteristic properties of natural rubber.

2. Molecular structure of both chain-ends of natural rubber

2.1 Initiating terminal of the rubber molecule

^{13}C-NMR spectroscopy of low-molecular weight natural rubber, obtained by fractionation, shows the signals corresponding to the *trans* methyl and methylene carbons, suggesting the presence of *trans*-isoprene units in the structure of NR (Figure 2) as similar to those of rubber from goldenrod leaves (Tanaka et al., 1983). However, the signals due to a dimethylallyl group are not detected in the case of *Hevea* rubber. The absence of *trans*-isoprene units in the *cis-trans* sequence suggests that the *trans*-isoprene units are in the *trans-trans* linkage, but not derived from isomerization of *cis*-isoprene units (Eng et al., 1994a).

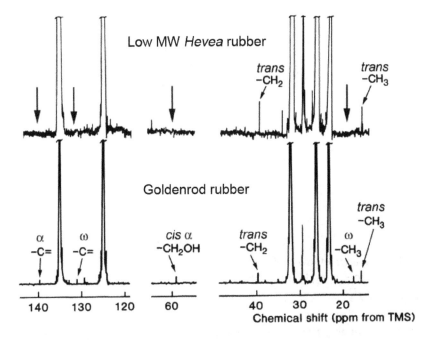

Fig. 2. ^{13}C-NMR spectra of low-molecular weight fraction of *Hevea* rubber and rubber from Goldenrod.

As shown in ^{1}H-NMR spectra (Figure 3), the *trans*-isoprene units in NR show the methyl proton signals corresponding to two-*trans*-polyprenol (Eng et al., 1994b). The absence of a dimethylallyl group in NR suggests that the initiating species for rubber formation in *Hevea* is a derivative of *trans-trans*-farnesyl diphosphate (FDP) modified at the dimethylallyl group or ordinary *trans-trans*-FDP, which is selectively modified after polymerization (Tanaka et al., 1996; Tanaka et al., 1997; Tanaka & Tangpakdee, 1997). The former assumption is consistent with the finding that the methyl proton of the *trans*-isoprene unit in the ω'-*trans-trans* group shows a similar chemical shift as that of the dimethylallyl-*trans-trans* arrangement (Eng et al., 1994b; Tanaka et al., 1996). This suggests that the ω'-group has a structure similar to that of the dimethylallyl group, and shows a similar magnetic shielding effect on the subsequent *trans*-isoprene unit.

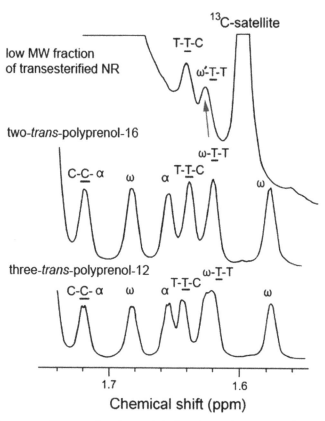

Fig. 3. ¹H-NMR spectra of low-molecular weight fractions of transesterified natural rubber, two-*trans*-polyprenol-16 and three-*trans*-polyprenol-12

The dimethylallyl group is not detected even in the low-molecular weight rubber isolated from seedlings (Tangpakdee et al., 1996), or even long-chain polyprenols in *Hevea* (Tangpakdee & Tanaka, 1998a). However, it is difficult to ignore the possibility that FDP directly initiates polymerization, because it was reported that FDP stimulates *in vitro* rubber formation upon incubation with isopentenyl diphosphate (IDP) and washed rubber particles (Archer & Audley, 1987; Audley & Archer, 1988; Madhavan et al., 1989; Cornish & Backhaus, 1990). Under the same conditions, [1-³H]neryl diphosphate was incorporated into rubber molecules (Audley & Archer, 1988). Recently, a high yield of rubber formation was reported for *in vitro* rubber synthesis by incubation of the bottom fraction of freshly tapped latex with IDP (Tangpakdee & Tanaka, 1997a). The resulting rubber showed no dimethylallyl group, whereas the rubber obtained in the presence of IDP and FDP clearly showed the dimethylallyl group and *trans*-isoprene units as shown in Figure 4 (Tangpakdee et al., 1997). This suggests that FDP will be the initiating species of rubber formation in *Hevea* tree. The fact that newly formed rubber contains no dimethylallyl group is additional evidence supporting the presence of ω'-*trans-trans* group at the initiating terminal arising from unidentified initiating species.

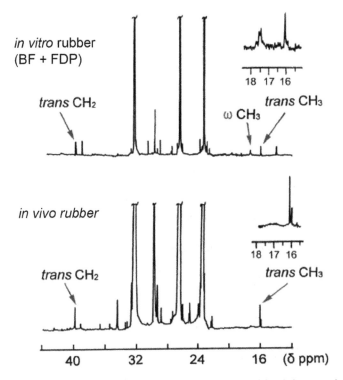

Fig. 4. ^{13}C-NMR spectra of *in vitro* NR formed on incubation of fresh bottom fraction (BF) and FDP (above) and *in vivo* rubber (below).

At the present, the ω–terminal group of low molecular weight and polyprenol fractions from NR was analyzed using high-resolution ^{13}C- and ^{1}H-NMR and 2D-COSY techniques (Mekkriengkrai, 2005). Very recently, the presence of the dimethylallyl group and two *trans*-isoprene units at the ω-terminal was observed in polyprenol from *Hevea* shootings and fresh bottom fraction as well as the lowest molecular weight fraction of washed rubber particles (WRP) obtained by washing the cream fraction from fresh latex with surfactant. This finding suggests that the initiating species of rubber biosynthesis is *trans, trans*-FDP as in the case of two-*trans* polyprenol. This ω–terminal was not detected in the high molecular weight fractions of WRP and low molecular weight fractions of the ordinary NR and DPNR, suggesting the occurrence of a modification at the dimethylallyl residue. Thus, it can be concluded that the structure of dimethylallyl group at the ω–terminal of the ordinary NR is modified by some enzymatic or chemical reactions. Figure 5 shows the presumed mechanism of ruber formation mediated by an enzyme based on hypothesis of Ogural *et al.* (1997). It has been postulated that the methyl group in dimethylallyl diphosphate is connected to a phenylalanine of the FQ (phenylalanine-glutamine) motif which is found not only in all prenyl diphosphate synthases but also in tRNA-dimethylallyl transferase, to hold this molecule in the direction necessary for a condensation reaction with IDP. This suggests that some reactions can proceed to modify the dimethylallyl group according to the removal of the ω–terminal from the pocket of the enzyme, which may result in the formation of various dimethylallyl group derivatives.

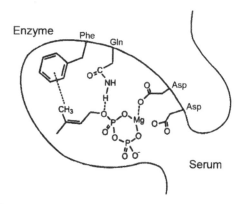

Fig. 5. Presumed role of enzyme on initiation and chain elongation steps

2.2 Terminating chain-end of rubber molecule

The primary alcohol or their fatty acid ester groups is a popular structure of the α-terminal group in many naturally occurring polyisoprenes such as the rubbers from *Lactarius* mushrooms and leaves of goldenrod and sunflower, even though rubber as well as polyprenols were presumed to be synthesized by the addition of IDP. This is owing to the fact that diphosphate group can be hydrolyzed to hydroxyl group or transesterified to fatty acid ester. However, signals corresponding to these structures at the α-terminal are not detected in the ^{13}C- and ^{1}H-NMR spectra of natural rubber. On the other hand, long-chain fatty acid ester groups are clearly observed, even after purification by deproteinization and acetone extraction (Eng et al., 1994a) (Figure 6).

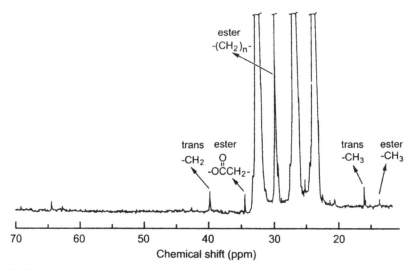

Fig. 6. ^{13}C-NMR spectrum (aliphatic region) of low-molecular weight fractions from deproteinized natural rubber

The long-chain fatty acid ester group was found to be about two moles per one rubber molecule, by ^{13}C-NMR and FTIR analysis (Eng et al., 1994a; Tangpakdee & Tanaka, 1997a; Tanaka, Y. et al., 1997). In addition, the ^{1}H-NMR spectrum of low-molecular weight rubber from 1-month-old *Hevea* seedlings shows the signals corresponding to $-CH_2OP$ and glyceride (Tangpakdee & Tanaka, 1997a). These groups can be removed by transesterification or saponification. These findings suggest that most of the chain-ends of NR consist of a phosphate group belonging long-chain fatty acid ester (Tangpakdee & Tanaka, 1997a; Eng et al., 1994a), which corresponds to phospholipids as follows (Figure 7):

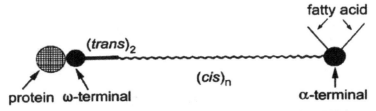

Fig. 7. Presumed structure of NR linear chain (Tangpakdee & Tanaka, 1997a)

The new information of structure of α-terminal group in NR was analyzed recently by selective decomposition of the branch-points followed by structural characterization by using high resolution ^{1}H-, ^{13}C- and ^{31}P-NMR, FTIR, and GPC techniques as well as diluted solution viscometry for highly purified NR (Tarachiwin et al., 2005a, 2005b). The selective decomposition of branch-points was carried out by using lipase, phosphatase, and phospholipases A₂, B, C, and D. The ^{1}H-NMR spectrum of acetone-extracted deproteinized NR (AE-DPNR) at 750 MHz (Figure 8) shows not only the signals derived from phospholipids, but also two small triplet signals due to mono- and diphosphate groups (Tarachiwin et al., 2005a). Figure 9 shows the ^{1}H-NMR spectra of AE-DPNR, lipase-treated AE-DPNR and phosphatase-treated DPNR. It is remarkable that these signals did not disappear even after lipase, phosphatase and phospholipase treatments of AE-DPNR. Lipase can decompose fatty acid ester of acylglycerol including phospholipids at C1 and C3 positions as shown in Scheme 1, while phosphatase decomposes only monophosphate ester linkage. The presence of mono- and diphosphate signals after lipase and phosphatase treatments clearly indicates that these phosphate groups are directly linked to rubber molecule.

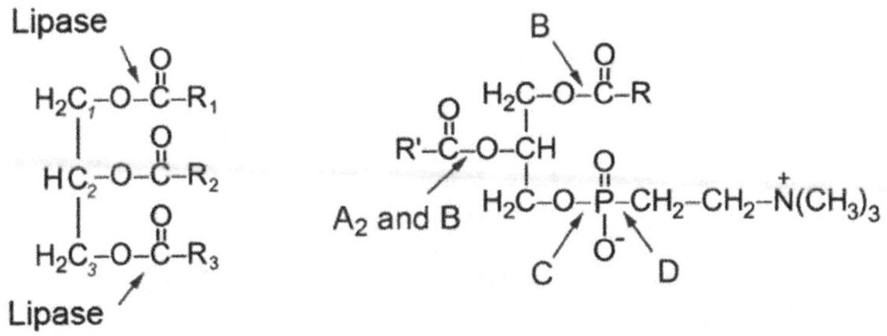

Scheme 1. Decomposition positions of lipase on the acylglycerol (left) and phospholipase on L-α-phosphatidylcholine (right)

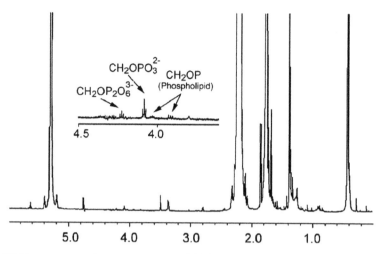

Fig. 8. ^{1}H-NMR spectrum of acetone-extracted DPNR

The phospholipids, which are the origin of long-chain fatty acid esters, are presumed to be linked to phosphate groups mainly by hydrogen bonding with a minor portion by ionic linkage (Tarachiwin et al., 2005a).

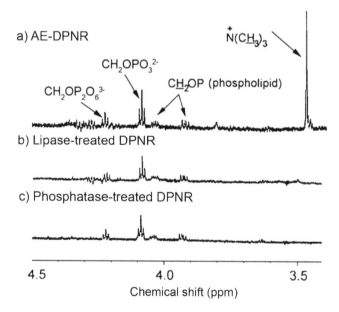

Fig. 9. ^{1}H-NMR spectra of acetone-extracted DPNR, lipase-treated DPNR and phosphatase-treated DPNR.

It is well known that four kinds of phospholipases can decompose selectively the linkages in a phospholipid. For example, the reaction site of L-α-phosphatidylcholine is shown in

Scheme 1. The treatment of DPNR latex with phospholipases A₂, B, and C resulted in the decrease of molecular weight and marked shift of the high-molecular weight peak to the low-molecular weight peak as well as narrowing the molecular weight distribution (MWD), while phospholipase D showed no change. This result demonstrates the presence of phospholipids in rubber molecules, which participate the formation of branch-points (Tarachiwin, 2004). The change of MWD and decrease in the molecular weight of DPNR after the treatment of lipase and phospholipases A₂, B, and C signify that the acylglycerol group and phosphate in phospholipids are directly concerned with the formation of branch-points in NR at the α-terminal.

It was reported that most phospholipids resist solubilization in polar and non-polar solvents by the formation of micelle and inverse micelle structures, respectively (Murari et al., 1982). The polar group in phospholipids has been reported to be participated in inter- and intramolecular hydrogen bonding that restricted the mobility of the phosphate group. (Alenius et al., 2002) These findings suggest that branch-points in NR are predominantly formed by hydrogen bonding between polar groups of phospholipids. However, the formation of branch-points by ionic linkage cannot be neglected. The formation of crosslinking by ionic linkages between negatively charges of phospholipids with divalent cation is plausible, since the process has been implicated in many membranes associated events (Hübner & Blume, 1998).

The presence of Mg^{2+} ions in FL-latex is expected to form ionic linkages between rubber molecules. The addition of ammonium sulfate ((NH_4)$_2SO_4$) in NR-latex for removing Mg^{2+} ions slightly decreased the molecular weight and polydispersity index. This indicates that Mg^{2+} ions have less effect on the branching formation than hydrogen bonding. Therefore, the α-terminal group of NR was postulated to consist of two kinds of functional group, i.e., monophosphate and diphosphate groups, which are linked with phospholipids *via* hydrogen bonding as a predominant linkage and some parts *via* ionic linkage as shown in Figure 10.

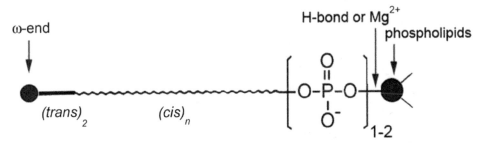

Fig. 10. Presumed structure for the α-terminal group for NR (Tarachiwin et al., 2005a)

2.3 Structural change of rubber molecule in rubber tree

Rubber tapped from the first opening mature tree, also known as virgin tree, contains a solvent-insoluble fraction formed *via* carbon–carbon linkages that are referred to as 'hard-gel', as high as 80–90% of whole solid rubber. This gel fraction showed the structure and cross-linking density corresponding to the rubber cured with peroxide. In contrast, the residual solvent-soluble fraction (the 'sol fraction') is of low molecular weight and formed by oxidative degradation. Both the ^{1}H- and ^{13}C-NMR spectra of the sol fraction of the rubber from virgin tree show prominent signals due to epoxide and hydroperoxide, as well as aldehyde (Tangpakdee & Tanaka, 1998b; Sakdapipanich et al., 1999a). It is remarkable that the ^{13}C-NMR signals due to methyl protons of *trans*-isoprene units in the initiating terminal were not detected in the sol fraction of virgin tree. The absence of both, the dimethylallyl group and the *trans*-isoprene units in the rubber from virgin tree, indicates that the oxidative degradation of rubber chains occurs by loss of *trans*-isoprene units at the initiating terminal.

2.4 Structure of branch-points, gel and storage hardening

It is well known that commercial high ammonia latex (HA-latex) increases the mechanical stability during storage, but it is always accompanied with the increase in the gel content as high as 60% after storage for a long period. The gel fraction in commercial HA-latex is assumed to be due to the reaction of branched chains to form three-dimensional chains. A slight increase in the gel content was also observed from DPNR-latex, although the rate is very slow compared to that of commercial HA-latex (Kawahara, 2002a). Accordingly, the gel formation in DPNR latex is not due to proteins, but by some reactions of the functional α-terminal group or by radical crosslinking reaction.

NR contains both soft-gel and hard-gel (Tangpakdee & Tanaka, 1997b). The content of soft-gel fraction in NR decreases by deproteinization with a proteolytic enzyme or can be partly decomposed in solution by the addition of small amounts of a polar solvent into a good solvent and almost completely solubilized by transesterification. These suggest that branch-points are originated mainly from functional groups at ω- and α-terminals. The molecular weight between the crosslinks (M_c) of the gel fraction in NR was 2-3 times of the number average molecular weight (M_n) value suggesting the presence of 2-3 rubber chains per crosslink, although the measurement is based on many assumptions.

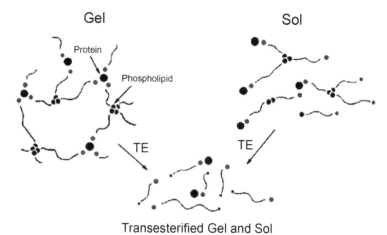

Fig. 11. Presumed structure of gel and sol phases in NR after transesterification

Based on these findings, it is possible to illustrate the structure of gel and sol phase in NR as shown in Figure 11. Two types of branch-points are presumed in rubber chain, the formation of branchings by both types of branch-points results in the formation of crosslinked chains. Here, one to two chains are assumed to be between the crosslink-points. The sol fraction may be a branched polymer having no three-dimensional branching. It is remarkable that all the branch-points containing ester linkages of fatty acid and/or phosphates can be decomposed by transesterification and the resulting rubber chains cannot form three-dimensional branched-chains because of only one type of the residual branch-points in the rubber chains.

It is noteworthy that a part of the gel fraction in commercial HA-latex cannot be solubilized by transesterification or saponification (Tarachiwin et al., 2003). This hard-gel has been presumed to be formed by radical reactions between rubber chains and tetramethylthiuram disulfide (TMTD), which is normally used as bactericide preservatives, in latex together with zinc oxide (ZnO). As shown in Figure 12, the addition of TMTD and ZnO into HA- and DPNR-latices resulted in a rapid increase in the gel content. The resulting gel fraction was hard-gel insoluble in toluene even after chemical or enzymatic treatments mentioned above. This suggests that TMTD and ZnO can be another factor to increase the gel content during storage of latex (Tarachiwin et al., 2003).

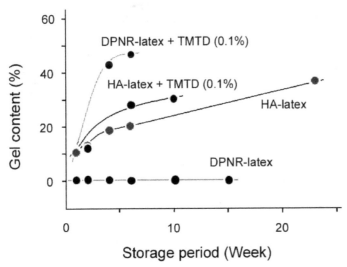

Fig. 12. Gel content in commercial HA-latex and DPNR-latex, prepared from HA-latex, with and without TMTD and ZnO (Tarachiwin et al., 2003)

Another example of hard-gel formation is observed for the rubber obtained from virgin *Hevea* trees, which have not been collected latex by tapping for a long period. It is known that the latex obtained from virgin *Hevea* trees shows very poor properties. This may be concerned with the fact that the rubber from virgin trees contained the gel fraction as high as 80% (Tangpakdee & Tanaka, 1998b). The gel fraction showed almost the same structure as the crosslinked rubber prepared from FL-latex in the presence of peroxide. This gel fraction cannot be solubilized by enzymatic or chemical reactions (Tangpakdee & Tanaka, 1998b). The M_c value of the gel fraction was about 3×10^3, showing that the gel is highly crosslinked rubber. It is remarkable that the sol fraction in the latex from virgin tress was very low-molecular weight rubber containing aldehyde and epoxide groups derived by oxidative degradation (Tangpakdee & Tanaka, 1998b). This finding indicates clearly the occurrence of radical reactions in rubber trees during storage as latex for a long period before taking out the latex by tapping. This was further confirmed by the analysis of the molecular weight and gel content after the first tapping of virgin *Hevea* trees as shown in Figure 13. The gel content decreased gradually by successive tapping after the first tapping and recovered to the same level as the ordinary FL-latex after 6 days. A similar tendency was observed for the molecular weight. The molecular weight of soluble rubber fraction increased to the ordinary values after tapping for 6 days. These findings indicate the occurrence of radical reactions on rubber chains in rubber tree during storage, to form carbon-carbon crosslinking and partly oxidative degradation products in laticiferous cells. This also suggests a possible role of rubber as a scavenger of hydroxyl radicals in latex (Tangpakdee & Tanaka, 1998b).

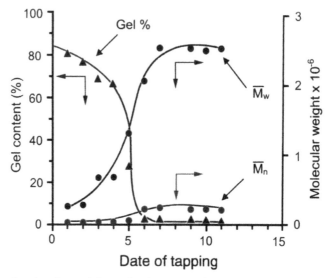

Fig. 13. Change of molecular weight and gel content of NR after the first tapping of virgin tree

Recently, as mentioned above, the structure of α-terminal group of NR was proposed to consist of two kinds of functional group, i.e., monophosphate and diphosphate groups, which are linked with phospholipids *via* hydrogen bonding as a predominant linkage and some parts *via* ionic linkage (Tarachiwin et al., 2005a, 2005b). Here, it is postulated that the branch-points in NR are formed by aggregation of the phospholipids which are linked to phosphate or diphosphate groups at the α-terminal. Phospholipids are presumed to aggregate together to form a micelle structure mainly *via* hydrogen bonding between polar groups in phospholipids molecules, as shown in Figure 14.

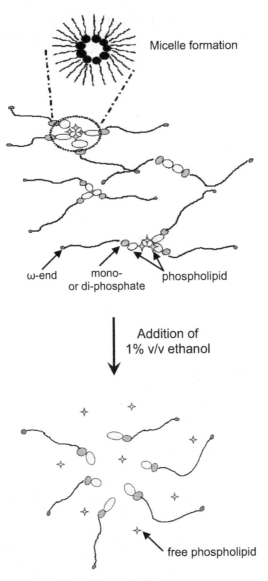

Fig. 14. Proposed structure of branch-points in NR

The storage-hardening (SH) phenomenon in solid NR has long been recognized to be a factor affecting the processing properties such as Mooney viscosity and Wallace plasticity during storage. These phenomena affect directly to the variability of NR in several aspects including processing properties (Sekhar et al., 1958, 1960). The change in the properties of NR during storage was presumed to be due to the cumulative effects of crosslinking and chain-scission, varying with the environmental conditions (Gan, 1997). It has been postulated that the hardening proceeds through the reactions between the rubber chains and so-called abnormal groups assumed to exist on rubber molecules such as epoxide (Sekhar et al., 1958), carbonyl (Sekhar et al., 1960; Subramaniam, 1976), and lactone (Gregory & Tan, 1976).

The physical properties during long-term storage as real condition for selected commercial Standard Thai Rubber (STR), i.e. STR XL, STR 5L and STR CV60 as high-graded NR, were investigated. Each zone of commercial NR after SH was also subjected to examination. STR 5L showed clearly increase in Mooney viscosity, MR_{30}, gel content and initial plasticity (P_0) which higher than those of STR XL. This result suggested that STR 5L showed the highest inconsistency in physical properties. The increasing in viscosity and gel content of STR 5L and STR XL samples suggest the occurrence of crosslink structure during storage. STR CV60, known as viscosity-stabilized NR sample, also showed increasing in Mooney viscosity, gel content, P_0, and high plasticity retention index (PRI) value during long storage. These findings indicate that SH occurred in the rubber samples even in carefully controlled production procedure. As for the different zones of samples, there is no clear relation about the gel content with respect to storage time, indicating that depth or positions of specimen in a certain rubber bale had not affected to the storage-hardening phenomenon.

In the past several years, branching and crosslinking formations in NR has been postulated to derive from chemical reactions of abnormal groups (Sekhar et al., 1958; Sekhar et al., 1960; Subramaniam, 1976; Gregory & Tan, 1976). However, it has been reported that these abnormal groups in NR are not major factors for branching and gel formations (Yunyongwattanakorn, 2005). The effect of non-rubber components on the gel formation and SH of NR after accelerated storage-hardening test (ASHT) in phosphorous pentoxide (P_2O_5) has been investigated (Yunyongwattanakorn et al., 2003).

SH phenomenon of solid NR is presumed to occur *via* reactions between some non-rubber components and abnormal groups in rubber chains. The main non-rubber constituents in NR are composed of proteins and lipids. The SH behavior under high and low humidity conditions using P_2O_5 and sodium hydroxide (NaOH) was analyzed for the various purified NR samples (Yunyongwattanakorn et al., 2003). The NR obtained from centrifuged fresh NR latex (CFNR) and deproteinized NR latex (DPNR) showed significant increase in the hardening plasticity index (P_H) value during storage under low humidity conditions, while that of the transesterified NR (TENR) and transesterified DPNR (DPTE-NR) was almost constant during storage, as shown in Figure 15. The low P_H value and no gel content in TENR was observed indicating that TENR had not much branch-points enough to form gel phase. The lack of branch-points in both transesterified rubber samples might be the main reason for the constant P_H value, even when using the strong drying agent. The above findings suggest that the fatty acid ester group plays the most importance role in the SH of rubber under low humidity conditions.

After keeping samples under high humidity conditions, the fresh NR (FNR), CFNR and DPNR showed decrease in the P_H value, while that of the TENR and DPTE-NR showed low P_H value and decreased with increasing the storage time, as shown in Figure 16. This implies that the rubber containing no fatty acid groups tended to cause auto-oxidation. It was also observed that the P_H value of all rubber samples decreased as the storage time increased, indicating that the SH was inhibited under high humidity.

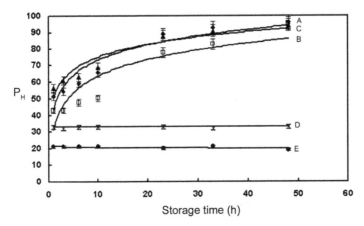

Fig. 15. Change in the hardening plasticity index (P_H) of the rubber samples after storage in P_2O_5 under vacuum; (A) FNR, (B) CFNR, (C) DPNR, (D) TENR and (E) DPTE-NR

The gel phase in rubber samples after SH is soft-gel originated by micelle formation between phospholipids molecules. This finding confirmed by the decrease in gel content of stored FNR after the addition of 2% v/v ethanol into rubber solution and complete decomposition of the gel fraction after transesterification (Yunyongwattanakorn, 2005).

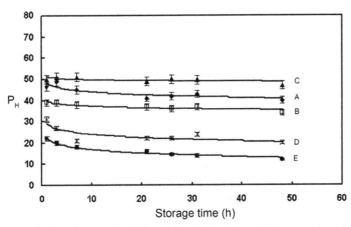

Fig. 16. Change in the hardening plasticity index (P_H) of the rubber samples after storage under high humidity under *vacuum*; (A) FNR, (B) CFNR, (C) DPNR, (D) TENR and (E) DPTE-NR

Significant increase in the plasticity, gel content and molecular weight between the crosslinks (M_c) when the samples were kept under low humidity condition suggested an important role of the humidity on gel formation during SH. The proposed structure of gel formation during ASHT based on all of findings mentioned above was schematically illustrated in Figure 17. The proteins and phospholipids at the chain-ends of rubber molecules may interact with water under ambient condition, thus the water may disturb the formation of branching points by hydrogen bonding.

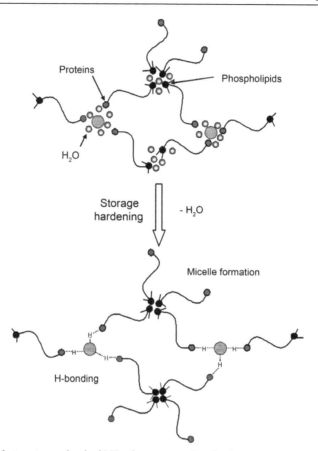

Fig. 17. Proposed structure of gel of NR after storage hardening

When water is removed from the rubber, with a drying agent such as phosphorus pentoxide or sodium hydroxide, proteins and phospholipids at the terminals of rubber chain may have a chance to form branching points by hydrogen bonding. Since both functional groups of NR are active, the gel fraction can be formed during storage due to the reaction between the functional groups of NR chains. Therefore, the formation of gel in NR during SH is presumed to compose of two types of branching points; the first one is expected to originate from phospholipids, which are associated to rubber chains and/or free phospholipid molecules. The phospholipids are associated together by the formation of micelle structure mainly *via* hydrogen bonding between polar groups in phospholipids molecules as reported (Tarachiwin et al., 2005a, 2005b). The other crosslinking point is due to proteins, which can be formed *via* hydrogen bonding.

3. Color substances and obnoxious odor in natural rubber

It is accepted that NR gives naturally occurring color, which restrict many applications such as light-color products. Therefore, characterization of color substances presenting in NR is very useful to develop the certain methodology to eliminate them completely or partly from

NR in the future. Recently, Sakdapipanich and co workers (2006) have tried to purify and characterize the color substances extracted from various fractions of *Hevea* rubber latex by certain methods, using high-resolution structural characterization techniques. It was found that the content of color substances extracted from fresh latex (FL), rubber cream, bottom fraction (BF), Frey Wyssling (FW) particles and STR 20 were different. Based on the high-resolution spectroscopic analyzes, it was found that the color substances extracted from NR were composed of carotenoids, tocotrienol esters, fatty alcohol esters, tocotrienols, unsaturated fatty acids, fatty alcohols, diglyceride and monoglyceride. The results will be useful for rubber-technologist to identify the origin to make obnoxious color in natural rubber, especially in some applications which are restricted by such the color.

3.1 Color substances in natural rubber

3.1.1 Enzymatic browning

Enzymatic browning is the discoloration resulting when monophonic compounds of plants or shellfish in the presence of atmospheric oxygen and polyphenol oxidase (PPO) are hydroxylated to *o*-diphenols and the latter are oxidized to *o*-quinones. PPO is also known and reported under various names (tyrosinase, phenolase, catechol oxidase, catecholase, monophenol oxidase, *o*-diphenol oxidase and *o*-phenolase) based on substrate specificity. Then, quinones may condense and react non-enzymatically with other phenolic compounds, amino acids, proteins or other cellular constituents to produce colored polymer or pigments (Iyidogan & Bayindirh, 2004), as shown in Figure 18 (Lee & Whitaker, 1995).

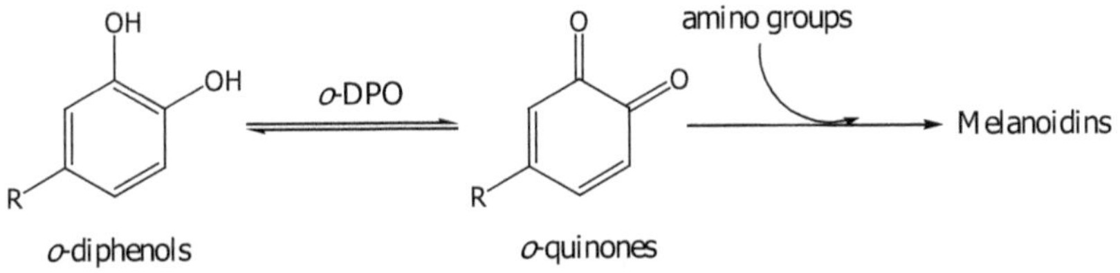

Fig. 18. Enzymatic browning of *o*-diphenols by *o*-diphenol oxidase (*o*-DPO)

PPO was earlier reported to be present in both lutoid and Frey-Wyssling particle (Wititsuwannakul et al., 2002). It was found that the latex PPO activity in lutiods was 5 to 34 folds higher than that of the Frey-Wyssling particles (Table 1).

| Rubber clone | PPO activity (nkat/ml latex)[a] | |
	Lutiod	Frey-Wyssling
RRIM 600	7.33	0.21
GT 1	9.21	0.65
KRS 21	4.34	0.77

[a] Minimal PPO activity was detected in C-serum and rubber fraction.

Table 1. Distribution of PPO activity in the ultracentrifuged fresh latex

3.1.2 Non-enzymatic browning

Non-enzymatic browning resulted from the following reactions is possibly concerned as the discolorations of NR.

Lipid oxidation

The important lipids involved in oxidation are the unsaturated fatty acid moieties, oleic, linoleic and linolenic. The rate of oxidation of these fatty acids increases with the degree of unsaturation. The mechanism of lipid oxidation is illustrated in Figure 19 (http://www.agsci.ubc.ca).

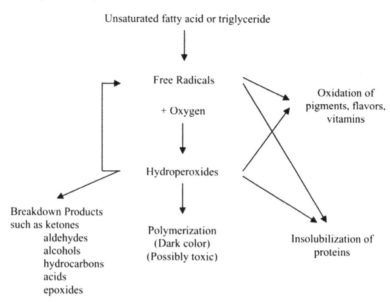

Fig. 19. Lipid oxidation mechanisms

Millard reaction

Millard reaction involves the reaction between carbonyl compounds (reducing sugars, aldehydes, ketones and lipid oxidation products) and amino compounds (lysine, glysine amine and ammonia proteins) to produce glycosyl-amino products, followed by Amadori rearrangement. An intermediate step involves dehydration and fragmentation of sugars, amino acids degradation, etc. A final step involves aldol condensation, polymerization and the formation of colored products.

3.1.3 Enzymatic browning prevention

The principles of browning prevention have not changed with time and are essentially the same as those applying to the inhibition of any tissue enzyme, i.e.:

1. Inhibition or inactivation of the enzyme
2. Elimination or transformation of the substrate(s)
3. Combination of both above

The examples of chemical agent for browning prevention are sulfiting agents, aromatic compounds, such as 4-hexylresorcinol, tropolone (2-hydroxy-2,4,6-cycloheptatrien-1-one), kojic acid (5-hydroxy-2-(hydroxymethyl)-γ-pyrone), etc., glucosidated substrates, proteolytic enzymes, carbohydrates, peptides, carbon monoxide, hypochlorite, and miscellaneous browning inhibitors. The enzymatic browning prevention can be also performed by physical treatments, i.e. blanching, ultrafiltration, sonification, supercritical carbon dioxide.

3.2 Obnoxious odor in natural rubber

Fulton W. S. (1993) has studied the problems of odor during rubber processing. It was found that coagulation and subsequent conversion of coagulum into bale rubber affects the smell of natural rubber. The field grade material tends to have a stronger smell than rubber prepared by the deliberately controlled coagulation of latex. The main constituents of the effluent gases from the rubber industry are low-molecular weight volatile fatty acid, which can be effectively removed by water scrubbers with efficiencies of 92-99%.

Isa Z. (1993) has studied how to control the mal-odor in Standard Malaysia Rubber (SMR) factories. The mal-odor from SMR factories is mainly attributed to the obnoxious volatile components, which are present in the exhaust gases discharged into the air through a chimney during the drying stage of SMR processing. The volatile compounds are originally produced from the microbial breakdown of the non-rubber components during the storage of scraps and cup lumps prior to processing. Before to the characterization by gas chromatography, the exhaust gases were collected by adsorption on charcoal adsorption tube. It was found that the volatile compounds in the exhaust gases were low molecular-weight volatile fatty acids such as acetic acid, propionic acid, butyric acid and valeric acid. The mal-odor can be reduced by a water-scrubber system.

Hoven V. P. and coworkers (2003) have determined the volatile organic components of various grades of solid rubber by gas chromatography (GC) and gas chromatography-mass spectroscopy (GC-MS) using direct sampling collection of head space technique. It was found that about fifty components, with molecular weight in the range of 40-200 amu, were identified. They can be classified into four groups, as follows: (1) compounds having low polarity; aliphatic and aromatic hydrocarbons; (2) compounds having moderate polarity; aldehydes, ketones; (3) compounds having high polarity; volatile fatty acids; (4) derivatives containing nitrogen or sulfur. The components discovered in all samples were ethylamine, benzylhydrazine and low molecular-weight volatile fatty acids, which are acetic acid, propionic acid, isobutyric acid, butyric acid, isovaleric acid and valeric acid. The obnoxious odor of all NR samples is mainly originated from low molecular-weight volatile fatty acids, whose quantity depends upon the odor intensity and the quality of NR latex. In 2004, Hoven V. P. and coworkers have incorporated odor-reducing substances into STR 20 and RSS 5 by physical mixing prior to vulcanization. It was found that according to the GC analysis and olfactometry test, the obnoxious odor from STR 20 and RSS 5 can be significantly reduced by physical mixing of chitosan, zeolite 13X and carbon black. Benzalkonium chloride and sodium dodecyl sulphate (SDS) did not exhibit the desired odor-reducing properties as a consequence of their thermal degradation during vulcanization. The ability to adsorb physically and/or chemically with the volatile fatty acids as well as their reinforcing effect indicates that chitosan and carbon black are strong candidates for odor reduction of NR.

Recently, Sakdapipanich and Insom (2006) have elucidated the mechanisms producing the components of obnoxious odor derived from various solid rubbers using a combination technique between head-space sampling and GC-MS. It was found that volatile components of STR XL were mainly comprised of hydrocarbons, which were probably derived from lipid oxidation of unsaturated fatty acids or triglycerides. In the case of STR 5L, the small amounts of volatile fatty acids derived from carbohydrate fermentation were detected. They were also liberated from STR 5, STR 20, cup lump and RSS. Finally, in the case of skim crumb rubber, sulfur-containing compounds derived from decomposition of proteins existing in NR were observed.

4. Skim rubber

Skim latex is a material resulting from the production of concentrated latex in the centrifugation process. After centrifuging the fresh field latex, 5-10% of total rubber, together with an enhanced proportion of the non-rubber constituents in the original latex, remains in the serum phase or skim latex (Bristow, 1990). The skim latex is composed of small rubber particles in range of 0.04 μm to 0.4 μm with the mean particle diameter of about 0.1 μm, while those of large rubber particles in concentrated latex is from 0.1 μm to 3 μm with a mean diameter of 1 μm, as shown in Figure 20. The skim rubber obtained from skim latex shows a unimodal distribution with a peak top between the high and low-molecular weight peaks in the MWD of ordinary rubber, centered around 1.0×10^6 g/mol, (Sakdapipanich et al., 2002a) as shown in Figure 21.

Recovery of skim rubber from skim latex

The residual rubber from skim latex is normally recovered by the addition of sulfuric acid. This method can separate skim rubber out from skim latex as coagulum like To-fu. Skim rubber contains 70-85% rubber component. It is known that there are a number of proteins contaminate in skim rubber. Normally, there are hydrocarbon, 5-10% acetone-soluble fatty materials and 10-20% proteins, compared with an average of 95% hydrocarbon, 3% fatty materials and 2% proteins in the case of smoked sheet prepared from fresh field latex. (Nithi-Uthai, 1998)

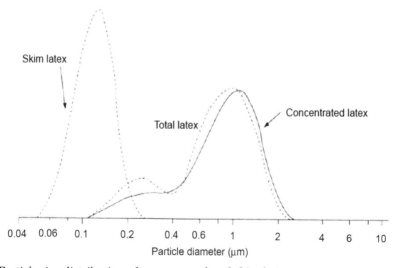

Fig. 20. Particle size distribution of concentrated and skim latices

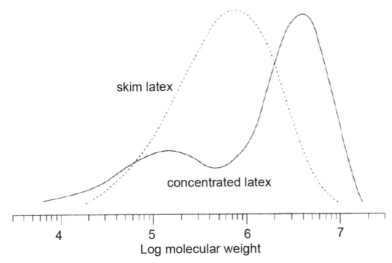

skim latex

concentrated latex

4   5   6   7

Log molecular weight

Fig. 21. Molecular weight distribution of the rubber from concentrated and skim latices

In addition to the sulfuric acid coagulation, there are several other methods such as

a. Auto-coagulation of skim latex by de-ammoniating to less than 0.1% and leaving for spontaneous coagulation within 5 days (Smith, 1969). The drawbacks of this method are that it required extensive coagulation-tank capacity to handle large quantities of the resulting coagulum and the obtained dry rubber foul smell.

b. Accelerated auto-coagulation of skim latex by partly de-ammonion of skim latex and addition of 0.1% di-octyl sodium sulfosuccinate and 1% $CaCl_2$ to skim latex (John & Weng, 1973). This treatment can coagulate the skim latex within 2 days.

c. The addition of enzyme to skim latex before coagulation with acid was also carried out. This process can reduce the nitrogen content in the skim latex to about 0.6-3.5% due to the decomposition of proteins by enzyme. Thus, after enzymolysis, the proteins attached to rubber particles are removed and the possible remaining stabilizer on the particle surface is fatty acid soaps (Nithi-Uthai, 1998; Morris, 1954).

d. Skim NR latex was recovered as concentrated skim latex by using deproteinization and salting-out techniques (DP/S) (Sakdapipanich et al., 2002b). The increase in skim rubber content is owing to the recovery of small rubber particles in the skim latex.

Structure of skim rubber

In the last decade, it was disclosed that skim rubber contains low or no ester content compared to concentrated rubber, i.e., 0.03 per rubber chain. This implies that the rubber molecules from skim rubber are not terminated by phospholipids as could be detected in the concentrated rubber molecules (Sakdapipanich et al., 1999b). Based on the previous results, it was observed that skim rubber showed insignificant increase in gel content during storage while the gel content of fresh and concentrated rubber increased up to 30% and 60%, respectively. Furthermore, in the case of concentrated rubber, the M_w and M_n values decreased after transesterification, the reaction which decomposes gel and branching in natural rubber, by about 20% and 35%, respectively. However, no significant change of these values was observed in skim rubber. This suggests that skim rubber composed of linear rubber molecules, which differ from concentrated rubber.

^{1}H-NMR spectra of fractionated skim rubber and 1-month old seedlings, as shown in Figure 22, revealed that skim rubber showed three major signals at 1.78, 2.10, and 5.18 ppm, which are assignable to -CH$_3$, -CH$_2$, and =CH of *cis*-1,4 isoprene units, respectively. The rubber from 1-month old seedlings showed additional signals around 4.0-4.3 ppm and 1.2 ppm, which have been assigned to terminal -CH$_2$O and long methylene sequence -(CH$_2$)$_n$- of fatty acids (Tangpakdee, 1998). The absence of both signals in fractionated skim rubber clearly indicates that skim rubber contains no phospholipid linked to rubber chain. Phospholipid groups including long-chain fatty acids were found to play an important role to form branched structure (Tarachiwin et al., 2005c). Thus, the ^{1}H-NMR results provide confirmation that skim rubber is composed of linear rubber molecules.

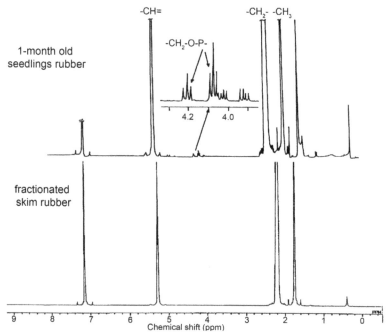

Fig. 22. ^{1}H-NMR spectra of (A) 1-month old seedlings and (B) fractionated skim rubber

Physical properties of skim rubber

The difference in rubber constituent in skim latex is the important parameter related to green strength of natural rubber. Figure 23 shows the green strength of concentarated, skim rubber, synthetic polyisoprene (IR) and transesterified-deproteinized rubber (Nawamawat, 2002). The concentrated rubber is composed of branched molecules, which linked together by hydrogen bonding *via* protein and ionic crosslinks cased by phospholipid groups. The latter branch points have been attributed to the high green strength of NR and induced the crystallization of the rubber on straining (Kawahara et al., 2002b). It was reported that the nitrogen and ester contents were dramatically decreased after deproteinization followed by transesterification of concentrated rubber, i.e., DPTE (Tangpakdee & Tanaka, 1997). DPTE composed of linear rubber molecules was found to show very low green strength comparable to that of concentrated rubber. Moreover, it was found that skim rubber showed similar green strength to DPTE as well as synthetic IR (Figure 23). Synthetic IR contains no functional terminal group, especially phospholipid group (Gregg & Macey, 1973), which is similar to that in the case of skim rubber. The Mooney viscosity and Mooney relaxation data of skim rubber were also found to be lower than that from concentrated rubber. Thus, this indicates that the skim rubber is softer and lower elasticity than concentrated rubber (Sakdapipanich et al., 2002). The low tensile strength and Mooney viscosity might lead to the benefit of skim rubber on the low energy consumption during processing, which are perfectly different from concentrated rubber.

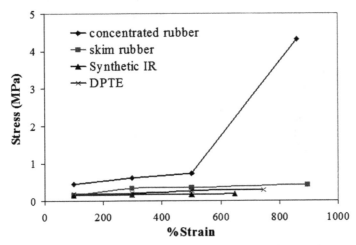

Fig. 23. Green strength of concentrated rubber, skim rubber, synthetic polyisoprene (IR) and DPTE

Application of skim rubber

In the previous work, the application of the skim rubber is used in many fields. For example, the small amounts of skim rubber act as cure-rate boosters, which can replace the secondary accelerators in some applications. In physical processing characteristics, the skim rubber is resembled compounded materials rather than elastomeric gum (Blackey, 1996). This is due to the preponderance of non-elastomeric substances in the material. The increases in hardness and state of curing correlate well with the nitrogen content, due to the physical effect of proteinous substances. The skim rubber can be controlled by a suitable choice of accelerator. About 20-25 parts of skim rubber can be blended with conventional rubber to give a high level of vulcanizate properties retention and reduced variability. The skim rubber was found to give better adhesion between a brassed metal and contiguous skim rubber (Schofeld, 1995). The rubber-metal adhesion and adhesion retention can be obtained by adding copper sulfide to the conventional rubber skim stock composition and followed by vulcanization to yield the end product. In addition, skim rubber was also studied for its use as urea encapsulant in the controlled release application (Tanunchai, 1999).

Recently, it was revealed the effective method using saponification reaction to remove the impurities, especially proteins and lipids, from skim rubber. It was also found that this highly purified skim rubber showed good solubility with no any gel formation, which is a merit on adhesive application (Nawamawat, 2002). The pressure-sensitive adhesive made from high-purified skim rubber also showed good tack and adhesion properties, good transparent and absence of protiens, which might cause allergy (Nawamawat, 2002). Thus, the highly-purified skim rubber can be further developed for more valuable as medical and surgical tape.

5. References

Alenius, H., Turijanmaa, K., Palosuo, T. (2002) *Occup. Environ. Med.* 59, 419

Archer, B. L., Audley, B. G. (1987) *Bot. J. Linnean Soc.* 94, 181–196.

Archer, B. L., Audley, B. G., McSweeney, G. P., Tang, C. H. (1969) *J. Rubb. Res. Inst. Malaya* 21, 560–569.

Arnold, A. R., Evans, P. (1991) *J. Natl. Rubb. Res.* 6, 75–86.

Audley, B. G., Archer, B. L. (1988) Biosynthesis of rubber, in: *Natural Rubber Science and Technology* (Roberts, A. D., Ed.), pp. 35–62. London: Oxford University Press.

Blackey D.C. (1966) High polymer lattices, vol. 1. Maclaren & Sons, London; 192.

Bristow G.M. (1990) J. nat. Rubb. Res., 5, 114-134.

Cornish, K., Backhaus, R. A. (1990) *Phytochemistry* 29, 3809–3813.

Crafts, R. C., Davey, J. E., McSweeney, G. P., Stephens, I. S. (1990) *J. Natl. Rubb. Res.* 5, 275–285.

Eng, A. H., Ejiri, S., Kawahara, S., Tanaka, Y. (1994a) *J. Appl. Polym. Sci. Appl. Polym. Symp.* 53, 5–14.

Eng, A. H., Kawahara, S., Tanaka, Y. (1994b) *Rubb. Chem. Technol.* 67, 159–168.

Fulton W. S. (1993) *Avoiding the Problems of Odour during Rubber Processing. Rubb. Devel.*, 46, 35-37.

Gan, S.N. (1997) *Trends in Polymer Science* 2, 69-82.

Gazeley, K. F., Gorton, A. D. T., Pendle, T. D. (1988) Latex concentrates: Properties and composition, in: *Natural Rubber Science and Technology* (Roberts, A. D., Ed.), pp. 63–140. London: Oxford University Press.

Gregg E. C. and Macey J. H. (1973), *Rubb. Chem. Technol.*, 46, 47.

Gregory, M.J. and Tan, A.S. (1976) *Proc. Int. Rubber Conf. 1975 Kuala Lumpur* 4, 28.

Hoven V. P., Rattanakarun K. and Tanaka Y. (2003) *Rubb. Chem. Technol.* 76, 1128.

Hoven V. P., Rattanakarun K., Tanaka Y. (2004) *J. Appl. Polym. Sci.* 92, 2253-2260.
       http://www.agsci.ubc.ca

Isa Z. (1993) *Bull. Rubb. Res. Inst., Malaysia,* 215, 56.

Iyidogan N., F.and Bayindirh A. (2004) J. Food Eng. 62, 299-304.

Jacob, J. L., d'Auzac, J., Prevot, J. C. (1993) *Clin. Rev. Allergy* 11, 325–337.

John C. K., Weng S. S. (1973) *J.Rubb. Res. Inst. Malaya.,* 23, 257.

Kawahara S., Isono Y., Sakdapipanich J. T., Tanaka Y. and Eng A. H. (2002a) *Rubb. Chem. Technol.* 75, 739-746.

Kawahara S., Kakubo T., Nishiyama N., and Tanaka Y. (2002b) *J. nat. Rubb. Res.* 2(3), 141.

Kawahara, S., Kakubo, T., Sakdapipanich, J. T., Isono, Y., Tanaka, Y. (2000) *Polymer* 41, 7483–7488.

Kawahara, S., Nishiyama, N., Kakubo, T., Tanaka, Y. (1996) *Rubb. Chem. Technol.* 69, 600–607.

Lee C. Y. and Whitaker J. R. (1995) *ACS Symp.* Series 600. Am. Chem. Soc., Washington DC. 8.

Madhavan, S., Greenblatt, G. A., Foster, M. A., Benedict, C. R. (1989) *Plant Physiol.* 89, 506–511.

Mekkriengkrai, D. (2005) [A Doctor thesis in Department of Chemistry]. Thailand; Faculty of Science, Mahidol University.

Mooibroek, H., Cornish, K. (2000) *Appl. Microbiol. Biotechnol.* 53, 355–365.

Morris J. E. (1954) Proc. 3rd Rubber Technology Conference. London., 13.

Nair, S. (1987) Characterisation of natural rubber for greater consistency, *Proceedings, International Rubber Conference 1987*, P. 2A/1–10, PRI, London.

Nakade, S., Kuga, A., Hayashi, M., Tanaka, Y. (1997) *J. Natl. Rubb. Res.* 12, 33–42.

Nawamawat, K. (2002) M.Sc. thesis, Mahidol University, Thailand.

Nishiyama, N., Kawahara, S., Kakubo, T., Eng, A. H., Tanaka, Y. (1996) *Rubb. Chem. Technol.* 69, 608–614.

Nithi-Uthai B., Nithi-Uthai P., Wititsuwannakul R., Promna J., Promnok J., and Boonrasi S. (1998) Proc. 7th Seminar on Elastomer. Bangkok, 1.

Ogura, K., Koyama, T., Sagami, H. (1997) Choleterol: its functions and metabolism in biology and medicine in subcellular biochemistry, p. 57-87. In Bittman, R. (ed.), Polyprenyldiphosphate synthase. Plenum Press, New York.

Pendle, T. D., Swinyard, P. E. (1991) *J. Natl. Rubb. Res.* 6, 1–11.

Murari R., Abd. El-Rahman M. M. A., Wedmid, Y., Parthasarathy S., Baumann, W. J. (1982) *J. Org. Chem.* 1982, 47, 2158-2163.

Sakaki, T., Hioki, Y., Kojima, M., Kuga, A., Tanaka, Y. (1996) *Nippon Gomu Kyokashi* 69, 553–556.

Sakdapipanich J. T., Nawamawat K., and Tanaka Y. (2002a) *Rubb. Chem. Technol,* 75, 179-185.

Sakdapipanich J. T., Nawamawat K., and Tanaka Y. (2002b) *J. Rubb. Res.,* 5, 1-10.

Sakdapipanich J. T., Suksujaritporn S., and Tanaka Y. (1999b) *J. Rubb. Res.,* 2, 160-168.

Sakdapipanich J. T., Tanaka, Y., Jacob, J. L., d'Auzac, J. (1999a) *Rubb. Chem. Technol.* 72, 299–307.

Sakdapipanich J.T. and Insom K. (2006) *Kautsch. Gummi Kunstst.* 7/8-06, 382-387.

Sakdapipanich J.T., Insom K. and Phuphewkeaw, N. (2006) *Rubber Chem. Technol.,* 80, 212-230

Schofeld U.S. patent 4,1995, 679.

Sekhar, B.C. (1958) *Rubb. Chem. Technol.* 31, 425-429.

Sekhar, B.C. (1960) *J. Polym. Sci.* 48, 133-137.

Sentheshanmuganathan, S. (1975) Some aspects of lipids chemistry of natural rubber, *Proceedings International Rubber Conference 1975*, 457–469.

Smith M. G. (1969) *J. Rubb. Res. Inst. Malaya.,* 23, 78.

Subramaniam, A. (1976) *Proc. Int. Rubber Conf. Kuala Lumpur 1975* 4, 3-11.

Tanaka, Y., Sato, H., Kageyu, A. (1983) *Rubb. Chem. Technol.* 56, 299-303.

Tanaka, Y., Eng, A. H., Ohya, N., Tangpakdee, J., Kawahara, S., Wititsuwannakul, R. (1996), *Phytochemistry* 41, 1501–1505.

Tanaka, Y., Kawahara, S., Tangpakdee, J. (1997) *Kautsch. Gummi Kunstst.* 50, 6–11.

Tanaka, Y., Tangpakdee, J. (1997b) *Curr. Polym. Res.,* 12–15.

Tangpakdee J. (1998) [A Doctor thesis in Department of Material systems engineering]. Japan; Faculty of Technology, Tokyo University of Agriculture and Technology.

Tangpakdee J. and Tanaka Y. (1997b) *Rubb. Chem. Technol.* 70, 707-713.

Tangpakdee J. and Tanaka Y. (1998) *J. Rubb. Res.* 1, 77-83.

Tangpakdee, J. and Tanaka, Y. (1997a) *J. nat. Rubb. Res.* 12, 112–119.

Tangpakdee, J., Tanaka, Y. (1998a) Long-chain polyprenols and rubber in young leaves of *Hevea brasiliensis, Phytochemistry* 48, 447–450.

Tangpakdee, J., Tanaka, Y. (1998b) *J. Rubb. Res.* 1, 77–83.

Tangpakdee, J., Tanaka, Y., Ogura, K., Koyama, T., Wititsuwannakul, R., Wititsuwannakul, D. (1997) *Phytochemistry* 45, 269–274.

Tangpakdee, J., Tanaka, Y., Wititsuwannakul, R., Chareonthiphakorn, N. (1996) *Phytochemistry* 42, 353–355.

Tanunchai T. (1999) [M.Sc. Thesis in Polymer Science and Technology]. Bangkok; Faculty of Graduate Studies, Mahidol University.

Tarachiwin L, Sakdapipanich J. T., Tanaka Y. (2005c), *Kautsch. Gummi Kunstst.* 58, 115-122.

Tarachiwin L., Sakdapipanich J. and Tanaka Y. (2003) *Rubb. Chem. Technol.* 76, 1177-1184.

Tarachiwin, L., Sakdapipanich, J.T., Ute, K., Kitayama, T., Bamba, T., Fukusaka, E., Kobayashi, A. and Tanaka, Y. (2005a) *Biomacromolecule* 6(4), 1851-1857.

Tarachiwin, L., Sakdapipanich, J.T., Ute, K., Kitayama, T. and Tanaka, Y. (2005b) *Biomacromolecule* 6(4), 1858-1863.

Tarachiwin, L. (2004) Ph.D thesis, Mahidol University, Thailand.

Verhaar, G. (1959) *Rubb. Chem. Technol.* 32, 1627–1659.

Wititsuwannakul D., Chareonthiphakorn N., Pace M., and Wititsuwannakul R. (2002) *Phytochemistry* 61, 115-121.

Yeang H.Y. Curr. Opin. Allergy Clin. (2004) *Immunol.* 4, 99-104.

Yeang H.Y., Arif S.A., Yusof F. Sunderasan E. (2002) *Sunderasan* 27, 32-45.

Yunyongwattanakorn, J. (2005) Ph.D thesis, Mahidol University, Thailand.

Yunyongwattanakorn, J., Tanaka, Y., Kawahara, S., Klinklai, W. and Sakdapipanich, J.T. (2003) *Rubb. Chem. Technol.* 76(5), 1228-1240.

Permissions

The contributors of this book come from diverse backgrounds, making this book a truly international effort. This book will bring forth new frontiers with its revolutionizing research information and detailed analysis of the nascent developments around the world.

We would like to thank Reda Helmy Sammour, for lending his expertise to make the book truly unique. He has played a crucial role in the development of this book. Without his invaluable contribution this book wouldn't have been possible. He has made vital efforts to compile up to date information on the varied aspects of this subject to make this book a valuable addition to the collection of many professionals and students.

This book was conceptualized with the vision of imparting up-to-date information and advanced data in this field. To ensure the same, a matchless editorial board was set up. Every individual on the board went through rigorous rounds of assessment to prove their worth. After which they invested a large part of their time researching and compiling the most relevant data for our readers. Conferences and sessions were held from time to time between the editorial board and the contributing authors to present the data in the most comprehensible form. The editorial team has worked tirelessly to provide valuable and valid information to help people across the globe.

Every chapter published in this book has been scrutinized by our experts. Their significance has been extensively debated. The topics covered herein carry significant findings which will fuel the growth of the discipline. They may even be implemented as practical applications or may be referred to as a beginning point for another development. Chapters in this book were first published by InTech; hereby published with permission under the Creative Commons Attribution License or equivalent.

The editorial board has been involved in producing this book since its inception. They have spent rigorous hours researching and exploring the diverse topics which have resulted in the successful publishing of this book. They have passed on their knowledge of decades through this book. To expedite this challenging task, the publisher supported the team at every step. A small team of assistant editors was also appointed to further simplify the editing procedure and attain best results for the readers.

Our editorial team has been hand-picked from every corner of the world. Their multi-ethnicity adds dynamic inputs to the discussions which result in innovative outcomes. These outcomes are then further discussed with the researchers and contributors who give their valuable feedback and opinion regarding the same. The feedback is then collaborated with the researches and they are edited in a comprehensive manner to aid the understanding of the subject.

Apart from the editorial board, the designing team has also invested a significant amount of their time in understanding the subject and creating the most relevant covers. They scrutinized every image to scout for the most suitable representation of the subject and create an appropriate cover for the book.

The publishing team has been involved in this book since its early stages. They were actively engaged in every process, be it collecting the data, connecting with the contributors or procuring relevant information. The team has been an ardent support to the editorial, designing and production team. Their endless efforts to recruit the best for this project, has resulted in the accomplishment of this book. They are a veteran in the field of academics and their pool of knowledge is as vast as their experience in printing. Their expertise and guidance has proved useful at every step. Their uncompromising quality standards have made this book an exceptional effort. Their encouragement from time to time has been an inspiration for everyone.

The publisher and the editorial board hope that this book will prove to be a valuable piece of knowledge for researchers, students, practitioners and scholars across the globe.

List of Contributors

Kenichi Iwata, Nik Noor Azlin binti Azlan and Toshio Omori
College of Systems Engineering and Science, Department of Bioscience and Engineering, Shibaura Institute of Technology, Japan

San San Yu
Department of Biotechnological Research, Ministry of Science and Technology, Myanmar

Irma Esthela Soria-Mercado and Graciela Guerra Rivas
Facultad de Ciencias Marinas, Universidad Autónoma de Baja California, Ensenada, BC, México

Luis Jesús Villarreal-Gómez
Centro de Ingeniería y Tecnología, Universidad Autónoma de Baja California, Tijuana, BC, México

Nahara E. Ayala Sánchez
Facultad de Ciencias, Universidad Autónoma de Baja California, Ensenada, BC, México

Sejeong Kook and Kiheon Choi
Duksung Women's University, Seoul, South Korea

Michael Greger
The Humane Society of the United States, USA

Ann Njoki Kingiri
African Centre for Technology Studies (ACTS), Off United Nations Crescent, Gigiri Nairobi, Kenya

Marian D. Quain, James Y. Asibuo, Ruth N. Prempeh and Elizabeth Y. Parkes
Council for Scientific and Industrial Research, Crops Research Institute, Kumasi, Ghana

Sadhana Talele
The University of Waikato, New Zealand

Kinga Turzo
University of Szeged, Faculty of Dentistry, Hungary

Taketo Wakai and Naoyuki Yamamoto
Microbiology & Fermentation Laboratory, Calpis., Ltd., Fuchinobe, Chuo-ku, Sagamihara-shi, Kanagawa, Japan

Georgia Papavasiliou, Sonja Sokic and Michael Turturro
Illinois Institute of Technology, Department of Biomedical Engineering, USA

Srinath Rao and Prabhavathi Patil
Department of Botany, Gulbarga University, Gulbarga Karnataka, India

Ji Gang Kim
National Institute of Horticultural and Herbal Science, Rural Development Administration, Suwon, Korea

Jitladda T. Sakdapipanich
Mahidol University, Thailand

Porntip Rojruthai
King Mongkut's University of Technology North Bangkok, Thailand

Printed in the USA
CPSIA information can be obtained
at www.ICGtesting.com
JSHW011432221024
72173JS00004B/773

9 781632 394675